WATER QUALITY AND ITS CONTROL

INTERNATIONAL ASSOCIATION FOR HYDRAULIC RESEARCH
ASSOCIATION INTERNATIONALE DE RECHERCHES HYDRAULIQUES

5

HYDRAULIC STRUCTURES DESIGN MANUAL

HYDRAULIC DESIGN CONSIDERATIONS

Hydraulic Structures Design Manuals:

1. J. Knauss (ed.), 1987. *Swirling flow problems at intakes.*
2. H.N.C. Breusers & A.J. Raudkivi, 1991. *Scouring.*
3. E. Naudascher, 1991. *Hydrodynamic forces.*
4. I.R. Wood (ed.), 1991. *Air entrainment in free-surface flows.*
5. M. Hino (ed.), 1994. *Water quality and its control.*
6. A.J. Raudkivi, 1993. *Sedimentation: Exclusion and removal of sediment from diverted water.*
7. E. Naudascher & D. Rockwell, 1994. *Flow-induced vibrations: An engineering guide.*

WATER QUALITY AND ITS CONTROL

Edited by:

MIKIO HINO

Department of Civil Engineering, Tokyo Institute of Technology, O-okayama, Meguro-ku, Tokyo, Japan
Presently: Faculty of Political Science, Chuo University, Hachiohji, Tokyo, Japan

A.A. BALKEMA / ROTTERDAM / BROOKFIELD / 1994

Photo on the binding: Advancing head of a density current on an inclined slope (1/60). (Photo by Y. Miyanaga).

Published by
A.A. Balkema, P.O. Box 1675, 3000 BR Rotterdam, Netherlands
A.A. Balkema Publishers, Old Post Road, Brookfield, VT 05036, USA

ISBN 90 5410 123 7

Printed in the Netherlands

Contents

Introduction

MIKIO HINO
Department of Civil Engineering, Tokyo Institute of Technology, O-okayama, Meguro-ku, Tokyo, Japan
Presently: Faculty of Political Science, Chuo University, Hachiohji, Tokyo, Japan

Hydraulic engineers have become progressively more and more involved with environmental problems in the latter half of the twentieth century. Originally, hydraulics developed as a branch of the physical sciences and engineering related to the water flow problems in civil and mechanical engineering works. During the first half of this century, hydraulics (classical hydraulics) became more firmly established on the basis of fluid mechanics.

The first encounter of hydraulicians with an environmental problem in a broader sense was with density stratified flows, such as the salt wedge intrusion at the river mouth. This topic was one of the main subjects at the 1953 IAHR Congress held at Minnesota, USA, the first large conference on hydraulics following World War II, and it had a strong impact on the development of hydraulic research during subsequent decades. Gradually the importance within hydraulics of density stratified flows (caused thermally, by salt water or by suspension of sediment) has increased from both practical and fundamental viewpoints.

Selective withdrawal, from a reservoir or lake, of warm water (for agricultural use) or cold water (for industrial use) has been studied both theoretically and experimentally.

In the 1960s, the rapid increase in the demand for electric power in the highly industrialized countries, on the one hand, and the technical evolution of constructing large steam turbines with greater and greater capacities, on the other, have brought about problem with the disposal of waste heat of water used to condense steam in turbine generators. The resultant increase in the temperature of ambient waters may have adverse effects on plants and animals (sea weed and fish or their larva) living in the sea and the rivers. Consequently, the region of raised temperature and its spread must be predicted acurately and then limited as narrowly as possible by installations of diffusers or other means. The research on waste heat disposal is classified in two categories; one is the dilution of the thermal plume in the ambient flow (near field problem), and the other is the diffusion of warm water over the flowing water in the cold region (far field problem). In this period the primary concerns were, nevertheless, with the physical aspects, although the new factor of heat had been introduced.

From the 1970s on, the importance of several factors other than just the physical ones, such as chemically toxic material, biological effects and ecological impacts had to be taken into account. A realization of the importance of the environment gained world wide acceptance; in 1972 the UN held the first UN Congress on the world environment in Sweden. With the evolution of these concerns, the role (or research field) of hydraulic engineers expanded greatly. At the same time, hydraulic engineers came to see the problems from a different and broader viewpoint of both hydraulics and physics and to contribute to their solutions. Extensive efforts have gone into the building of various kinds of numerical models for the prediction of thermal as well as ecological processes in reservoirs and lakes, and for the disposal of waste heat from thermal power generation.

During these periods, the activities of hydraulic engineers were greatly stimulated and assisted by the advent of electronic computers; without them most of the accomplishments during this period would have been impossible. The other significant weapon for hydraulic research during this period was the advancement in experimental instrumentation for measuring, recording and analysing data. These two were enhanced by advances and developments in electronics.

In the 1990s, the fields of activity for hydraulic engineers will grow ever wider. The environmental problem has expanded to global scale; it now includes (1) the prediction on the effects of temperature rise due to the greenhouse effect caused by the increased CO_2 gas concentration, (2) counter-measures for the sea level rise due to the melting of glaciers, (3) those for floods and landslides caused by heavy rainfall in some parts of the earth, (4) at the other extreme drought and desertification in other regions caused by changes in the global climate, and (5) acid rainfall due to industrialization, motorization and/or deforestation.

This volume of the IAHR Hydraulic Design Manual Series contains, the theories, design formulas and engineering practices dealing with these broad problems in environmental hydraulics. The coordinator invited as contributors researchers with extensive experience with these problems from all over the world. For two reasons about half of the contributors are from Japan. On the one hand, a few of prominent researchers in the other countries invited initially were unable to respond, perhaps because of their busy schedules. On the other hand, Japan, a comparative small and densely populated country, has been rapidly industrializing since the early 1960s and it has suffered severely from various kinds of public nuisances, atmospheric pollution and water pollution. Because of the diligent endeavor of scientists and engineers from a variety of fields, many of these environmental problems have been resolved.

The authors who have contributed to this manual have donated their precious time, and their efforts will surely be appreciated by all that use the results of their work.

Lastly, all involved regret the many delays in completing this volume; although some contributors completed their parts very early, others were not able to do so (However, the whole manuscripts have been completed by April, 1990). The coordinator of the manual would like to thank to Professor Naudascher for giving him this opportunity and for his continual encouragement. He is also grateful to Professor McNown for his kind and devoted endeavor in the painstaking work of editing.

Water quality in lakes and reservoirs

Thermal stratification: Their prediction and control 1

YOICHI MIYANAGA
Hydraulics Department, Central Research Institute of Electric Power Industry, Abiko, Abikoshi, Japan

MASAHIRO TANAKA
Hydraulic Engineering Department, Kajima Technical Research Institute, Tobita Kyu, Chofu-Shi, Tokyo, Japan

MIKIO HINO
Department of Civil Engineering, Tokyo Institute of Technology, O-okayama, Meguro-ku, Tokyo, Japan
Presently: Faculty of Political Science, Chuo University, Hachiohji, Tokyo, Japan

1.1 INTRODUCTION

Water quality changes in lakes and reservoirs occur as the result of both material (nutrient, sediment or toxic) transport, and biochemical processes, mostly in a density stratified environment. Since the density gradient associated with the thermocline prevents the vertical transport of momentum, suspended materials, solutes, nutrients and biological species, it influences the dynamics of flow and the process of bio-chemical activities in a lake. In order to predict water quality change and to take appropriate measures, one often uses numerical models that take these processes into account. Hydraulic measures to control water quality such as selective withdrawal and artificial mixing can be applied to achieve various objectives, and they are of practical importance in the management of lakes and reservoirs. This section describes: (1) Predictive models for temperature variation and stratification in lakes and reservoirs; (2) Physics of density currents; (3) Theory and design procedures for selective withdrawal; (4) Artificial mixing as a measure for water quality control.

1.2 THERMAL STRATIFICATION PROCESS

Hydraulic and thermal characteristics of lakes and reservoirs are subject to heat and momentum transport between the atmosphere and the water body, to tributary inflows, to artificial intake and pumping, to the Coriolis force, etc.

One important phenomenon that affects both hydraulic and thermal characteristics of lakes and reservoirs is the occurrence of a thermocline where the vertical profile of temperature changes sharply. The thermocline limits the vertical transport of heat, momentum and material within the lake, and thus it separates the hydraulic, thermal and ecological characteristics of the lake into two parts. Its depth varies temporally owing to wind-induced mixing and heat exchange at the lake surface.

Thermal stratification is classified into seasonal stratification and diurnal stratification, according to the two cycles of solar radiation. Imberger & Hamblin (1982) suggested that the seasonal thermocline is an accumulation of past diurnal mixing events. In lakes located in middle latitudes, the usual depth of the seasonal thermocline is from 5 to 15 m with temperature differences of 10 to 20°C. The seasonal thermocline is formed in spring or early summer and remains until fall. In contrast, the diurnal stratification is so faint and unstable that it can be destroyed in a day by a mild wind with a speed of a few m/s. Therefore the diurnal stratification seems to have little relation to practical problems except that it is the seed of the seasonal stratification. However, in a turbid lake in which the water is not clear such as an eutrophic lake, a sharp thermocline can form, and the diurnal thermocline then sometimes remains for a few days. In such circumstances, the diurnal stratification modifies the flow pattern and affects dispersion processes in the lake (Ishikawa & Tanaka, 1990).

The characteristics of the seasonal stratification and the diurnal one are very different. If a lake is shallower than the usual depth of the seasonal thermocline, the lake is called 'hydraulically shallow' or simply a shallow lake. In a 'deep lake' the seasonal thermocline forms and remains for half a year. In a 'deep artificial reservoir', a special case of the 'deep lake', operations such as withdrawal from or pumping into a reservoir also affect significantly the hydraulic and thermal characteristics. Research work on the stratification in lakes and reservoirs has been mainly directed toward the seasonal stratification. In studies of the evolution of the seasonal thermal structure, individual diurnal excursions are averaged out and only the long term behavior is investigated.

In a shallow lake, a physical or numerical model for non-stratified water has been adopted in investigations of the hydraulic processes because the diurnal stratification is weak, e.g. in a 'shallow lake' the density profile of flow is approximately uniform over the entire depth, because wind-induced mixing extends to the lake bottom. Recently, however, the diurnal stratification has been studied in the field. Imberger and his coworkers (Imberger, 1985, Spigel et al., 1986) described the details of the mixing processes of the diurnal stratification and its energetics in their extensive work on lake dynamics. Ishikawa and his coworkers (Tanaka & Ishikawa, 1989; Ishikawa et al., 1989) investigated the diurnal stratification and its effects on wind-induced currents and water qualities in a shallow eutrophic lake.

1.2.1 *Types of layer mixing*

Wind mixing in a lake depends on wind strength, size of lake and stratification. Spigel and Imberger (1980) classified the types of layer mixing into four regimes

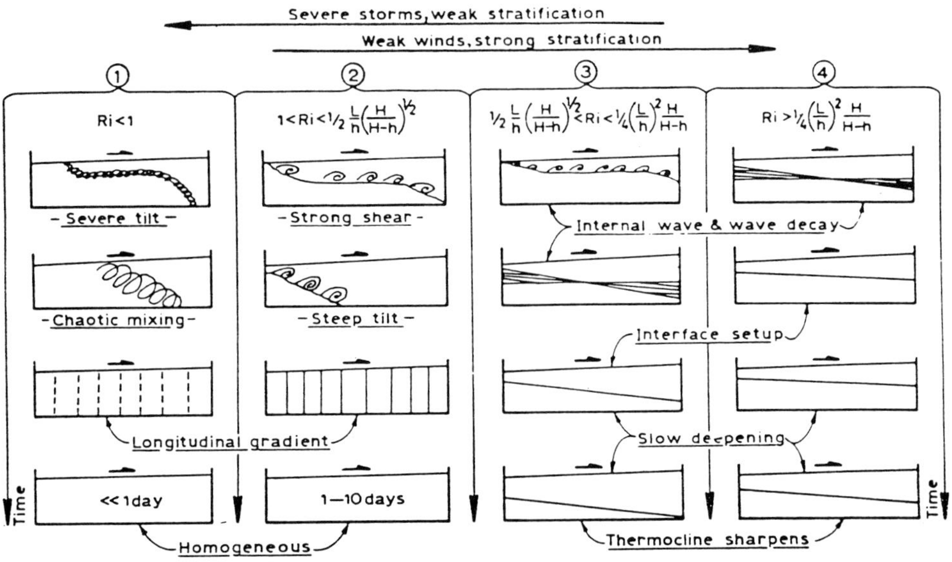

Figure 1.1. Overall features of the four regimes for a two layer basin (from Spigel & Imberger, 1980)

as shown in Figure 1.1; they used a mixed layer Richardson number and the lake's aspect ratio (the mixed layer thickness to length of lake).

The mechanisms whereby the thermal stratification in lakes and reservoirs is shifted vertically by wind are classified as DIM (Dynamic Instability Model) and TEM (Turbulent Erosion Model), according to the mixing process that occurs at the interface (Price et al., 1978; Spigel & Imberger, 1980). In the former, the mixing of the stratified interface is caused by the shear instability of wind-driven currents; in the latter, the mixing is caused by migration of turbulence produced at the water surface by wind.

DIM type entrainment is expected to be applicable in cases for which the wind set-up of the interface is not fully developed, while TEM is for the case in which the set-up is completed (Kranenburg, 1985). A criterion for these two types of mixing can be described by means of a modified Wedderburn number (W_d), which is the ratio of the time for DIM type entrainment to erode the stratification,

$$T_e = \frac{\sqrt{\varepsilon g h}}{U_*^2} H$$

to the time needed for the interfacial set-up to be developed

$$T_s = \frac{L}{\sqrt{\varepsilon g h}}$$

A criterion on the types of stratification is given as (Ishikawa & Tanaka, 1990)

$$W_d = \frac{\varepsilon g h}{U_*^2} \cdot \frac{H}{L} \begin{cases} < 1 \text{ (DIM)} \\ \gg 1 \text{ (TEM)} \end{cases} \tag{1.1}$$

where: ($\varepsilon = (\rho_{lower} - \rho_{upper})/\rho_{lower}$) is the relative density difference between upper and lower layers, h = thermocline depth, $U_* = \sqrt{\tau_0/\rho}$ = shear velocity of wind at the water surface (τ_0= wind shear stress on the surface), H = mean depth of lake or reservoir, L = horizontal scale of lake or reservoir.

1.2.2 *Entrainment law*

The hypolimnetic water is entrained into the epilimnion through the mixing at a thermocline. The process is expressed in terms of the entrainment velocity W_e

$$W_e = \frac{dh}{dt} \tag{1.2}$$

or alternatively in terms of the entrainment coefficient E

$$E = \frac{W_e}{V} = \frac{1}{V}\frac{dh}{dt} \tag{1.3}$$

where V is a reference velocity such as friction velocity U_* or flow velocity U_0. The idea of an entrainment coefficient was first used explicitly by Morton et al. (1956) for the treatment of vertical plumes. The concept was soon extended to turbulent entrainment at the interface of a two-layer stratified flows by Ellison and Turner (1959). Since then, a variety of experiments and theories have been presented.

It is now clear that the mechanism, and thus the entrainment formula should be divided into the two types mentioned in the preceding section. The entrainment coefficient is given as a function of Richardson number R_{io}^*, depending on the types of mixing, as

$$E \propto \begin{cases} R_{io}^{*\,-1} & \text{(TEM type)} \\ R_{io}^{*\,-1/2} & \text{(DIM type)} \end{cases} \tag{1.4}$$

where

$$R_{io}^* = \varepsilon g h / U_*^2 \tag{1.5}$$

The TEM type formula was derived from the concept of energy balance by Kraus and Turner (1967), and the DIM type formula was discussed first by Pollard et al. (1973) and later by Price (1979) and Kundu (1981). The entrainment of the DIM type has been shown to be larger by two orders of magnitude than that of the TEM type (Kranenburg, 1984, 1985).

Table 1.1. Empirical criteria for lake stratification

Parameter	Criterion	Reference
Fetch (F)	Thermocline depth (m) = $4[F(km)]^{1/2}$	Ragotzkie (1978)
$\alpha \left[\frac{\text{annual inflow}}{\text{lake volume}}\right]$	$\alpha < 10$ stratified $\alpha > 20$ homogeneous	Aki & Shirasuna (1977)
average densimetric Froude number (F_d)	$F_d < 0.01$ stratified $F_d > 0.01$ homogeneous	Iwasa et al. (1975)

1.2.3 *Empirical criteria of thermocline formation*

Empirical criteria summarized in Table 1.1 are useful in determining whether a thermocline will occur, without the need for solving the momentum and heat equations. The parameters used for the classification are the fetch for natural lakes, where the effects of inflow and outflow are negligible, the ratio of annual inflow volume to lake volume and the densimetric Froude number for artificial reservoirs.

1.3 NUMERICAL OR MATHEMATICAL MODELS FOR PREDICTION OF THERMAL STRATIFICATION

Water quality models of lakes and reservoirs are reviewed in considerable detail in Chapter 2. Further information is contained in Imberger and Patterson (1981, 1990). Thus, only the one-dimensional model that is in most common use is described in this section. Mathematical prediction models for thermal stratification in lakes and reservoirs can be divided into two groups; convection/diffusion models and integral models.

1.3.1 *Convection/diffusion models*

The characteristics of the temperature profile in a lake are predicted simply by using a one-dimensional (vertical) diffusion model that takes into account the absorption of the insolation by water (Dake & Harleman, 1969). The one-dimensional diffusion model may be refined by considering the role of convection and introducing the concept of 'effective diffusion' that represents eddy diffusion, adding to molecular diffusion (Orlob & Selna, 1970; Huber et al., 1972). The one-dimensional model has been improved for this case by incorporating density current theory by Huber et al. (1972), Iwasa et al. (1975) and Aki & Shirasuna (1977).

Diffusion model. The heat budget equation for the one-dimensional diffusion model is written as follows:

$$\frac{\partial T}{\partial t} = \frac{1}{\rho c}\frac{\partial \phi}{\partial z} + \frac{\partial}{\partial z}\left(K_z \frac{\partial T}{\partial z}\right) \tag{1.6}$$

where T = temperature, z = vertical distance, t = time, K_z = vertical diffusion coefficient, ϕ = heat flux of internally absorbed radiation, ρ = density of water and c = specific heat of water.

Convection/diffusion model. For reservoirs that are subject to significant inflow and outflow, estimate of convective heat transport is important. The heat budget equation for the one-dimensional convection/diffusion model is written as follows (Harleman, 1982):

$$\frac{\partial T}{\partial t} + \frac{1}{A}\frac{\partial}{\partial z}(Q_v T) = \frac{1}{A}\frac{\partial}{\partial z}\left(AK_z \frac{\partial T}{\partial z}\right) + \frac{B}{A}(u_i T_i - u_0 T) + \frac{1}{\rho c}\frac{\partial \phi}{\partial z} \tag{1.7}$$

where A = horizontal area, B = width, u_i = horizontal velocity of inflow, u_0 = horizontal velocity of outflow, T_i = inflow water temperature. The vertical flow rate Q_v is given by

$$Q_v(z, t) = B\int_0^z u_i(z, t)\mathrm{d}z - B\int_0^z u_0(z, t)\mathrm{d}z \tag{1.8}$$

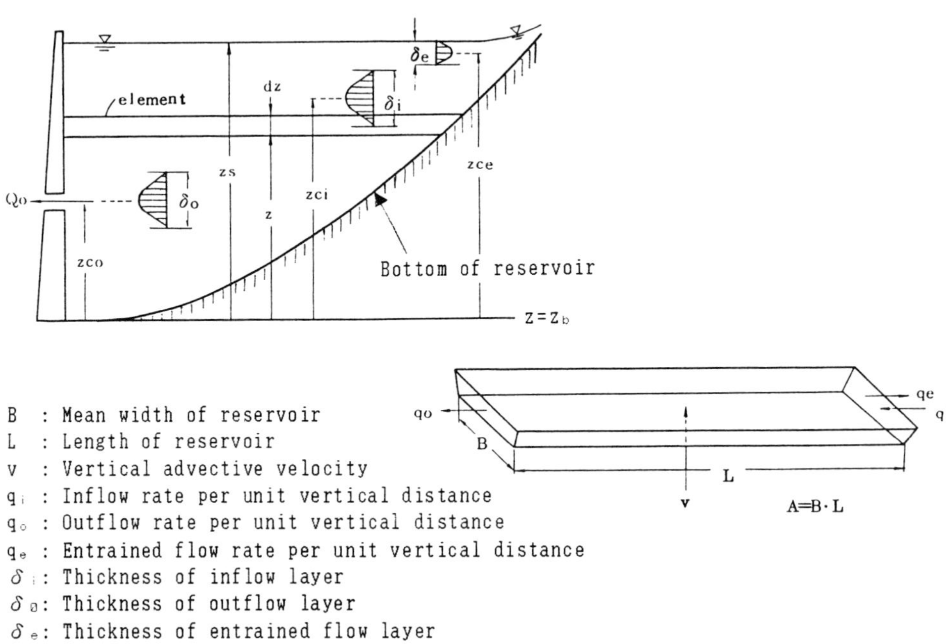

Figure 1.2. Schematic representation of one dimensional convection/diffusion model

The eddy diffusivity is given by

$$K_z = K_{z0}(1 + \alpha R_i)^{-1} \tag{1.9}$$

where K_{z_0} = eddy diffusion coefficient in the absence of stratification, α = a constant, and R_i = *a gradient Richardson number,*

$$R_i = \left(-\frac{g}{\rho}\frac{\partial \rho}{\partial z}\right)\left(\frac{\partial U}{\partial z}\right)^{-2} \tag{1.10}$$

Figure 1.2 shows the concept of a one-dimensional convection/diffusion model. This model can be used to predict the temperature profile by calculating heat exchange at the surface, absorption of insolation, and convective and diffusive transport of the heat.

1.3.2 *Integral models*

Integral or mixed-layer models were proposed by Kraus & Turner (1967), and followed by Stefan & Ford (1975), Sherman et al. (1978), Ford & Stefan (1980), Murota et al. (1985), Spigel et al. (1986), Imberger & Patterson (1990) and Bloss & Harleman (1979).

Integral (energy) models use the turbulent kinetic energy supplied directly or indirectly by the wind shear to work against gravity at the bottom of a mixed layer to entrain denser water into the mixed layer. The vertical one-dimensional turbulent energy budget for a column of fluid in lakes is considered. The increase in the mean potential energy of this column is assumed equal to the work done against gravity by the turbulent velocity fluctuations as presented originally by Kraus & Turner (1967). The terms in the equation can be parameterized based on the physical consideration as shown in Figure 1.3 to obtain the general entrainment law (Imberger et al., 1978),

$$\frac{1}{q_t}\frac{\mathrm{d}h}{\mathrm{d}t} = \frac{C_k - \dfrac{C_L R_S^3}{(R_i + R_s)^2}}{R_i(1 + \dfrac{C_T}{R_i} - C_s F^2)} \tag{1.11}$$

in which h = the thickness of the mixed layer, q_t = the intensity of the turbulent velocity fluctuations, R_i (= $\alpha \Delta T gh/q_t^2$) = the turbulent Richardson number, R_s (= Nh/q_t) = a type of gradient Richardson number, N $\left(= (-g/\rho\ \partial\rho/\partial z)^{1/2}\right)$ = buoyancy or Brunt-Väisälä frequency of the quiescent fluid just below the thermocline or mixed layer base, F $\left(= (\upsilon \cdot \upsilon/\alpha\Delta T gh)^{1/2}\right)$ = the layer internal Froude number, υ = the mean velocity of mixed layer, and C_k, C_L and C_s are dimensionless coefficients (Table 1.2).

A heat balance equation is necessary to close the integral model:

$$h\frac{\mathrm{d}T}{\mathrm{d}t} = \Delta T\frac{\mathrm{d}h}{\mathrm{d}t} - (\overline{w'T'})_{z=h} + \phi \tag{1.12}$$

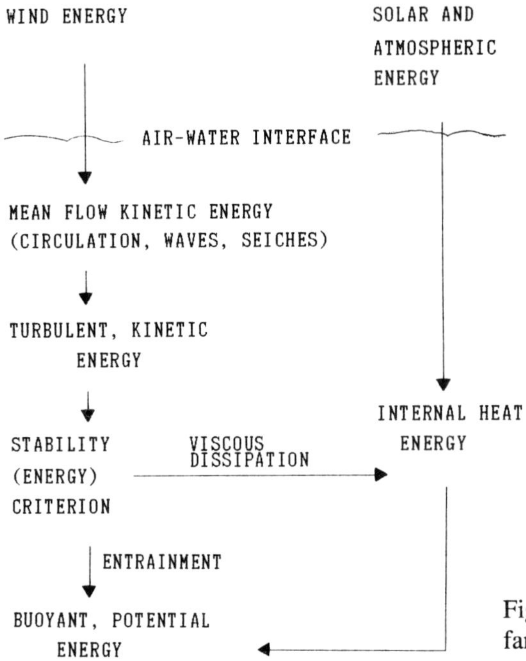

Figure 1.3. Energy flux diagram (from Stefan & Ford, 1975)

Table 1.2. Coefficients chosen by various authors (from Sherman et al., 1978)

Author	η	C_k	C_T	C_s	A	C_L
Zeman & Tennekes (1977)	2.2	0.5	3.6	$0.5R_s$	$\frac{q}{N}$ $\frac{q^2}{g\alpha\Delta T}$	0.024
Niiler (1975)	1.3	1.0	0	1	0	0
Mahrt & Lenschow (1976)	0	0.2	0	1	0	0
Kraus & Turner (1967)	1.4	1	0	0	0	0
Linden (1975)	1	$C_{Ko}(s/h)^{9/2}$*	0	0	1 cm fixed	0.01
Alexander & Kim (1976)	1	exponential	0	0	0	0
Imberger et al. (1977)	$\frac{1.1(1 + h/5)}{(1 + h/20)}$	$\frac{1.0}{(1 + h/5)^3}$	0	0	0	0
Best fit to experiments reviewed in Table 1.2.	1.8	0.38	2.25	0.3	$\frac{q}{N + g\alpha\Delta T/q}$	0.04

*s = stroke; C_{Ko} = constant for particular experiment.

where T = mean temperature in the mixed layer, ΔT = temperature difference between the mixed layer and lower layer, $-\overline{w'T'}$ = turbulent temperature flux. If h is defined as the depth of mixed layer base, the second term on the right hand side in the above equation can be neglected.

1.3.3 *Features of convection/diffusion models and integral models*

In applying the diffusion model, difficulties are encountered in estimating with sufficient accuracy the diffusion coefficient; i.e. the turbulent diffusion process is strongly suppressed at the thermocline or within a density stratification, and, on the contrary, is greatly enhanced by the mixing induced by large eddies due to the K-H (Kelvin-Helmholtz) instability phenomenon. If one employs instead the integral model or layer model, one has difficulty in applying the entrainment law. At this stage, however, the integral model is recommended in practical applications.

Some kinds of turbulence model (such as k-ε and LES, Large Eddy Simulation) have been utilized to simulate the mixed layer dynamics. But the numerical models based on the turbulence model require large amount of computer time and the closure assumptions would not always be valid. Rodi (1987) gave summaries of some kinds of the turbulence models.

1.4 PREDICTION OF DENSITY CURRENTS IN RESERVOIRS

If a fluid of a given density (cold river flow or flood flows with high concentration of sediment) enters into an almost stagnant fluid of slightly different density, it can

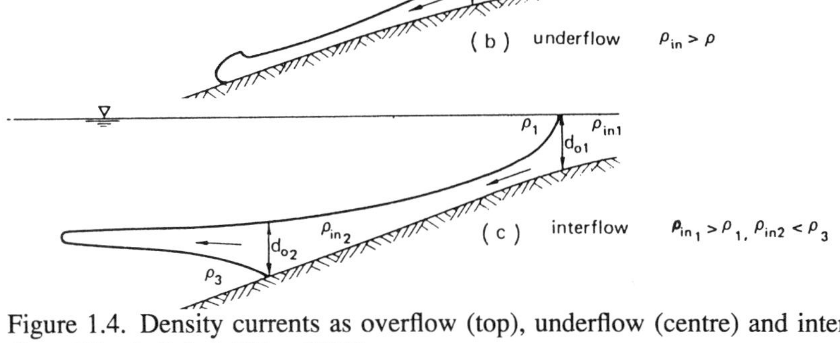

Figure 1.4. Density currents as overflow (top), underflow (centre) and interflow (bottom) (from Wunderlich & Elder, 1973)

undergo three different types of behavior, depending on the density difference, as shown in Figure 1.4 (Wunderlich & Elder, 1973). Such a flow in reservoirs and lakes is induced by the difference in specific weight and called by names such as density current, gravity current, underflow, or turbidity current (if it is caused by suspended sediment). The phenomena of density currents can be encountered in various fields, and it occurs in the natural environment and in laboratories (Turner, 1973; Simpson, 1987).

1.4.1 *Engineering implication of density currents*

A proper knowledge of the hydraulic behaviour of the density current through the reservoir is useful for (1) the proper instalation or operation of the opening gates to intake the water with desired temperature and water quality or to withdraw the turbid and muddy water from the reservoir, (2) the prediction and management of the water quality in lakes and reservoirs, and (3) the determination of the loss of storage capacity as a result of settling of suspended sediment.

1.4.2 *Plunging phenomenon*

When a stream flow enters the water body, it plunges underneath or rises buoyantly and spreads over it; which of these will occur depending on the density difference at a control section called the 'plunge point'. The density difference is due to either temperature, dissolved substances or the suspended solid content between river flow and ambient water in a reservoir. If an inflow has a density greater than that of the reservoir water, the inflow plunges under the free surface and forms an underflow (Task Committee, ASCE, 1963). The overflow, the counterpart to the plunging, occurs if an inflow with a lighter density separates from the bottom and remains on the surface of the reservoir.

The plunging phenomenon is important in reservoir water quality management and modeling because inflows that continue down along the bottom of a reservoir enhance the suspension of sediment and transport the material into various parts of stratified reservoirs. A steady plunging flow is shown in Figure 1.5. The flow will change from a homogeneous open channel flow to a gradually varied flow at the plunge point. Intensive mixing occurs at this point, changing both the volume and the density of the inflow. The temperature or density of the flowing water will be a weighted average of the temperature of the river inflow and that of ambient water current.

The plunging process is predicted theoretically on the basis of the momentum equation, including the effects of the shear stresses at the bottom and interface and of the dilution by the entrainment of the ambient fluid. Approximate solutions have been given by Singh & Shah (1971), Savage & Brimberg (1975), Hebbert et al. (1979) and Fukuoka et al. (1980). The most general solution of the momentum equation is that given by Akiyama & Stefan (1984), who applied the Ellison &

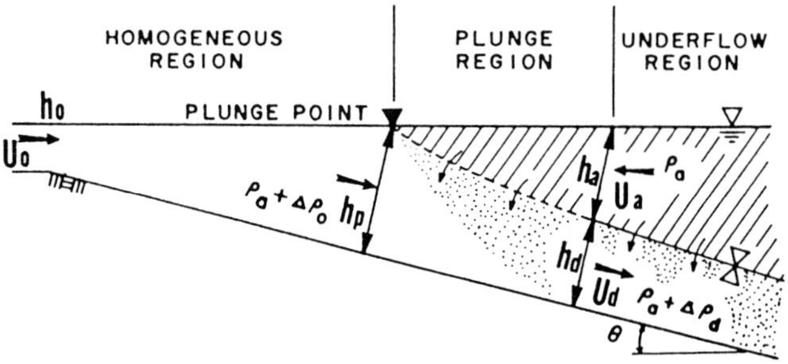

Figure 1.5a. Flow parameters of plunging flow

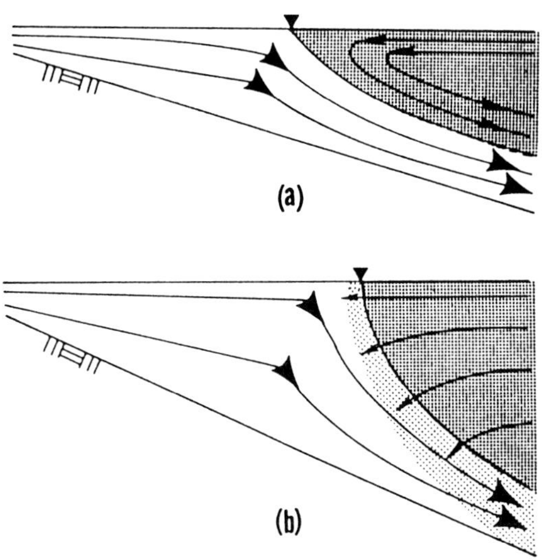

Figure 1.5b. Extreme cases of plunging flow: a) nonmixing on mild slope; and b) mixing on steep slope (from Akiyama & Stefan, 1984)

Turner (1959) model for slowly varied density currents with entrainment.

The densimetric Froude number at the plunge point, F_p, which characterizes the plunge depth h_p, is defined by Equation (1.13):

$$F_p = \left(\frac{q^2}{\varepsilon g h_p^3}\right)^{1/2} \tag{1.13}$$

where q = inflow flux per unit width, $\varepsilon(=\Delta\rho/\rho)$ = relative density difference between inflow and reservoir water. A simplified theory on the steady two-layered flow of two invisid fluids on a horizontal bed yields, from the momentum and mass conservation laws, the relation on the plunge depth, h_p, or the densimetric Froude number, F_p, as

$$h_p = \left(\frac{4q^2}{\varepsilon g}\right)^{1/3} \quad \text{or} \quad F_p = \frac{1}{2} \tag{1.14}$$

Actually, F_p is not a constant but is rather a parameter that depends on the other

Table 1.3. Summary of estimation for densimetric Froude number F_p at plunge point

Author	Estimation of F_p	Comment
Singh & Shah (1971)	$\left[1.3 + 1.85\left(\frac{q^2}{\varepsilon g}\right)^{-1/3}\right]^{-3/2}$	Experimental formula
Savage & Brimberg (1974)	$\frac{2.05}{1 + f_i/f_b}\left(\frac{I}{f_b}\right)^{0.478}$	Numerical formula
Fukuoka & Fukushima (1980)	$\left[\frac{1 - h_2^2 + k_2 I(1 + h_2)}{k_3 - 1/(1 - k_1 I)}\right]^{1/2}$	Theoretical study, h_2 is obtained from Benjamin's theory
Akiyama & Stefan (1984)	$\left[\frac{1+x}{2} + \frac{\gamma}{4} + \sqrt{\left(\frac{1+x}{2} + \frac{\gamma}{4}\right)^2 - \frac{x}{1+\gamma}}\right]^{-3/2} x^{1/2}$ $x = \alpha_2 I/(f_i + f_b)$ for mild slope $x = \alpha_1$ for steep slope	Theoretical study

Note: q = inflow rate per unit span, ε = density difference relative to ambient water, g = gravitational acceleration, I = bed slope, f_i = friction coefficient at the interface, f_b = friction coefficient at the bed, h_2 = underflow depth normalized by plunge depth, k_1, k_2, k_3 = theoretical coefficients, γ = rate of entrainment to inflow volume flux, α_1, α_2 = theoretical coefficients.

hydraulic properties. Table 1.3 summarizes various experimental and theoretical studies that evaluated F_p in Equation (1.13). Comparisons of the theories with a wide range of experimental (laboratory and field) data are contained in Akiyama & Stefan (1984).

1.4.3 *Progression of density current*

Non-uniform flow. After plunging, the density current or underflow progresses down along the bottom of reservoir until it finds a water layer for which the density is equal to that of the mixed inflow. The features of density current flows are like open-channel flows, but they differ essentially because of the reduced buoyancy and the entrainment of the stagnant fluid lying above them. Neglecting the entrainment, the basic equation of motion of gradually varied two-layered flow is derived (Schijf & Schönfeld, 1953; Turner, 1973) as

$$\frac{\mathrm{d}h}{\mathrm{d}x} = \frac{(\theta - C_D F^2)}{1 - F^2} \tag{1.15}$$

where h is the depth of underflow, x is the downslope distance, C_D is the drag coefficient ($\tau = C_D U^2$), θ is the angle of bottom slope, and the Froude number is defined

$$F = \frac{q}{(\varepsilon g h^3)^{1/2}} = \frac{U}{(\varepsilon g h)^{1/2}} \tag{1.16}$$

More general treatment which considers the entrainment of ambient water has been presented by Ellison & Turner (1959), extending the theory of plumes based on the momentum and mass consevation equations as

$$\frac{dh}{dx} = \frac{(2 - \tfrac{1}{2}\sigma_1 R_{io})E - \sigma_2 R_{io} \tan\theta + C_D}{1 - \sigma_1 R_{io}} \tag{1.17}$$

$$\frac{h}{3R_{io}}\frac{dR_{io}}{dx} = \frac{(1 + \tfrac{1}{2}\sigma_1 R_{io})E - \sigma_2 R_{io} \tan\theta + C_D}{1 - \sigma_1 R_{io}} \tag{1.18}$$

where σ_1 and σ_2 are profile constants defined by integrating across the plume,

$$\sigma_1 = \frac{1}{h^2 C_0}\int_0^\infty zC(z)dz, \quad \sigma_2 = \frac{1}{hC_0}\int_0^\infty C(z)dz \tag{1.19}$$

where C_0 is the average concentration and R_{io} is an overall Richardson number defined

$$R_{io} = \frac{\varepsilon g h \cos\theta}{U^2} = \frac{\varepsilon g h^3 \cos\theta}{q^2} = F^{-2} \tag{1.20}$$

where U (the mean velocity of the underflow) and h (the depth) are defined by integral

$$Uh = \int_0^\infty u(z)dz, \quad U^2 h = \int_0^\infty u^2(z)dz \tag{1.21}$$

The rate of entrainment, E, is a function of the Richardson number ($E = f_n(R_{io})$) and it depends strongly on the slope of bottom. Turner (1973) has shown that for density currents with no bottom friction and donimated by the turbulent entrainment

$$E = R_{io} \tan\theta \tag{1.22}$$

He also indicated that if the similarity of profiles is assumable, gravity currents will spread linearly with distance x, their mean velocity will be constant and proportional to $B^{1/3}$ (where $B = \varepsilon g h U$; the buoyancy flux per unit width), and the density difference will be inversely proportional to x.

The underflow depth h developed along the slope and the dilution, $\Delta Q/Q$, may be approximately estimated (for a triangular underflow; Hebbert et al., 1979)

$$h = \frac{6}{5}Ex + h_0, \quad \frac{\Delta Q}{Q} = \left(\frac{h}{h_0}\right)^{5/3} - 1 \tag{1.23}$$

where h_0 is the initial underflow depth, (after the plunge has been negotiated),

$$h_0 = \left(\frac{2}{F_0^2 \tan^2\theta} \cdot \frac{Q^2}{\varepsilon g}\right)^{1/5} \tag{1.24}$$

where $F_0^2 = u_0^2/\varepsilon g d_0$, d_0 = the hydraulic depth at the section of h_0.

Uniform flow. After the completion of plunging, an underflow approaches a uniform flow, since the weight of the underflow layer is balanced by the frictional forces, once if the entrainment becomes very small and therefore negligible. The

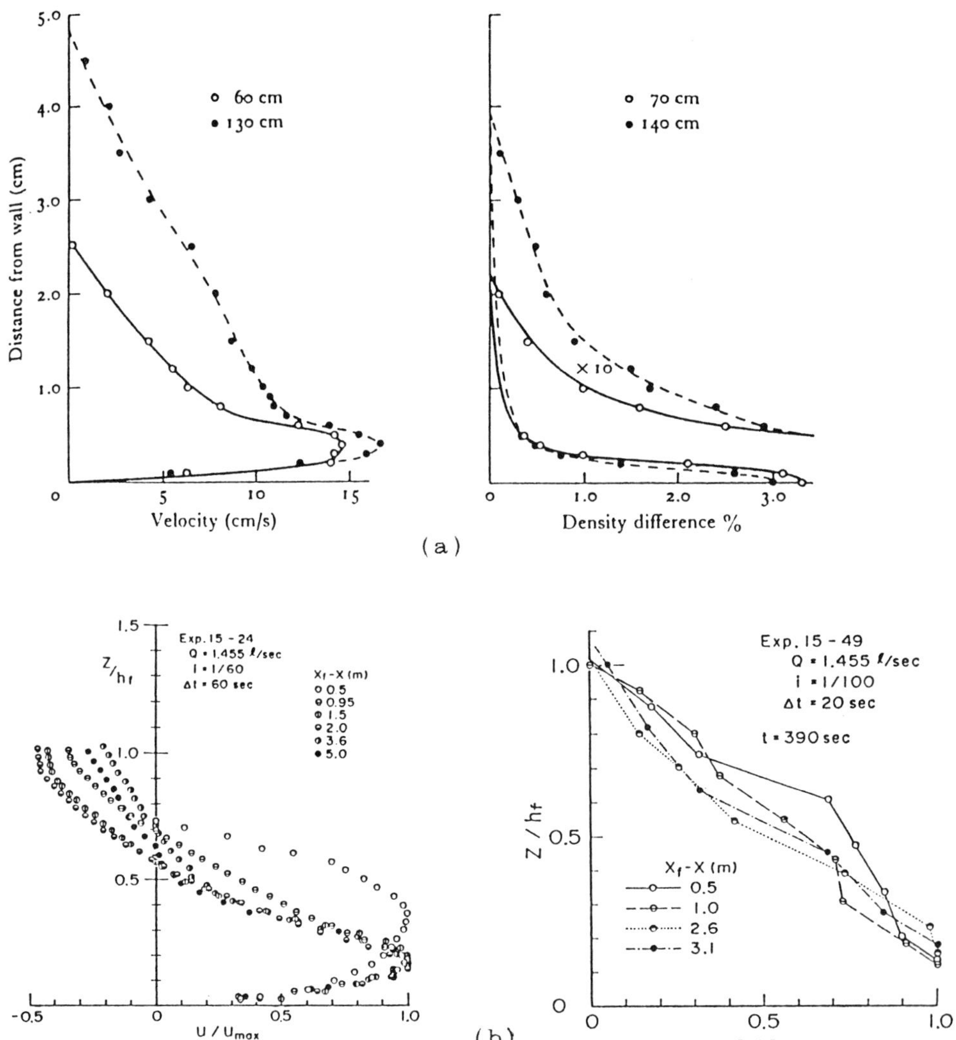

Figure 1.6. Velocity and concentration distributions in the uniform flow of density current, a) from Ellison & Turner, 1959; b) from Ashida et al., 1981)

actual profiles of velocity and concentration (or density) show some variation in the vertical direction. For simplicity, however, they are considered to be uniform. The velocity of a uniform flow becomes

$$U = \left[\frac{\sigma_2}{C_D}\varepsilon g h \sin\theta\right]^{1/2} \qquad (1.25)$$

where σ_2 indicates a profile constant defined by Johnson as above and h is the uniform flow depth.

Profiles of velocity and density distribution. Ippen & Harleman (1952) investigated a stationary, two-dimensional laminar underflow in an infinite homogeneous fluid. Stefan (1973) extended their theory to include the case of a finite ambient fluid.

For turbulent underflows, Hinze (1960) derived a formula of velocity distribution for the assumed density distributions, and Ashida & Egashira (1975) presented a model based on the mixing length theory. From this solution, the bed shear stress can be determined, on the basis of the assumption of similarity for the velocity profiles of the density current.

1.4.4 *Density current head*

The region of uniform flow for a density current is preceded by a frontal region known as the nose where mixing and entrainment of the ambient fluid occur. Von Karman (1940) argued theoretically that the angle at the nose of a density current is $\pi/3$ for an inviscid fluid. A non-dimensionalized shape of density current head (z/h_f vs. x/h_f, where h_f is the maximum height of the head, the subscript f refers to the head) has been obtained theoretically for a two-layer inviscid fluid by Benjamin (1968) and for flows with sufficiently large Reynolds number ($R_f = U_f h_f/\nu > 10^3$, U_f = velocity of the head), experimentally (Figure 1.7) by Ippen and Harleman (1952), Keulegan (1958), Middleton (1966), Hirano & Hadano (1981) and Altinakar et al. (1990). Because of the no-slip condition at the bed of a real fluid, the foremost point of the head (i.e. nose) is elevated.

The maximum height of the head is twice the depth of the uniform flow, and its velocity is less than that of the uniform flow. The maximum height of the head (and the height of underflow depth) increases proportionally as it progresses downwards due to the supply of volume from the uniform flow that follows the head and by the entrainment of ambient fluid (Theory: Hebbert et al., 1979; Hirano & Hadano, 1981; Experiment: Fukuoka et al., 1977, Britter & Linden, 1980; Hirano & Hadano 1981; Altinakar et al., 1990).

Irrigular fluctuation of the rate of increase in velocity and height of the head which may be caused by the merging of vortices within the head has been reported by Fukuoka et al. (1978), Hirano & Hadano (1981) and Altinakar et al. (1990).

With increase in the slope angle, the rate of change in h_f increases

$$\frac{d}{dx}\left(h_f\right) = 0.23\theta \tag{1.26}$$

where the slope angle θ is in radians.

Britter & Linden (1980) showed on the basis of their experiment that the

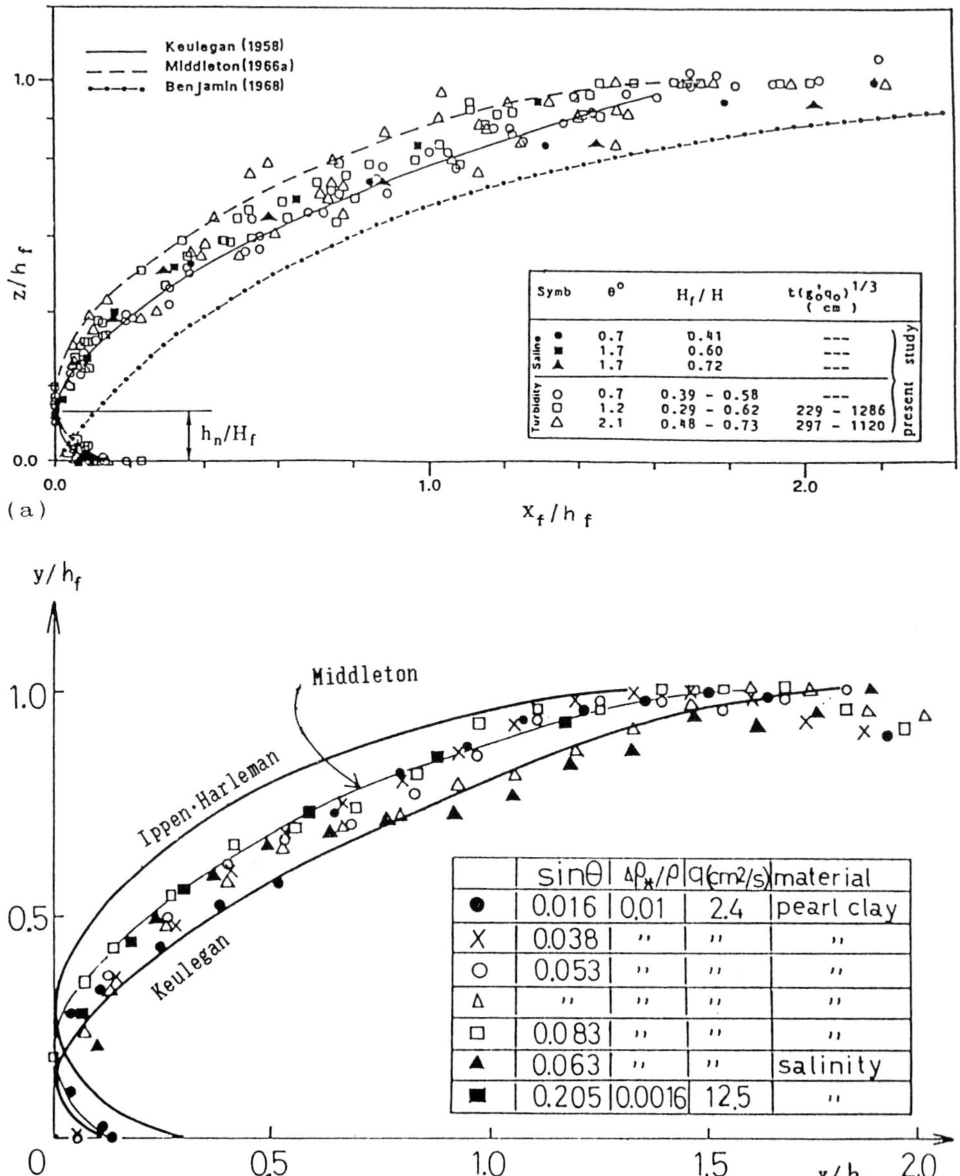

Figure 1.7. Shape of the density current head: a) from Altinakar et al. 1987; b) from Hirano & Hadano, 1981)

velocity of the head U_f is

$$\frac{U_f}{(\varepsilon g q)^{1/3}} = 1.5 \pm 0.2 \tag{1.27}$$

where q = discharge per unit width of turbid flow. Equation (1.27) is valid for slope angles between 5-90°. Altinakar et al. (1987) pointed out that the left-hand side of Equation (1.27), i.e. the value of the dimensionless head velocity approaches to the ratio of the depth of the turbidity current to the total depth.

The head velocity for a conservative (no-entrainment) density current is given as a function of the initial buoyancy flux $B_0 = \varepsilon g q_0$, the slope angle θ and the Reynolds number *Re*. For sufficiently large Reynolds numbers (> 10^3), the head velocity U_f is expressed as (Britter & Linden, 1980; Faust et al., 1983):

$$U_f = (\varepsilon g q_0)^{1/3} f_n(\theta) \tag{1.28}$$

If the density current entrains liquid from the ambient fluid, the velocity of the advancing head changes depending on the initial value of the densimetric Froude number F_0 and the entrainment coefficient E. The fact that the velocity of head increases if the value F_0 is small, and vice versa, and tends asymptotically to a constant value is shown theoretically and experimentally by Hirano & Hadano (1981). From the tendency of the velocity of head, the asymptotic behaviour of the head is derived (Tamai, 1980):

$$\frac{h_f}{h_{f0}} \propto \frac{x}{x_0}, \quad \frac{U_f}{U_0} \sim \text{const}, \quad \frac{\Delta\rho}{(\Delta\rho)_0} \propto \left(\frac{x}{x_0}\right)^{-1} \tag{1.29}$$

where the variables with subscript 0 means the values at a reference location. These relations are quite the same as those of the underflow mentioned on Page 15.

1.4.5 *Intrusion*

After reaching a level in the ambient water with a density equal to its own, the inflow penetrates horizontally as an intruding plume. The physics of an intruding layer is discussed by, for instance, Wu (1969), Maxworthy (1972), Masuda & Nagata (1974), Manins (1976), Kao (1977), Toda & Nagata (1980), Amen & Maxworthy (1980), Fukuoka & Fukushima (1980), Fukuoka et al. (1980) and Faust et al. (1983).

For the non-viscous case, a general formula on the steady propagation speed of the intruding plume in which the ambient and the current are both linearly stratified is derived by Kao (1977) as

$$U = \left[(\beta - \beta_d)\frac{h_d^2}{4}\right]^{1/2} \tag{1.30}$$

where

$$\beta = -\frac{1}{\rho_0}\frac{d\rho_a}{dz}, \qquad \beta_d = -\frac{1}{\rho_0}\frac{d\rho_d}{dz} \tag{1.31}$$

and $\rho_a(z)$ = ambient density, $\rho_0 = \rho_a(0)$, ρ_d = density of intruding plume, h_d = intruding plume depth.

For the case of a current of uniform density ρ_d flowing along the interface of two layers of uniform densities ρ_1 and ρ_2, the propagation speed is

$$U = \left(2\frac{\Delta\rho}{\rho_2} rgh_d\right)^{1/2} \tag{1.32}$$

where $r = (\rho_2 - \rho_d)/(\rho_2 - \rho_1)$.

From the dimensional analysis, the velocity of the nose and the thickness of the intruding layer are given as (Manins, 1976).

$$U = (qN)^{1/2} f(Re, \tau, \psi) \propto (qN)^{1/2} \tag{1.33}$$

$$h_d = \left(\frac{q}{N}\right)^{1/2} g(Re, \tau, \psi) \propto \left(\frac{q}{N}\right)^{1/2} \tag{1.34}$$

where

$$N = \left(-\frac{g}{\rho_0}\frac{\partial\rho}{\partial z}\right)^{1/2} = (\beta g)^{1/2} \tag{1.35}$$

N means the environmental buoyancy frequency or Brunt-Väisälä frequency, q is the source volume flux per unit breadth, the intrusion Reynolds number is defined as $Re = q/\upsilon$, $\tau = Nt$ is the buoyancy time and ψ $(= h_d/H)$ is the ratio of the intrusion layer thickness to the depth of the ambient fluid. For a steady state range and for large Reynolds numbers, the above expressions reduce to the right hand side expressions. Both these relations ($U \propto (qN)^{1/2}$ and $h_d \propto (q/N)^{1/2}$) are derivable by the transformation from Kao's theory when β_d reduces to zero. From the above relations, (1.33) and (1.34), it is evident that the internal Froude number Fr $(= U/Nh_d)$ is of order unity. For the estimation of the layer thickness of selective withdrawal which will be described in the next section, the formula on the intrusion layer thickness can be applied.

For a strong stratification ($Ri \gg 1$) in viscosity-buoyancy range, the intruding length L is related with the elapsed time as (Maxworthy, 1972)

$$L = C\left(-\frac{1}{\rho_0}\frac{\partial\rho}{\partial z}\right)^{1/6} Q^{2/3} t^{5/6} \tag{1.36}$$

The relation is verified experimentally by Masuda & Nagata (1974) and Tamai (1980, from reanalysis of the data by Fukuoka et al. (1978)), but Manins (1976) could not confirm the relation.

At the other end of the reservoir, e.g. the upstream face of a dam, the intruding layer or underflow is reflected and it propagates upstream as a surge with a thickness twice that the intruding layer had initially. The shapes of the propagating density underflow at 2-min. intervals are shown, from the numerical simulation

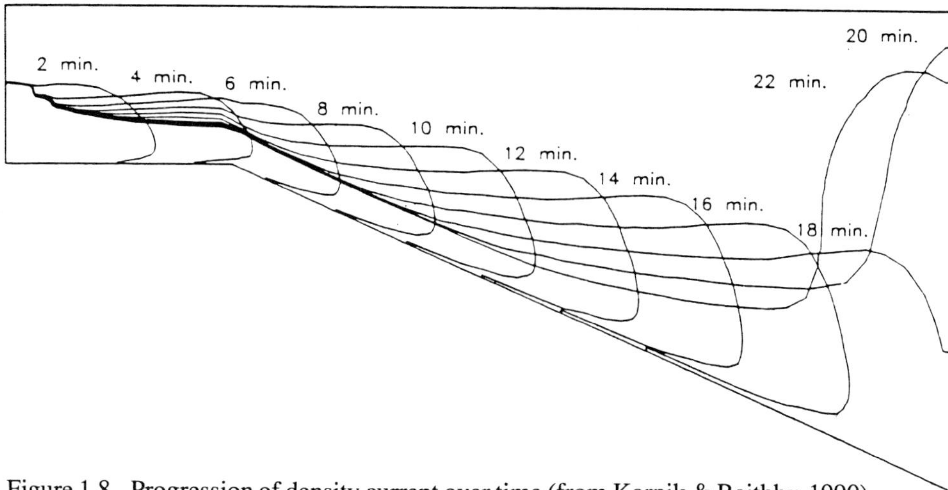

Figure 1.8. Progression of density current over time (from Karpik & Raithby, 1990)

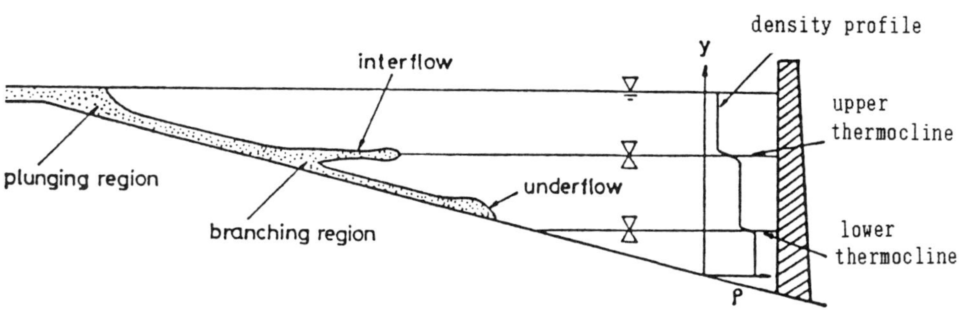

Figure 1.9. Classification of turbidity currents in a reservoir (Fukuoka & Fukushima, 1980)

by the laterally averaged hydrodynamic model (LAHM) of Karpik & Raithby (1990), in Figure 1.8 where the vertical scales have been expanded for clarity.

The actual behaviour of the density current is more complicated owing to the thermal stratification in the reservoir and the density profile of the turbidity current itself. In particular, the flow bifurcates at the elevation of the thermocline in the reservoir. Patterns of density currents in the reservoir are classified as shown in Figure 1.9 by Fukuoka & Fukushima (1980) who proposed an analytical model for each flow pattern.

1.4.6 *Density currents into lake and diverging channel*

In practical situations, the phenomenon must be treated as a three-dimensional one; that is a narrow river flows into a much wider and almost stagnant body of water (in the case of lakes) or into a diverging channel (in the case of reservoirs

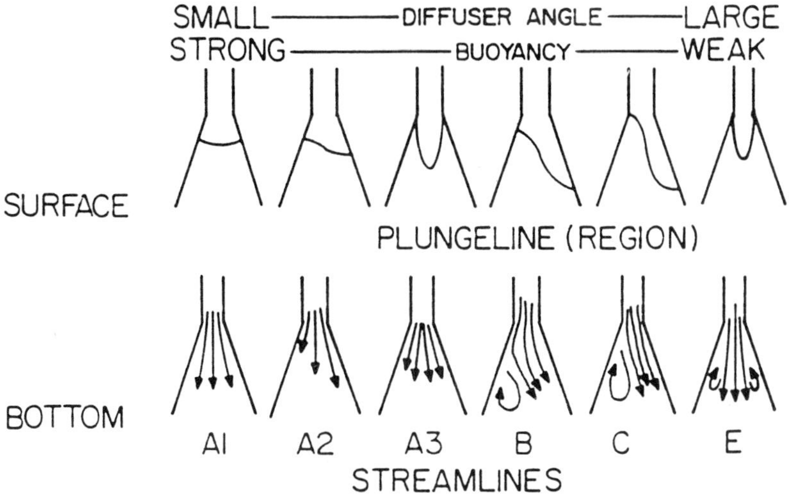

Figure 1.10. Principal regimes of negatively buoyant flow into horizontal divering channel (from Johnson et al., 1987a)

formed by the construction of a dam on a river). Experimental and analytical studies of this more complicated case have been performed for the plunging in lakes by Hauenstein & Dracos (1984) and for the case of diverging channel by Fukuoka et al. (1983), Hirano & Hadano (1983), Akiyama & Stefan (1988) and Johnson et al. (1987a,b, 1989), (Figure 1.10).

Numerical simulations of density currents in reservoirs have been carried out, [3-D models: Spraggs & Street (1975) and Waldrop & Tatom (1976), 2-D models: Edinger & Buchak (1979), Norton et al. (1973), Karpik & Raithby (1990), Watanabe & Fukushima (1990) and Yoon et al. (1990)].

1.4.7 *Sediment transport by turbidity currents*

For sediment transport, the motion of the head of the turbidity current is of primary importance. Field data available on the sediment transport due to the underflow are limited. Howard (1953) showed from the observations in Lake Mead, USA, a reservoir extending some 120 km from Hoover dam on the Colorado river, that appreciable underflow occurred although the relative density difference was low enough as $\Delta\rho/\rho = 0.0005$. In that case, the suspended particles consisted mainly of wash load for which the concentration by weight was only approximately 0.08%, and the sediment transport rate was estimated to be 1.5×10^4 tons/day.

Ashida & Egashira (1975) studied the turbidity current generated in Kurobe Reservoir in August 1969 using their model; the conditions were a depth of 1.2 m, a maximum velocity of 2.7 m/s and a friction velocity at the bed of 11 cm/s. They concluded that this turbidity current played an important role in the sediment transport in the reservoir.

Lambert & Giovanoli (1987) observed a turbidity current in the Rhone River Delta in Lake Geneva in June 1985. The peak velocity observed was 0.9 m/s at 4.5 m above the bottom of the lake, where the total depth was 150 m. They concluded that this underflow had caused a landslide of accumulated deposits in the Rhone River Delta.

Observation of density currents on the Nurek hydrostation on the Vaksh river, USSR was reported by Mukhamedov et al. (1972).

1.5 SELECTIVE WITHDRAWAL

1.5.1 *Purpose*

Selective withdrawal was originally developed as a measure for withdrawing water at a desired temperature, and it was accomplished by selecting the proper intake level. It is also applied in such other types of water quality control as (1) reduction of nutrient or sediment accumulation in a reservoir by selecting an intake at the level with a high concentration of nutrient or sediment and (2) intake of water with low turbidity. In Japan, turbidity persistence in reservoirs caused by floods and landslides has created serious environmental problems, the fine clay particles supplied from the watershed during a flood remain in suspension in a reservoir for a long time.

Selective withdrawal has been used at many reservoirs, and its operation has been refined both theoretically and empirically. The design technology is by now well supported by field experience.

1.5.2 *Theory on selective withdrawal*

Selective withdrawal from stratified fluids

The flow to a line sink or a point sink in a stratified fluid has been solved analytically for the two-dimensional case (for line sinks) by Yih (1958), Kao (1970), Imberger (1972), Imberger et al. (1976) and Ashida & Egashira (1978), and for the axisymmetrical case (for point sinks) by Hino & Onishi (1967, 1969b, 1970), Lawrence (1980) and Ivey & Blake (1985). The flow regimes are divided into two categories; Inertial-buoyancy and viscous-buoyancy regimes (Imberger, 1972).

From these theories, the thickness of the withdrawal layer can be determined for the region of inertia -buoyancy balance in terms of the intake parameters (F_c and G_c) as follows:

$$\text{line sink:} \quad \delta = Fc^{-\frac{1}{2}} \left(\frac{q}{\sqrt{\beta g}} \right)^{\frac{1}{2}} \tag{1.37}$$

$$\text{point sink:} \quad \delta = Gc^{-\frac{1}{3}} \left(\frac{Q}{\theta\sqrt{\beta g}} \right)^{\frac{1}{3}} \tag{1.38}$$

where δ = withdrawal layer thickness, q = withdrawal discharge per unit width, Q = withdrawal discharge, Fc = critical Froude number for the two-dimensional case, Gc = critical Froude number for the axisymmetrical case, β ($= -1/\rho \, d\rho/dz$) = gradient of density stratification and θ = horizontal angle of water intake ($\theta = 2\pi$, if the intake withdraws water axisymmetrically from all direction; $\theta = \pi$, if the intake is installed on a wall). In practical cases of selective withdrawal, the formula for point sinks, Equation (1.38), is effective.

For a withdrawal from the bottom of the lake, the quantities q (or Q) and δ must be changed to $2q$ (or $2Q$) and 2δ, respectively.

The theoretical predictions of the critical values in Equations (1.37) and (1.38) are

$$\text{link sink:} \quad F_c = \frac{1}{2\pi}$$

$$\text{point sink} \quad G_c = 0.04$$

Figure 1.11 shows a comparison between measurements and theoretical predic-

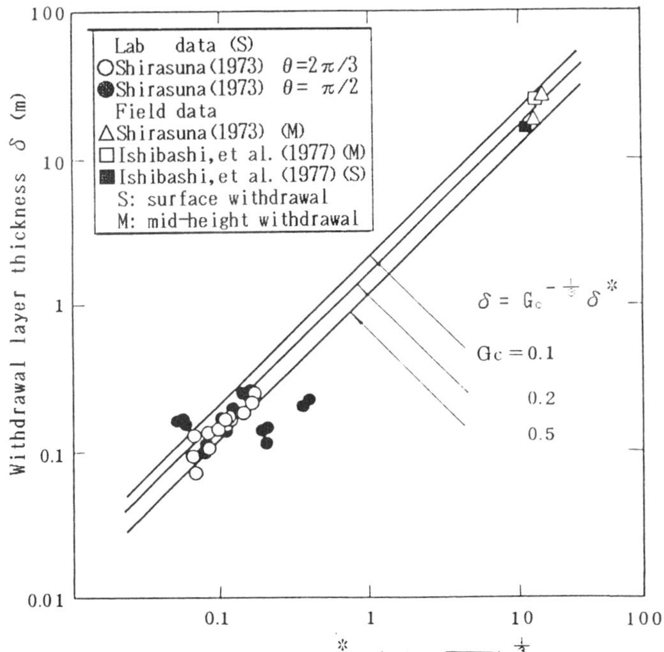

Figure 1.11. Withdrawal layer thickness δ as a function of flow parameters and the critical Froude number G_c for the axisymmetric case (Miyanaga, 1988)

Table 1.4. Values of critical Froude number G_c for selective withdrawal based on experiments

Withdrawal condition	Froude Number*	Reference
Point sink at bottom	$G_{cb} = 0.16$	Hino & Furusawa (1969)
Point sink at mid-height	$G_{cm} = 0.04$	Hino & Furusawa (1969)
Point sink at bottom	$G_{cb} = 0.324$	Aki & Shirasuna (1977)
Point sink at mid-height	$G_{cm} = 0.134$	Aki & Shirasuna (1977)

* Theoretically, $G_{cm} = G_{cb}/4$

tions for the thickness of a withdrawal layer. They are in a comparatively good agreement. The theoretical value of Gc is 0.04 for withdrawal from the middle layer, and it compares well with values found in practice. However, Aki & Shirasuna (1977) on the basis of field and laboratory experiments, recommended the values of G_c listed in Table 1.4. These are larger than the ones obtained from laboratory experiments by Hino & Furusawa (1969). The differences may be caused partly by the different definitions of withdrawal layer in these experiments and partly by the effect of intake geometry. Onishi et al. (1986) investigated the effect of intake geometry and proposed an evaluation method in terms of R/δ and H/δ (where R is radius of intake and H is height of intake) as shown in Figure 1.12.

The design criteria for the withdrawal layer half-thickness $\delta_{1/2}$ recommended by Imberger (1972), Imberger et al. (1976) and Hocking et al. (1988) are as follows; for the case of inertia-buoyancy balance:

$$\text{line sink:} \quad \delta_{1/2} = 2\left(\frac{q}{N}\right)^{1/2} \quad \left(q/(NL^2\upsilon^2)^{1/3} > 1\right) \tag{1.39}$$

$$\text{point sink:} \quad \delta_{1/2} = \left(\frac{Q}{N}\right)^{1/3} \quad (Q/\upsilon L > 1) \tag{1.40}$$

or for the case of viscosity buoyancy balance,

$$\text{line sink:} \quad \delta_{1/2} = 2.8\upsilon^{1/3}\left(\frac{L}{N}\right)^{1/3} \tag{1.41}$$

$$\text{point sink:} \quad \delta_{1/2} = 2.9(\upsilon D)^{1/6}\left(\frac{L}{N}\right)^{1/3} \tag{1.42}$$

where L is the length of the reservoir, D is the diffusivity and N is the buoyancy frequency.

For the steady-state axisymmetric withdrawal, Ivey & Blake (1985) introduced a transition parameter denoted by

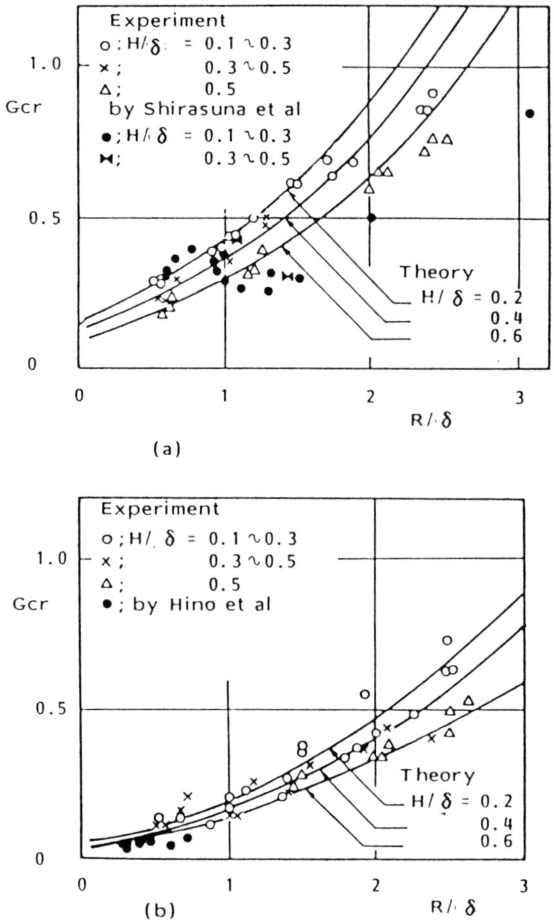

Figure 1.12. Effect of height (H) and radius (R) of intake on the critical densimetric Froude number (Onishi et al., 1986). a) Upper layer withdrawal; b) Middle layer withdrawal

$$S = \left(Q^2 N / \upsilon^3\right)^{1/15} \tag{1.43}$$

and they showed that the regime is in the inertial-buoyancy balance for $S > 3$ and the layer is viscous-convective if $S < 2$. They discuss further the classification of the flow regime.

The formula by Hino & Onishi (1969a), Equation (1.38), reduces to $\delta = 1.58(Q/N)^{1/3}$ which compares well with the formula recommended by Ivey & Blake:

$$\delta_{1/2} = 0.71\,(Q/N)^{1/3} \tag{1.44}$$

Selective withdrawal from two-layered fluids

Selective withdrawal from two-layered fluids has been discussed by Craya (1949), Faust et al. (1983), Harleman (1965), Harleman et al. (1958), Jirka (1979), Jirka & Katavola (1979) and Senshu (1967).

For the case of the withdrawal from discontinously stratified fluids (two-layered fluids), the upper layer is drawn down at the lower layer depth *d*. Prior to drawdown, the withdrawn fluid comes entirely from the lower layer. The critical conditions for the axisymmetric drawdown from a point sink are given as follows:

Inertia-buoyancy balance:

$$d = c_1 \left(\frac{Q^2}{\varepsilon g} \right)^{1/5} \qquad (S > S_{crit}) \tag{1.45}$$

Viscous-buoyancy balance:

$$d = c_2 \left(\frac{\upsilon Q}{\varepsilon \; g} \right)^{1/4} \qquad (S < S_{crit}) \tag{1.46}$$

where the constants of proportionality c_1 and c_2 are $c_1 = 0.69$ (theoretical value given by Craya, 1949) $c_1 = 0.81$ (experimental value by Harleman et al, 1958) $c_2 = 2.1$ (Ivey & Blake, 1985); and the transition criterion S is defined by:

$$S = \left(Q^3 \varepsilon g / \upsilon^5 \right)^{1/20} \tag{1.47}$$

and

$$S_{crit} = 2 \sim 5.$$

1.5.3 *Design procedure*

Figure 1.13 is a design flow chart for selective withdrawal. Water quality

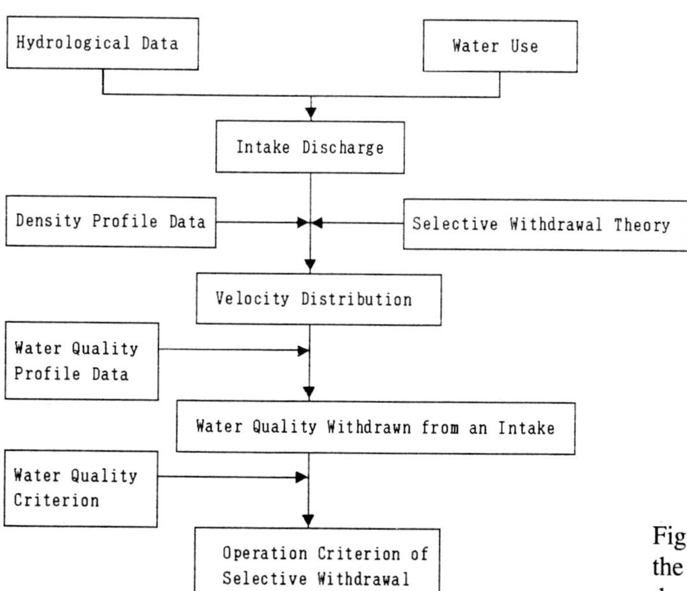

Figure 1.13. Flow chart for the design of a selective withdrawal system

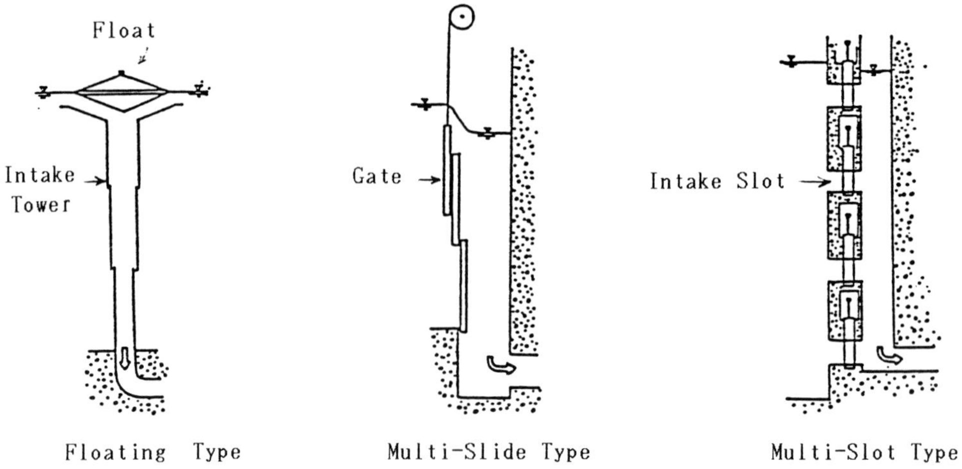

Figure 1.14. Typical designs of selective withdrawal systems used in Japan

withdrawn from an intake is determined by the distribution profiles of the water quality and velocity. Water quality distribution profile must be measured directly or predicted numerically. The intake velocity distribution is obtained from the theory as described in the preceding section.

Figure 1.14 shows designs of three typical selective withdrawal systems that are used in Japan. The floating type maintains surface withdrawal because it follows the fluctuations of the water-level. It is used mostly for avoiding colder water releases during the irrigation season. The multi-slide type is designed to allow one to select arbitrarily an intake level by adjustment of the gates. An old intake is often replaced by this type whenever existing intake structures are being reconstructed. Although the multi-slot type has several fixed levels of intake, it cannot provide an arbitrary intake level. Nevertheless, it has the advantage that it is the simplest to maintain.

1.6 ARTIFICIAL MIXING AND DESTRATIFICATION

1.6.1 *Purpose*

Artificial mixing for destratification (destruction of thermal stratification by hydraulic measures) has various applications: (1) controlling algal growth in the water column, (2) eliminating hypolimnetic oxygen depletion and preventing the release of nutrients and heavy metals (iron and manganese) from sediment soils under anaerobic conditions by supplying air into the deeper parts of lakes so as to keep the reservoir water aerobic and (3) preventing ice-formation both for fish protection and for protection of withdrawal structure (from the pressure exerted by ice slab). Extensive field experiments have been performed since the 1950s on the improvement of water quality in lakes and reservoirs by artificial mixing.

Both mechanical mixers and rising plumes of bubbles from compressed air systems can be used as mixing devices. Compressed air systems have been used more extensively than mechanical devices because of their high efficiency of mixing and simplicity of operation. An air-bullet type of device called 'air-lifter' has been proposed; it discharges an air bullet intermittently through a long vertical shaft and operates efficiently (Kojima, 1982, 1988). The method of hypolimnetic aeration is also effective (Bernhardt, 1967; Bernhardt & Wilhelms, 1975; Fast & Lorenzen, 1976). These facilities have been installed in various eutrophic reservoirs with a considerable degree of success for such water quality improvements as the elimination of mold odor, suppression of algal growth and red tide.

1.6.2 *Theory on air-bubble plume*

The flow induced by an air-bubble plume was analyzed initially for pneumatic breakwaters by Kurihara (1954, 1965), followed by Bulson (1961) and, more generally, by Kobus (1968), Cederwall & Ditmars (1970) and Goossens & Smith (1975). These theories are developed based on the theory of the forced plume (Rouse et al., 1952 and Morton et al., 1956), treat the bubble plume as a buoyant fluid and allow one to solve the momentum equation on the assumption of similarity of velocity distribution in a plume.

The solution for the ratio of water volume flux to air discharge is as follows:

$$\frac{Q_w}{Q_a} = f_1 \sigma b \left[\pi P_{atm} ln \left\{ \frac{H_*}{(H_* - z)} \right\} / \{ 2 f_2 \rho u_b Q_a \} \right]^{1/2} \tag{1.48}$$

$$H_* = H + \frac{P_{atm}}{\rho g} \tag{1.49}$$

where Q_w = water volume flux, Q_a = air volume flux, σ = ratio of the width of momentum spreading to the width of air-bubble plume, b = width of air-bubble plume, P_{atm} = atmospheric pressure, z = height above the air outlet, ρ = density of water, u_b = average rising speed of air-bubbles, f_1, f_2 = coefficients defined from the function of velocity distribution, H = depth of air outlet.

Figure 1.15 shows that the results of theory correspond well with those of experiment (Kobus, 1968).

Analysis and model experiment on the air-bullet type of device have been made by Asaeda & Nakai (1989). A numerical simulation of destratification by aeration or flood has been given by Iwasa et al. (1977, 1986). A method of destratification by propeller (or mixer) and its 2-dimensional numerical simulation were discussed by Busnaina et al., (1981). A low-energy propeller positioned just below the water surface provides a downward-directed jet of water and thereby locally mixes the interface.

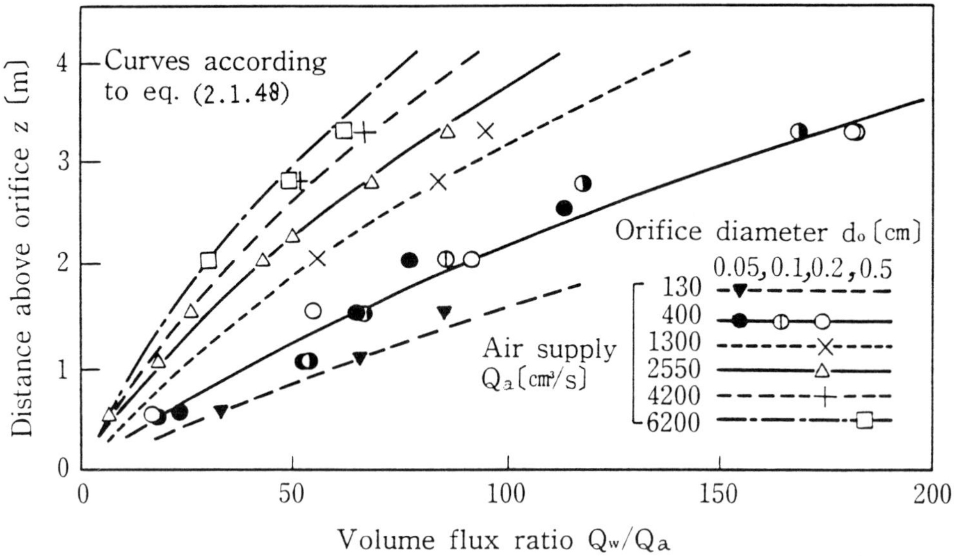

Figure 1.15. Volume flux ratio for a single orifice (Kobus, 1968)

1.6.3 *Principle of algal growth control by air-bubble plume mixing*

Algal growth control by mixing has been explained theoretically by Lorenzen & Mitchell (1973). They included the effects of light limitation and nutrient depletion. The photosynthesis necessary for algal growth is hindered through dispersion of algae below the photic layer, and the algal growth rate is suppressed by lowering the water temperature in the upper layer through mixing with the hypolimnetic cool water. This theory provided the relationship between the mixed depth and the algal biomass shown in Figure 1.16.

1.6.4 *Design procedure*

The US Environmental Protection Agency (EPA) published a manual (Lorenzen & Fast, 1977) describing the design of two major artificial mixing methods – destratification and hypolimnion aeration. The basic design flow chart is given in Figure 1.17. Air flows required to mix lakes of various depths can be determined from Table 1.5. The mixed depth required to control algal growth is determined from Figure 1.16 based on theory. The effectiveness of artificial mixing depends on such factors as the levels at which the mixing devices are placed, the period of operation and both hydrological and meteorological conditions. In order to take these factors into account, one must analyze by means of a numerical model the flows caused by inflow, outflow and the artificial mixing. The US EPA manual proposes a numerical model based on the Kobus theory.

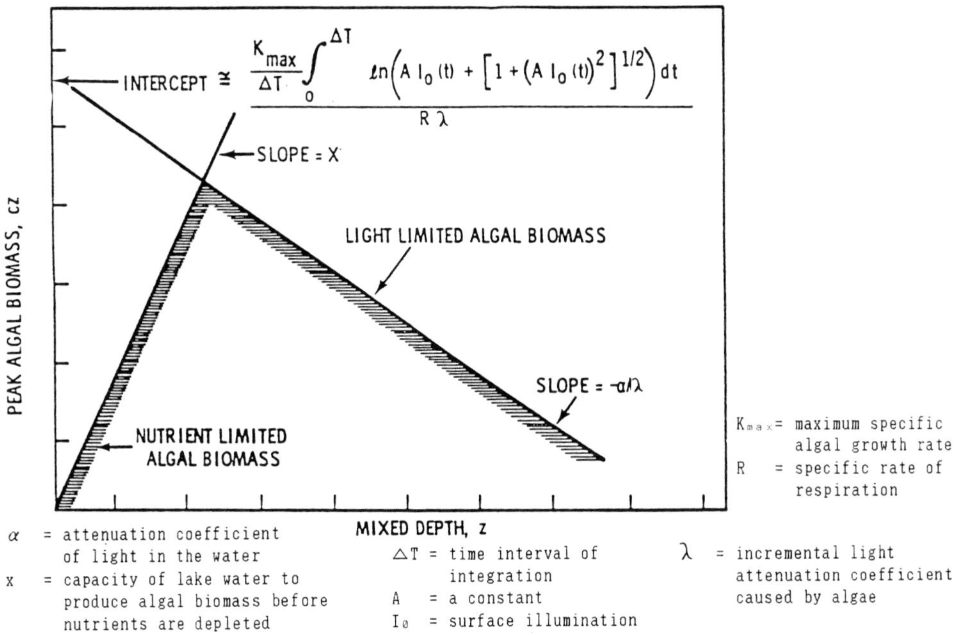

Figure 1.16. Theory on algal growth control by Lorenzen & Mitchell (1973)

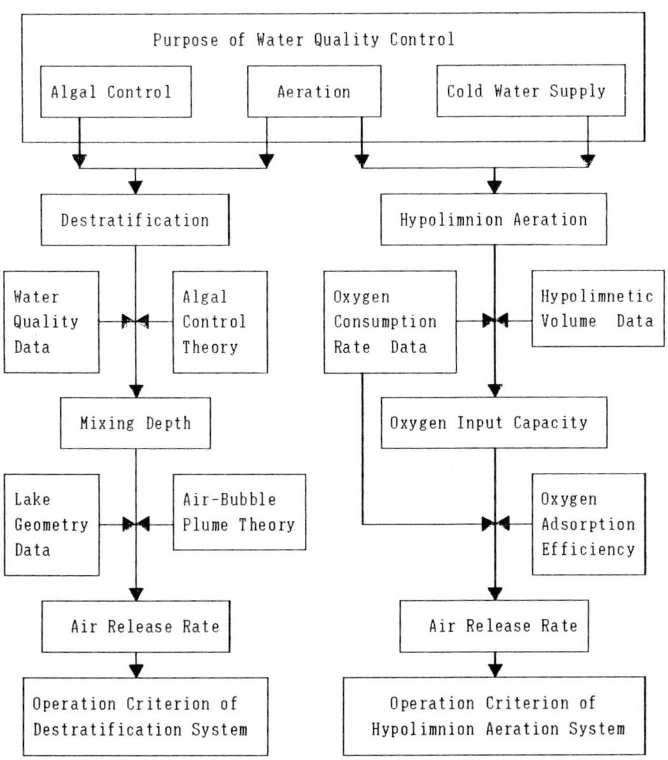

Figure 1.17. Flow chart for the design of artificial mixing systems

Table 1.5. Approximate rates of air flow required to mix lakes of various depths

Lake	Avg. depth (m)	Max. depth (m)	Surface area (m^2)	Volume (m^3)	Air required (SCMM)*
1	3.0	6.4	6.94×10^5	2.05×10^6	7
2	3.0	6.4	1.38×10^6	4.11×10^6	17
3	3.0	6.4	2.78×10^6	8.18×10^6	34
4	3.0	6.4	6.94×10^6	2.04×10^7	68
5	7.9	15.5	1.59×10^6	1.25×10^7	17
6	7.9	15.5	3.18×10^6	2.49×10^7	34
7	7.9	15.5	7.94×10^6	6.26×10^7	68
8	7.9	15.5	1.99×10^7	1.56×10^8	142
9	15.2	30.5	3.96×10^6	5.95×10^7	34
10	15.2	30.5	9.94×10^6	1.48×10^8	68
11	15.2	30.5	2.47×10^7	3.71×10^8	136
12	15.2	30.5	9.94×10^7	1.48×10^9	> 340
13	28.3	61.0	1.92×10^7	5.43×10^8	142
14	28.3	61.0	7.69×10^7	2.17×10^9	283

*SCMM = Standard cubic meters per minute.

However, the effects of stratification of the ambient fluid and of other significant flows that may occur are not taken into account in the model.

A one-dimensional convection/diffusion model as described in Section 1.3.1 can be developed for the analysis of the effect of mixing by aeration; it incorporates Equation (1.48) for the calculation of the water flow entrained by an air bubble plumes (Stefan & Ford, 1975; Matsunashi & Miyanaga, 1988). Figure 1.18 presents the concept of their model.

In order to design the facility, one must first determine the maximum height of water flow induced by the air bubbles. This height is determined from Equation (1.50) for a stratified ambient fluid. If the entrained water volume extends to the surface, it determines the installation depth of the mixing device:

$$\frac{(z_{\max} - z_i)}{z_i} = \frac{F_i^2}{(4\sigma^2\sigma_d^2 f_1^3/f_2 - F_i^2)} \tag{1.50}$$

where

$$F_i = \frac{Q_w}{\left(A_i\sqrt{\Delta\rho g z_i/\rho_0}\right)} \tag{1.51}$$

and $z_{\max}$ = maximum height to which the entrained water will rise, z_i = height of the thermocline, σ_d = ratio of the width of density-difference profile to the width of air-bubble plume, Q_w = water volume flux at z_i, A_i (= $\pi b^2(z_i)/4$) = horizontal cross sectional area of air-bubble plume at z_i, $\Delta\rho$ = density difference across thermocline in the ambient fluid, ρ_0 = reference density of water. Equation (1.50) is compared with experiments in Figure 1.19.

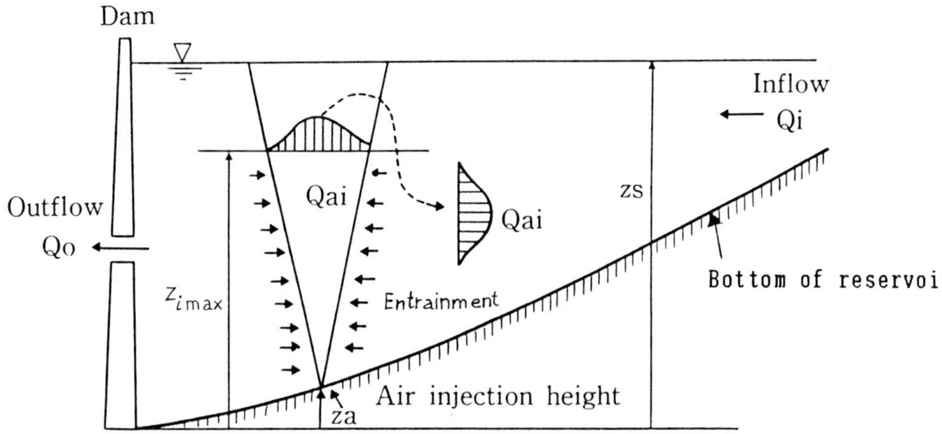

Figure 1.18. Schematic representation of the model (Matsunashi & Miyanaga, 1988)

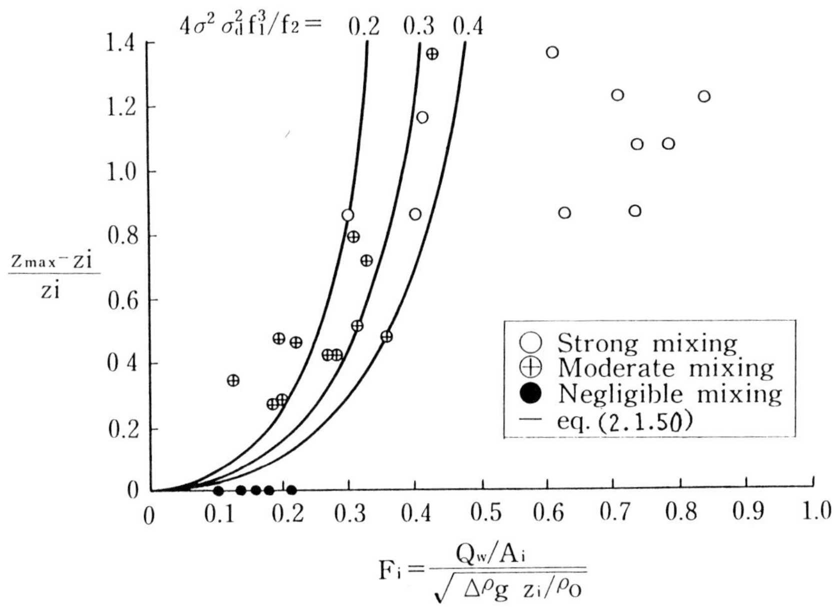

Figure 1.19. Relationship between maximum plume height z_{max} and densimetric Froude number F_i

Figure 1.20 shows several aeration devices. The multi-diffuser system has a simple structure and is easily installed. It is used for destratification. Also a way of preventing a diffuser pipe from being covered by sediment is sometimes needed. Air-lift pipe systems are divided into two types: diffuser and air-chamber. The former is used for hypolimnion aeration and the latter for destratification.

Examples of destratification systems in use are summarized in Table 1.6. The

Table 1.6. Examples of destratification systems (Lorenzen and Fast, 1977)

Lake	Location	Air release depth (m)	Mean depth (m)	Volume (10^6 m^3)	Area (10^6 m^2)	Power (HP)	Q_{air} (SCMM)*	$\frac{Q_{air}}{V}$ $\left(\frac{SCMM}{10^6 m^3}\right)$	$\frac{Q_{air}}{A}$ $\left(\frac{SCMM}{10^6 m^2}\right)$	ΔT (°C) Surface-bottom	Comment
Clines Pond	Corvallis, Oregon	5	2	0.0028	0.0014	1/4	0.028	10	20	~1	Good mixing
University Lake	North Carolina	9	3	2.6	0.81	3	0.4	0.15	0.49	1-2	O_2 depletion
Eufaula Reservoir	Oklahoma	28	16	703.0	43.5		34	0.05	0.78	9	Central pool 125 psig 1-2°C
Kezar Lake	New Hampshire	8	3	2.2	0.71		3	1.3	4.2	< 1	15 psig, good mixing
Lake Roberts	New Mexico	9	4	1.2	0.28		2.8	2.3	10	< 1	
El Capitan	California 1970	21	10	18.0	1.84	50	6.1	0.34	3.3	2.5	Did not mix algae
	California 1971	28	10	21.0	2.22	50	6.1	0.29	2.7	2.5	
Test Lake 1	Boltz, Kentucky	19	9	3.6	0.39		3.2	0.89	8.2		
Test Lake 2	Falmouth, Kentucky	13	6	5.5	0.91		3.3	0.59	3.6		
Section Four Lake	Michigan	18	10	0.11	0.011	20	2.2	20	200	0	Compressor run only 8 hrs a day
Ottoville Quarry	Ohio 1970	17	9	0.062	0.0071	3/4	0.91	14.5	128	1	Non-continuous compressor operation
	Ohio 1971	17	9	0.062	0.0071	2	2.2	35	310	2	
Casitas Reservoir	California 1972	43-49	26	248.9	9.61	150	18	0.07	1.9	4**	Air injected about 25-30 m off the bottom
	California 1973	43-49	28	294.8	10.6	150	18	0.06	1.7	7**	

Buchanan Lake	Ontario, Canada	13	5	0.42	0.086	2	0.28	0.67	3.3	10	
Valen Lake	Ontario, Canada	5	2	1.2	0.61	2	0.28	0.24	0.46	0.5	
Parvin Lake	Colorado	10	4	0.85	0.19		2.1	2.5	11	3	Helixor[R]
Lake Wohlford	California	14	6	3.09	0.53	50	5.9	1.9	11	< 1	Compressor run 9 hrs a day
Cox Hollow	Wisconsin	9	4	1.48	0.39	7.5	2	1.4	5.1	0-3	Aero-hydraulic cannons[R]
Wahnbach Reservoir	W.Germany 1961-62	43	19	41.6	2.14	15	2	0.05	0.93	5-6	
	W.Germany 1964	43	19	41.6	2.14	50	5.9	0.14	2.8	1	
Hyrum Reservoir	Utah	21	11	16.6	1.77	25	2.8	0.17	1.6	2-4	Did not mix algae
Allatoona	Georgia	43	9	457.8	48.5	300	28	0.06	0.58	4-8	

*SCMM = Standard cubic meters per minute.
**Temperature at surface and 36 m depth.

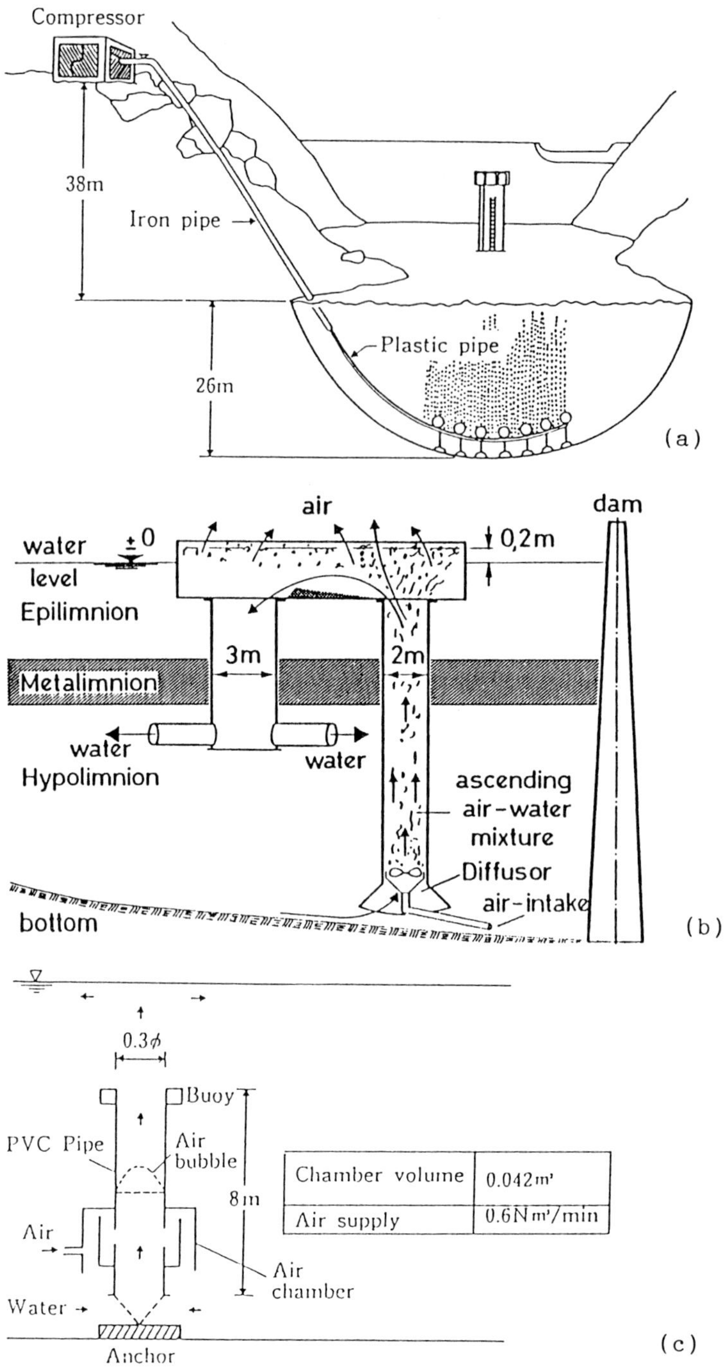

Figure 1.20. Examples of devices for artificial mixing by aeration: a) Multi-diffuser system (Lorenzen & Fast, 1977); b) Air-lift pipe system: diffuser type (Bernhardt & Wilhelms, 1975); c) Air-lift pipe system: air-chamber type (Kojima, 1982; Saeki, 1974; Kaiyo-Kogyo Co. & Toa Kensetsu Kogyo)

amount of air supplied, either per unit volume (Q_a/V) or per unit area (Q_a/A) of lake, varies widely. Those systems that provided good mixing generally used more than one standard cubic meter of air per minute (SCMM) per million cubic meters of lake volume, and more than 5 SCMM per million square meters of surface area.

REFERENCES

Aki, S. & T. Shirasuna 1977. Numerical simulation of temperature and turbidity in reservoirs, *17th Congress of IAHR, Baden-Baden*, 2: 41-48.

Akiyama, J. & H.G. Stefan 1984. Plunging flow into a reservoir. Theory, *J. Hydr. Engrg.*, 110: 484-499.

Akiyama, J. & H.G. Stefan 1988. Turbidity current simulation in a diverging channel, *Water Res. Res.* 24: 579-587.

Altinakar, M.S., W.H. Graf & E.J. Hopfinger 1987. Hydraulics of the head of a turbidity current, Topics in Lake and Reservoir Hydraulics, *Proc. Technical Session C1, 22nd Cong. IAHR, Lausanne*, 50-53.

Altinakar, M.S., W.H. Graf & E.J. Hopfinger 1990. Weakly depositing turbidity current on a small slope, *J. Hydraulic Res.*, 28(1): 55-80.

Amen, R. & T. Maxworthy 1980. The gravitational collapse of a mixed region into a linearly stratified fluid, *J. Fluid. Mech.*, 96(1): 65-80.

Asaeda, T. & M. Nakai 1989. Analysis on the hypolimnetic heavier water lifting system using air bullets, *Proc. JSCE*, 411(II-12): 91-98 (in Japanese).

Ashida, K. & S. Egashira 1975. Basic study on turbidity currents, *Proc. JSCE*, 237: 37-50 (in Japanese).

Ashida, K. & S. Egashira 1978. Velocity profiles of a stably stratified density currents, *Bull.*, 21-B2: 455-465. Disaster Prevention Res. Inst., Kyoto Univ. (in Japanese).

Ashida, K., S. Egashira & H. Nakagawa 1981. Behaviour of turbid water mass released into two dimensional reservoir, *Bull.*, 24-B2: 265-282. Disaster Prevention Res. Inst., Kyoto Univ. (in Japanese).

Benjamin, T.B. 1968. Gravity currents and related phenomena, *J. Fluid Mech.*, 31: 209-248.

Bell, H.S. 1942. Density currents as agents for transporting sediments, *J. of Geology*, L5.

Bernhardt, H. 1967. Aeration of Wahnbach Reservoir without changing the temperature profile. *Journal American Water Works Association*, 59(8): 943-964.

Bernhardt, H. & A. Wilhelms 1975. Hypolymnetic aeration as a means of controlling redox processes on the bottom of the eutrophic reservoir. *Internat. Verein. Limnol.*, 19: 1957-1959. Stuttgart.

Bloss, S. & D.R.F. Harleman 1979. Effect of wind-induced mixing on the seasonal thermocline in lakes and reservoirs. Rept. No.249, M.I.T.

Britter, R.E. & P.F. Linden 1980. The motion of the front of a gravity current travelling down an incline, *J. Fluid Mech.*, 99: 531-543.

Britter, R.E. & J.E. Simpson 1978. Experiments on the dynamics of gravity current head, *J. Fluid Mech.*, 88: 223-240.

Bulson, P.S. 1961. Currents produced by an air curtain in deep water, *Dock and Harbour Authority*, 42(487): 15-22.

Busnaina, A.H., D.G. Lilley & P.M. Moretti 1981. Prediction of local destratification of lakes, *J. Hydr. Div., Proc. ASCE*, 107(HY3): 259-272.

Cederwall, K. & J.D. Ditmars 1970. Analysis of air bubble plumes. CIT, Keck Lab. of Hydraulics and Water Resources, Rep. No. KH-R-24.

Clasen, J. & H. Bernhardt 1983. In-situ-Trübungsmessungen in der Wahnbachtalsperre, *GWF-Wasser Abwasser,* 124: 575-581.

Craya, A. 1949. Recherches theoriques sur l'ecoulement de couches superposees de fluides de densites differentes. La Houille Blanche, Jan-Fev, 45-55.

Dake, J.M.K. & D.R.F. Harleman 1969. Thermal stratification in lakes. Analytical and laboratory studies, *Water Resour. Res.*, 5(2): 484-495.

Darden, R.B., J. Imberger & H.B. Fischer 1975. Jet discharge into a stratified reservoir, *J. of Hydr. Div., Proc. ASCE*, 101: 1121-1220.

Debler, W.R. 1959. Stratified flow into a line sink, *J. Eng.Mech. Div., ASCE,* 85(EM3): 51-65.

Edinger, J.R. & E.M. Buchak 1979. A hydrodynamic two-dimensional reservoir model. Development and test application to Sutton Reservoir, Elk River, West Virginia. Prepared for US Army Engineer Division, Ohio River.

Ellison, T.H. & J.S. Turner 1959. Turbulent entrainment in stratified flows, *J. Fluid Mech.*, 6: 423-448.

Fast, A.W. & M.W. Lorenzen 1976. Synoptic survey of hypolimnetic aeration, *J. of Environ. Eng. Division, Proc. ASCE*, 102(EE6): 1161-1173.

Faust, K.M., R.A. Denton & E.J. Plate 1983. Schichtungs Effekte in Fluss-stauhaltungen. Laborversuche (Aspects of stratified flow in river reservoirs. Laboratory results), Wasserwirtschaft, 73: 386-392.

Ford, D.E. & H.G. Stefan 1980. Thermal predictions using integral energy model, *J. Hydr. Div., Proc.ASCE*, 106(HY1): 39-55.

Fukuoka, S., K. Mizumura & T. Kanoh 1978. Fundamental study on dynamics of the head of gravity currents, *Proc. JSCE*, 274: 41-55 (in Japanese).

Fukuoka, S. & Y. Fukushima 1980. Analytical study on density currents bifurcating in a two dimensional stratified reservoir, *Proc. JSCE*, 294: 73-84 (in Japanese).

Fukuoka, S., Y. Fukushima & K. Nakamura 1980. Study on the plunge depth and interface form of density currents in a two-dimensional reservoir, *Proc. JSCE*, 302: 55-65 (in Japanese).

Fukuoka, S., M. Tsunoda & K. Inaba 1983. Internal surge in two and three dimensional reservoirs, *Proc. 27th Japan Conf. on Hydr. Eng., JSCE*, 109-116.

Garcia, M. & G. Parker 1989. Experiments on hydraulic jumps in turbidity currents near a canyon-fan transition. *Science*, 245: 393-396.

Graf, W.H. 1983. The behaviour of silt-laden current. *International Water Power & Dam Construction*, 35(9): 33-38.

Grover, N.C. & C.L. Howard 1938. The passage of turbid water through Lake Mead. *Trans. ASCE*, 103.

Goossens, H.J. & J.M. Smith 1975. The hydrodynamics of unconstrained bubble columns for mixing lakes and reservoirs. *Chemie Ingenieur Technik*, 47(22): 951.

Harleman, D.R.F. 1961. Stratified flow. In V. Streeter (ed.), *Handbook of Fluid Dynamics.* McGrow-Hill.

Harleman, D.R.F. 1982. Hydrothermal analysis of lakes and reservoirs. *J. of Hydr. Division, Proc. of ASCE*, 108: 302-325.

Harleman, D.R.F. & R.A. Elder 1965. Withdrawal from two-layered stratified flows. *J. Hydr. Div., Proc. ASCE*, 91: 43-58.

Harleman, D.R.F., R.S. Gooch & A.T. Ippen 1958. Submerged sluice control of stratified flow. *J. Hydr. Div., Proc. ASCE*, 84.

Hauenstein, W. & Th. Dracos 1984. Investigation of plunging density currents generated by inflows in lakes. *J. of Hydraulic Res.*, 22(3): 157-179.

Hebbert, B., J. Imberger, I. Loh & J. Patterson 1979. Collie river underflow into the Wellington reservoir *J. Hydr. Div., Proc. ASCE*, 105: 533-545.

Hino, M. & M. Furusawa 1969. Experiments on selective withdrawal from stratified fluid. *16th Congress of Coastal Engrg. in Japan*, 231-236 (in Japanese).

Hino, M. & S. Onishi 1967. Consideration on the flow of selective withdrawal from the deep layer of linearly stratified fluid. *Proc. 14th Japanese Conf. on Coastal Eng., JSCE*, 246-252 (in Japanese).

Hino, M. & S. Onishi 1969a. Effect of a point sink location on flow regime of stratified fluid. *Proc. JSCE*, 163: 39-48 (in Japanese).

Hino, M. & S. Onishi 1969b. Investigation of axisymmetrical stratified flow towards a point sink. *Coastal Eng. in Japan, JSCE*, 12: 179-187.

Hino, M. & S. Onishi 1970. Axisymmetric flow of stratified fluid into a cylindrical sink. *Trans. of JSCE*, 2(I): 151-155.

Hinze, J.O. 1960. On the hydrodynamics of turbidity currents. *Geologie Em Munbouw*, 39: 18-25.

Hirano, M. & K. Hadano 1981. Characteristics of motion of density current head. *Proc. JSCE*, 314: 67-73 (in Japanese).

Hirano, M. & Hadano, K. 1983. Motion of the head of a density current advancing through a gradually widening channel. *Proc. JSCE*, 332: 47-54.

Hocking, G.C., B.S. Sherman & J.C. Patterson 1988. Algorithm for selective withdrawal from stratified reservoir. *J. Hydr. Eng., ASCE*, 114: 767-719.

Howard, C.S. 1953. Density currents in Lake Mead. *Proc. Minnesota Int. Hydr. Convention, Univ. Minnesota*, 355-368.

Huber, W.C., D.R.F. Harleman & P.J. Ryan 1972. Temperature prediction in stratified reservoirs. *J. of Hydr. Div., Proc. ASCE*, 98: 645-666.

Imberger, J. 1972. Two-dimensional sink flow of a stratified fluid contained in a duct. *J. Fluid Mech.*, 53(2): 329-349.

Imberger, J. 1985. The diurnal mixed layer. *Limnol. Oceanogr.*, 30(4): 737-770.

Imberger, J. & P.F. Hamblin 1982. Dynamics of lakes, reservoirs, and cooling ponds, *Ann. Rev. Fluid Mech.*, 14: 153-187.

Imberger, J., R. Thompson & C. Fandry 1976. Selective withdrawal from a finite rectangular tank. *J. Fluid Mech.*, 78(3): 489-512.

Imberger, J. & J.C. Patterson 1981. A dynamic reservoir simulation model-DYRESM 5, 310-361. In *Transport Models for Inland and Coastal Waters*. Academic Press.

Imberger, J. & J.C. Patterson 1990. Physical Limnology. In J.W. Hutchinson & T.Y.Wu (eds), *Advances in Applied Mechanics*, 127: 303-475. Academic Press.

Imberger, J., J. Patterson, B. Hebbert & I. Loh 1978. Dynamics of reservoirs of medium size. *J. Hydr. Div., Proc. ASCE*, 104: 725-743.

Ippen, A.T. & D.R.F. Harleman 1952. *Steady state charateristics of subsurface flow*. US Natl. Bur. Standards Circ. 521.

Ishibashi, T. et al. 1977. Field observation on flow and heat budget in a stratified reservoir. *Proc. 21st Japanese Conf. on Hydraulics, JSCE*, 1-8 (in Japanese).

Ishikawa, T. & M. Tanaka 1990a. Diurnal stratification and its effects on wind-induced current. *Int. Conf. on Physical Modelling and Dispersion*, 13A: 25-30.

Ishikawa, T. & M. Tanaka 1990b. Theoretical study on the DIM-type entrainment, *Proc. JSCE*, 417 (II-13): 99-108.

Ishikawa, T., M. Tanaka & M. Koseki 1989. On the influence of diurnal stratification on water quality in a shallow lake. *Proc. JSCE*, 411(II-12): 247-254 (in Japanese).

Ivey, G.N. & S. Blake 1985. Axisymmetric withdrawal and inflow in a density-stratified container. *J. of Fluid Mec.*, 161: 115-137.

Ivey, G. & J. Imberger 1978. Field investigation of selective withdrawal. *J. Hydr. Div., Proc. ASCE*, 104: 1225-1237.

Iwasa, Y., M. Noguchi & A. Kojima 1975. Prediction of thermally stratified flow in a reservoir. *Disaster Prevention Res. Inst. Annuals*, 18B: 565-573. Kyoto Univ. (in Japanese).

Iwasa, Y., N. Matsuo & M. Endo 1977. Destratification process of a reservoir resulting from flood flows. *Disaster Prevention Res. Inst. Annuals*, 20B: 259-270. Kyoto Univ. (in Japanese)

Iwasa, Y. & N. Matsuo 1986. Numerical simulation of flow dynamics and water quality induced by air bubble plume. *Proc. of 5th Symp. Multiphase Flow, Tokyo*, 265-280.

Jirka, G.H. 1979. Supercritical withdrawal from two-layered fluid systems, Part I. Two-dimensional skimmer wall. *J. Hydr. Res.*, 17: 43-51.

Jirka, G.H. & D.S. Katavola 1979. Supercritical withdrawal from two-layered fluid systems, Part 2. Three-dimensional flow into round intake. *J. Hydr. Res.*, 17: 53-62.

Johnson, T.R., G.J. Farrell,, C.R. Ellis & H.G. Stefan 1987a. Negatively buoyant flow in diverging channel I – Flow regimes. *J. Hydr. Eng., ASCE*, 113: 716-730.

Johnson, T.R., C.R. Ellis, G.R. Farrell & H.G. Stefan 1987b. Negatively buoyant flow in diverging channel II– 3-D flow field description. *J. Hydr. Eng., ASCE*, 113: 731-742.

Johnson, T.R., C.R. Ellis & H.G. Stefan 1989. Negatively buoyant flow in diverging channel IV – Entrainment and dilution *J. Hydr. Eng., ASCE*, 115(4): 437-456.

Kao, T.W. 1970. Free-streamline theory for inviscid stratified flow into a line sink. *Phys. Fluids*, 13(3): 558-564.

Kao, T.W. 1977. Density currents and their applications. *J. Hydr. Div., Proc. ASCE*, 103: 543-555.

Karpik, S.R. & G.D. Raithby 1990. Laterally averaged hydrodynamics model for reservoir predictions. *J. Hydr. Eng., ASCE*, 116(6): 783-798.

Keulegan, G.H. 1958. Twelfth progress report on model laws for density currents. *The motion of saline fronts in still water*. US Natl. Bur. Standards Rep. 5831.

Kikkawa, H. & K. Yamamoto 1971. The study on behaviour of reservoir water, *Proc. JSCE*, 186: 39-54 (in Japanese).

Kobus, H.E. 1968. Analysis of the flow induced by air-bubble systems. *Coastal Engrg. Conference, London*, 2: 1016-1031.

Kojima, S. 1982. Water quality improvement in eutrophic lakes by air lifting shaft – (I) Principle and effects, *Yosui to Haisui*, 24(1): 25-31 (in Japanese).

Kojima, S. 1988. Artificial circulation of lake water as a counter-measure to eutrophication – Its principle and results. *Japan J. of Water Treat. Biol.*, 24(1) (in Japanese).

Kranenburg, C. 1984. Wind-induced entrainment in a stably stratified fluid *J. Fluid. Mech.*, 145: 253-273.

Kranenburg, C. 1985. Mixed-layer deepening in lakes after wind set up. *J. Hyd. Engng., ASCE*, 111(9): 1279-1297.

Kraus, E.B. & J.S. Turner 1967. A one-dimensional model of the seasonal thermocline. *Tellus*, 19(1): 98-106.

Kundu, P.K. 1981. Self-similarity in stress-driven entrainment experiment. *J. Geophy. Res.*, 86: 1979-1988.

Kurihara, M. 1954. On the study on pneumatic breakwater. Report of Res. Inst. on Appl. Mech., Kyushu Univ., 5: 17-41.

Kurihara, M. 1965. On the study of a pneumatic breakwater in Japan. *Coastal Engrg. in Japan*, 8.

Lambert, A. & F. Giovanoli 1987. Records of channelized turbidity currents in Lake Geneva (Rhone Delta) – Topics in lake and reservoir hydraulics. *Proc. Technical Session C1, 22nd Cong. IAHR, Lausanne*, 54-55.

Lawrence, G.A. 1980. Selective withdrawal through a point sink. *Proc. 2nd Intl. Symp. on Stratified Flows, Trondheim*, 411-425. Tapin.

Lorenzen, M.W. & A.W. Fast 1977. A guide to aeration/ circulation techniques for lake management. EPA-600/3/ 77/004, U.S.EPA, Corvallis.

Lorenzen, M.W. & R. Mitchell 1973. Theoretical effects of artificial destratification on algal production in impoundments. *Environ. Sci. Tech.*, 7(10): 939-944.

Manins, P.C. 1976. Mixed-region collapse in a stratified fluid. *J. Fluid Mech.*, 77(1): 177-183.

Manins, P.C. 1976. Intrusion into a stratified fluid. *J. Fluid Mech.*, 74(3): 547-560.

Masuda, A. & Y. Nagata 1974. Water wedge advancing along the interface between two homogenous layers. *J. Oceanogr. Soc. Japan*, 30: 39-47.

Matsunashi, S. & Y. Miyanaga 1988. *The Application of simulation model for the flow induced by air bubble plume in thermally stratified reservoirs*. Central Res. Inst. Electric Power Industry, No.U88029 (in Japanese).

Maxworthy, T. 1972. Experimental and theoretical studies of horizontal jets in a stratified fluid. *Int. Symp. Stratified Flows, Novosibirsk*, Communication 17.

Middleton, G.V. 1966. Experiment on density and turbidity currents, I. *Canadian J. Earth Sciences*, 3: 523-546.

Middleton, G.V. 1967. Experiment on density and turbidity currents, III. *Canadian J. Earth Sciences*, 4.

Morton, B.R., G.I. Taylor & J.S. Turner 1956. Turbulent gravitational convection from maintained and instantaneous sources. *Proc. Roy. Soc. London, A*, 234: 1-23.

Mukhamedov, A.M., N.P. Kulesh & Y.S. Mukhamedov 1972. Origin and movement of the bottom current in the reservoir of the Nurek power station and its effect on the silting of the reservoir. *Proc. Int. Symp. on Stratified Flow, Novosibirsk, USSR*, 541-551.

Murota, A., K. Michioku & S. Hazama 1985. Prediction of thermal structure in reservoirs by bulk mixed layer model. *Proc. 29th Japanese Conf. on Hydraulics, JSCE*, 395-400 (in Japanese).

Notron, W.R., I.P. King & G.T. Orlob 1973. *A finite element model for Lower Granite reservoir*. Prepared for Walla Walla District, US Army Corps of Engineers.

Onishi, S., K. Imamura & S. Haraguchi 1986. Study on selective withdrawal in reservoirs. *Proc. JSCE*, 369: 79-87 (in Japanese).

Orlob, G.T. & L.G. Selna 1970. Temperature variations in deep reservoirs. *J. Hydr. Div., Proc. ASCE*, 96(HY2): 391-410.

Pao, H.P. & T.W. Kao 1974. Dynamics of establishment of selective withdrawal of a stratified fluid from a line sink, Part 1. *J. Fluid Mech.*, 65: 657-688.

Plapp, J.E. & J.P. Mitchell 1960. A hydrodynamic theory of turbidity currents. *J. Geophy. Res.*, 65: 983-992.
Pollard, R.T., P.B. Rhines & R.O. Thompson 1973. The deepening of the wind mixed layer, *Geophy. Fluid Dyn.*, 3: 381-404.
Prandtl, L. 1952. *Essentials of Fluid Dynamics*. Blackie and Son, Ltd, Glasgow.
Price, J.F. 1979. On the scaling of stress-driven entrainment experiments, *J. Fluid Mech.*, 90: 509-529.
Price, J.F., C.N.K. Mooers & J.C. Leer 1978. Observation and simulation of storm-induced mixed-layer deepening. *J. Phys. Oceanogr.*, 8: 582-599.
Ragotzkie, R.A. 1978. Heat exchange in lakes, In A. Lerman (ed.), *Lakes*. Springer Verlag.
Roberts, P.I.W. & P.R. Matthews 1984. Dynamics of jets in two-layer stratified fluids. *J. Hydr. Eng., ASCE*, 110:1201-1217.
Rodi, W. 1987. Examples of calculation methods for flow and mixing in stratified fluids. *J. Geophys. Res.*, 92(C5): 5305-5328.
Rouse, H., C.S. Yih & H.W. Humphreys 1952. Gravitational convection from a boundary source. *Tellus*, 4: 201-210.
Savage, S.B. & J. Brimberg 1975. Analysis of plunging phenomena in water reservoirs. *J. Hydr. Res.*, 13: 187-205.
Schijf, J.B. & J.C. Schönfeld 1953. Theoretical considerations on the motion of salt and fresh water. *Proc. Minnesota International Convention, IAHR*, 321.
Senshu, S. 1967. Some problems concerning the design of cooling water from deep layer. *Proc. 14th Japan Conf. Coast. Eng.*, 253-260.
Sherman, F.S., J. Imberger & G.M. Corcos 1978. Turbulence and mixing in stably stratified waters. *Ann. Rev. Fluid Mech.*, 10: 267-288.
Shirasuna, T. 1973. A hydraulic measure for turbidity problems in reservoirs. *Hydroelectric Power*, 126: 3-25 (in Japanese).
Simpson, J.E. 1987. Gravity currents. In *The environment and the laboratory*. Ellis Horward Ltd.
Singh, B. & C.R. Shah 1971. Plunging phenomenon of density currents in reservoirs, *La Houille Blanche*, no. 1: 59-64.
Spigel, R.H. & B. Farrant 1984. Selective withdrawal through a point sink and pycnoline formation in linearly stratified flow, *J. Hydr. Res.*, 22: 35-51.
Spigel, R.H. & J. Imberger 1980. The classification of mixed-layer dynamics in lakes of small to medium size. *J. Phys. Oceanogr.*, 10: 1104-1121.
Spigel, R.H., J. Imberger & K.N. Rayner 1986. Modeling the diurnal mixed layer. *Limnol. Oceanogr.*, 31(3): 533-556.
Spraggs, L.D. & R.L. Street 1975. *Three-dimensional simulation of thermally-influenced hydrodynamic flows*. Report TR-190, Dept. of Civ. Engrg., Stanford Univ., Stanford, Calif.
Stefan, H.G. 1973. High concentration turbidity currents in reservoirs. *Proc. 15th Cong. IAHR, Istanbul*, 1: 341-352.
Stefan, H. & D.E. Ford 1975. Temperature dynamics in dimictic lakes. *J. of Hydr. Div., Proc. ASCE*, 101(HY1): 97-114.
Stefan, H.G. & T.R. Johnson 1989. Negatively buoyant flow in diverging channel – III. Onset of underflow. *J. Hydr. Eng., ASCE*, 115(4): 423-436.
Tamai, N. 1980. *Hydraulics of Density Currents*, Gihoodo Pub Co.
Tanaka, M. & T. Ishikawa 1989. On the influence of diurnal thermal stratification on wind-driven current. *Proc. JSCE*, 405(II-11): 63-72 (in Japanese).

Task Committee on preparation of Sedimentation Manual 1963 . Sediment transportation mechanics – Density current. *J. Hydr. Div., Proc. ASCE*, 89(HY5): 77-87.

Toda, I. & Y. Nagata 1980. End wall effects in experiments on water intrusion along the interface between two homogeneous layers. *J. of Oceanographical Soc. of Japan*, 36: 15-22.

Turner, J.S. 1973. *Buoyancy Effects in Fluids*. Cambridge Univ. Press.

von Karman, T. 1940. The engineer grapples with nonlinear problems. *Bull. Am. Math. Soc.*, 46: 615-683.

Waldrop, W.R. & F.B. Tatom 1976. Analysis of the thermal effluent from the Gallatin Steam Plant during low river flows. Report no.33-30, Tennessee Valley Authority.

Watanabe, M. & Y. Fukushima 1990. Numerical analysis of density underflow by the k-ε turbulent model. *Proc. 45th JSCE Annual Convention, Part II*, 200: 452-453 (in Japanese).

Wilkinson, D.L. & I.R. Wood 1972. Some observations on the motion of the head of a density current. *J. Hydraul. Res.* 10: 305-324.

Wood, I.R. 1978. Selective withdrawal from two-layered fluid. *J. Hydr. Div., ASCE*, HY12: 1647-1659.

Wu, J. 1969. Mixed region collapse with internal wave generation in a density-stratified medium. *J. Fluid Mech.*, 35(3): 531-544.

Wunderlich, W.O. & R. Elder 1973. Mechanics of flow through Man-Made Lakes. Am. Geophy. Union; Geophysical Monograph 17. Man-Made Lakes, Washington, USA.

Yih, C.S. 1958. On the flow of a stratified fluid. *Proc. 3rd US Nat. Cong. Appl. Mech.*, 857-861.

Yih, C.S. 1965. *Dynamics of Nonhomogeneous Fluids*. Macmillan.

Yoon, T.H., C.Y.Chun & W.S. Yook 1990. Two dimensional density flow into a reservoir with sloping bottom. *Proc. of 7th Congress, The Asian and Pacific Regional Division, IAHR, Beijin*, III: 293-299.

Zic, K., H.G. Stefan & G. Ellis 1990. Laboratory study of water destratification by a bubble plume. *J. Hydr. Res.*, 30: 7-27.

Lake and reservoir eutrophication: Prediction and protection 2

HEINZ G. STEFAN
St. Anthony Falls Hydraulic Laboratory, Department of Civil and Mineral Engineering, University of Minnesota, Minneapolis, Minnesota, USA

2.1 INTRODUCTION

Hydraulic engineers have made great contributions to civilizations by creating water reservoirs that provide water, flood protection, and electric power when needed. The recognition for these contributions has sometimes been overshadowed by bad water quality. Only in recent years has information from limnology, aquatic biology, ecology, and water chemistry been synthesized with hydrodynamics to develop methods and models for water quality estimation and forecast applicable to reservoirs and lakes.

This chapter addresses lake and reservoir water quality from the point of view of its predictability by modeling. Particular emphasis is given to the very common problem of eutrophication and measures to prevent or reverse it.

2.2 DEFINITION AND CAUSES OF EUTROPHICATION

2.2.1 *Definitions*

Eutrophication* refers to the process of high primary biological production caused by excessive availability of nutrients in a lake or reservoir, and all its consequences. Primary biological production refers to the production of aquatic plant mass (e.g. phytoplankton, periphyton, macrophytes) by photosynthesis. The euphotic layer is the water layer below the water surface in which light intensity is sufficient and light wave length is appropriate (~ 400 to 700 nanometers) for photosynthesis.

Detailed treatments of photosynthesis and plant growth in lakes can be found in the extensive literature on limnology, including books by Goldman (1966),

*Eutrophic means literally well-fed or over-fed.

Hutchinson (1957), and Wetzel (1975); including journals such as *Limnology and Oceanography* and *Hydrobiologia* and *Water Research* and proceedings of symposia and conferences (e.g. Likens et al., 1979; Lorenzen, 1979). This body of literature, particularly on chemical and biological limnology, is not usually consulted by hydraulic engineers, but it is necessary for an understanding of eutrophication and its control. To this we should add books dealing specifically with eutrophication and its control (e.g. NAS, 1969; OECE, 1982; World Health Organization, 1981; Cooke et al., 1986; DNR, 1974; US EPA, 1979, 1980b, 1983, 1984; and NALMS, 1985, 1986, 1987; Sas, 1989; Ryoling & Rast, 1989).

Limnologists classify lakes and reservoirs in terms of their primary biological productivity (eutrophy). Productivity can be measured by the standing crop of plant mass, usually the floating unattached plants called phytoplankton. Plant mass is measured per volume of water in units of (1) total dry weight (g/l), (2) carbon contained in the plants (mg/l), (3) chlorophyll-a contained in the plants (μgchla/l), or (4) oxygen demand necessary to decompose (re-mineralize) the plant mass (mg O_2/l). The stoichiometric ratios between these four different measures are unfortunately not constant but vary over certain ranges.

Chl-a and carbon are the most commonly used measures of plant concentration in lakes and reservoirs. The basic classes of eutrophy given by limnologists are oligotrophic, mesotrophic, and eutrophic. Oligotrophic lakes have low biological productivity and are characterized by low algal concentrations (Table 2.1) and high water clarity and transparancy. Eutrophic lakes have undesirable high productivity, while mesotrophic lakes rank in between. Algal blooms are common and obvious in eutrophic lakes and reservoirs. In a eutrophic lake algae can be a severe nuisance. Some species of algae add color and taste to the water, other species release toxic substances. Algal mass also can clog water intakes and severely impact water uses.

2.2.2 *Causes of eutrophication*

Progressive eutrophication is a long-term trend in many lakes and reservoirs, typically over a time scale of centuries. However, manmade changes to the watershed of a lake or reservoir may accelerate eutrophication such that the quality of water changes significantly in just a few years (cultural eutrophication).

Eutrophication requires nutrient availability which can support a large plant population. The sources of these nutrients are the atmosphere, lake sediments, or the watershed. Atmospheric loading is significant for some large lakes such as the North American Great Lakes. Sediment nutrient recycling can be significant in shallow lakes, but the nutrients must reach the surface where light is available for photosynthesis. Intermittently, weakly temperature-stratified lakes transport a higher proportion of sediment nutrients to the euphotic zone than lakes with a strong and stable stratification. New reservoirs may lack a nutrient-rich sediment layer, but some lakes have sediments that contribute as much phosphorus to eutrophication as the inflow from the watershed.

Often the cause of eutrophication lies within the watershed of a lake. This leads to the recognition of a lake or reservoir as part of a larger ecological entity. As Wetzel (1975) states, 'The lake ecosystem is intimately coupled with the land surrounding it in its drainage area and its running (lotic) waters that transport, and metabolize enroute, components of the land to the lake'. In most watersheds, human activity is the dominant cause of change. These changes tend to increase the rate of nutrient inflow to a lake and, consequently, the rate of change in the eutrophication process.

2.3 LAKES AND RESERVOIRS AS ECOSYSTEMS

Most reservoirs and lakes contain large numbers of living organisms ranging from viruses and bacteria to fish and mammals and birds. These interact in complex and diverse ways both with each other and with chemicals. The interactions between the many organic and nonorganic components can be described in terms of an ecosystem that comprises abiotic components (such as nutrients), producers (such as photosynthetic plants), consumers (such as fish), and decomposers (such as bacteria) (Fig. 2.1) (see also: Thornton et al., 1990; Stefan, 1981).

At least 19 elements (all of which can be found dissolved in water) are essential for life, although some of these are needed in minute traces. These basic elements are known collectively as nutrients. Five of them are required in large amounts:

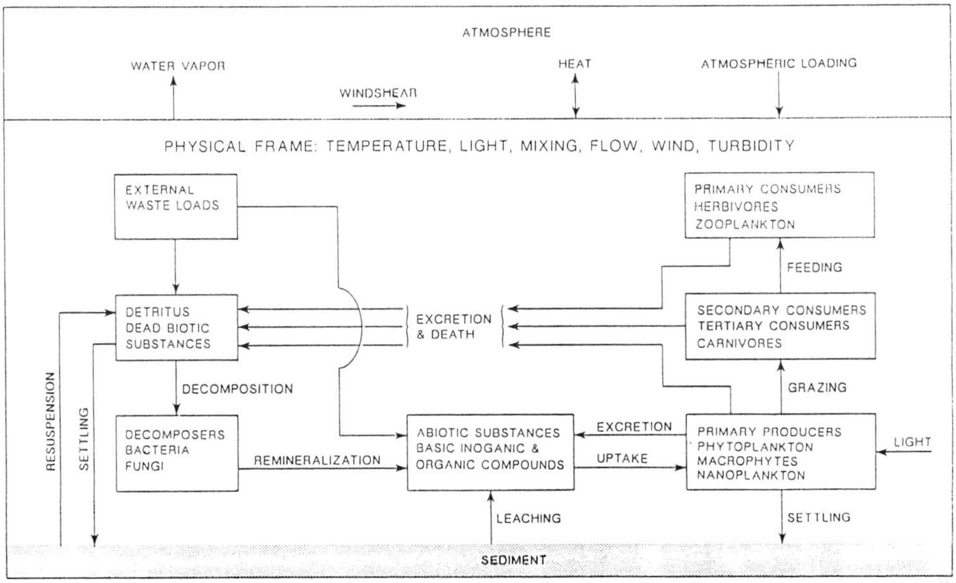

Figure 2.1. Material flow in lake/reservoir ecosystem

carbon, hydrogen, oxygen, nitrogen, and phosphorus. These are needed in proportions of approximately 106:181:45:16:1. The first three (C, H, O) are readily available either in water (H_2O) or dissolved carbon dioxide (CO_2) and thus never become a long-term limiting factor in aquatic plant growth. However, the concentrations of nitrogen and phosphorus dissolved in natural waters are much lower, and it is usually one of these two elements that provides the limiting factor for growth of the most common nuisance algae (Likens, 1972). The relative importance of these two for lentic freshwater bodies has been in dispute for a long time. Phosphorus is often utilized in the form of phosphates or orthophosphates as a fertilizer, in detergents and in industry. It is present in wastewater in large quantities, and only partially removed by wastewater treatment. Excess phosphorus in natural waters can lead to prolific plant growth.

Aquatic plants occur in many sizes and several degrees of complexity. The smaller and flowerless photosynthetic plants are probably of greater interest in lake and reservoir management. The most important of these belong to the group encompassing the algae. There are four main types of algae, three of which are classified in terms of their color. There are seasonal cycles of different algal species in lakes, and different species have different nutrient requirements.

Zooplankton (e.g. daphnia) are small animals that feed on selective species of algae as their food source. They are the 'primary consumers' in the food web while the algae are primary producers. Attempts have been made to control algal growth by increasing certain species of zooplankton by 'biomanipulation'. This methodology is elegant but still under development (Shapiro et al., 1982).

Bacteria constitute an extremely important group of nonphotosynthetic organisms. They hold a unique role as decomposers; some (aerobes) utilize free oxygen to break down the organic matter of dead plants and animals, while others (anaerobes) use oxygen found in salts for the same purpose. Several bacteria types are responsible for diseases; these are known as pathogens. Three common water-borne bacterial diseases are typhoid, Asiatic cholera, and bacillary dysentery.

The species of fish present in a body of water depend, in part, on the degree of eutrophication. Fish that prefer oligotrophic lakes include many of the more highly prized fish such as trout, grayling, and salmon.

There are significant differences between natural lakes and reservoirs. Reservoirs usually have greater drainage basin areas, making control of inflow of material very difficult, especially if the drainage basin has several different municipal or agricultural discharges from politically separate units. Also, reservoirs usually have shorter water residence time, and a greater areal water load. Portions of the reservoir bed get exposed to the atmosphere, which changes sediment chemistry, and water withdrawal can be from selective layers.

The annual temperature cycle is one of the most significant determinants of the physical, chemical, and biological interactions of a lake or reservoir. The significance of thermal stratification to the eutrophication process and to reservoir restoration lies in separation of the upper and lower lake zones during the summer through differences in water density.

In eutrophic reservoirs the high content of organic matter in the reservoir sediments, originating from the production of algae and macrophytes as well as from inflow via streams, provides an excellent source of energy for bacteria, fungi, protozoa, and some invertebrate animals. The respiration of such organisms consumes the oxygen that was stored in the hypolimnion at the time of thermal stratification in the spring and releases some of the elements bound in the organic matter. There may also be extensive chemical oxygen demand. As the oxygen is reduced, at a rate called the lake's oxygen deficit rate, the ability of the sediments to retain elements such as phosphorus in iron-hydroxy complexes is altered and those elements are released to hypolimnetic water. As summer progresses, phosphorus, nitrogen, and other nutrients increase in concentration in the hypolimnion and frequently are transposed vertically into the lighter warm epilimnion where they subsidize algal growth. This entire process is called internal nutrient loading.

2.4 INDICATORS OF EUTROPHICATION

Eutrophication cannot be measured in a single, unique, unambiguous way. Current methods measure symptoms (e.g. standing crop of algae, Secchi depth, primary productivity) or causes (e.g. total phosphorus concentrations, phosphorous loadings), or consequences (e.g. low dissolved oxygen levels in lake hypolimnion) (see examples in Tables 2.1 and 2.2).

Table 2.1. Trophic levels

	Chla (μg/l)
Oligotrophic	< ~4
Mesotrophic	~4 to ~10
Eutrophic	> 10

Table 2.2. General relationship of lake productivity to average concentrations of epilimnetic total phosphorus (after Wetzel, 1975)

General level of productivity	Change in alkalinity in epilimnion during summer (mval/l)	Total phosphorus (μg/l)
Ultra-oligotrophic	< 0.2	< 5
Oligo-mesotrophic	0.6	5-10
Meso-eutrophic	0.6-1.0	10-30
Eutrophic		30-100
Hypereutrophic	> 1.0	> 100

Modified from Vollenweider, 1968

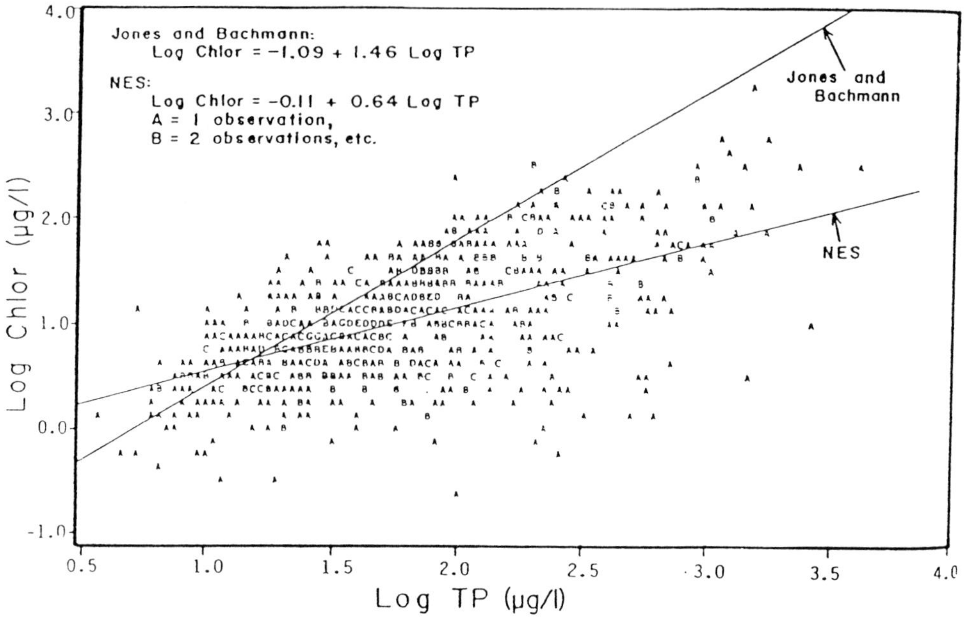

Figure 2.2. Relationship of ambient summer chlorophyll-a to total phosphorus in 757 US lakes samples by NES. Jones & Bachmann's (1976) regression line is given for comparison (from Lambou et al., 1982)

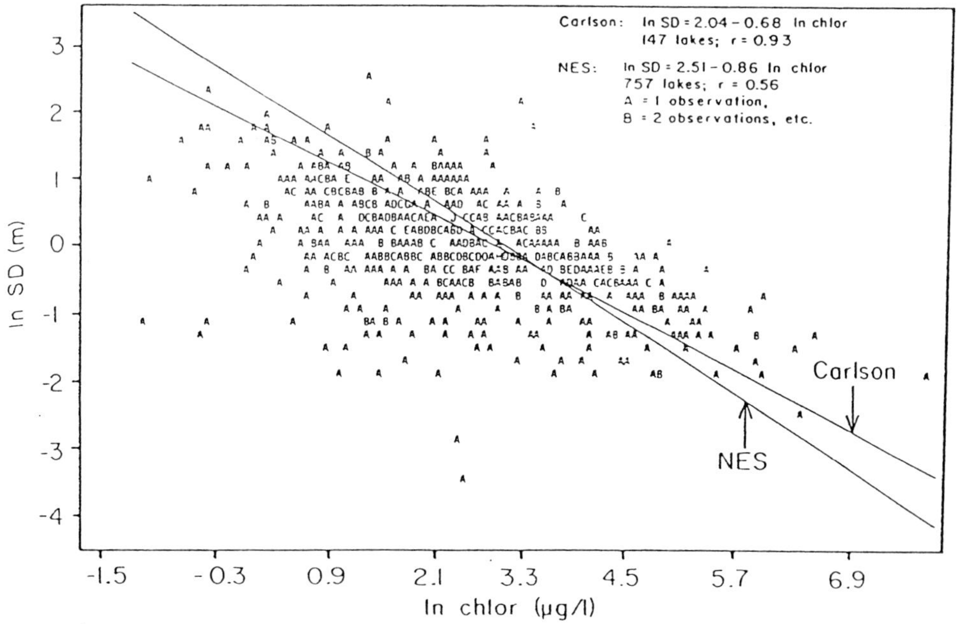

Figure 2.3. Relationship of summer Secchi disk depth to Chlorophyll-a in 757 US lakes sampled by NES. Carlson's (1977) regression lines given for comparison (from Lambou et al., 1982)

The relationships between chlorophyll-a, total phosphorus, Secchi disk depth, and trophic state were examined using data on 757 US lakes collected by the US National Eutrophication Survey (Lambou et al., 1982; see Figs. 2.2 and 2.3). By comparing predicted Secchi disk depths with observed summer Secchi disk depths, it was determined that in many lakes nonchlorophyll-related light attenuation is important in controlling the amount of chlorophyll-a produced per unit of total phosphorus.

Single and multivariable indices which use biological manifestations rather than primary nutrients have been proposed to classify lakes and reservoirs in terms of their trophic status. The trophic state measurements and indices are more successful in ranking lakes against total phosphorus than chlorophyll-a. The use of many of the commonly employed trophic state measurements, which assume or imply that there is a simple relationship between total phosphorus or Secchi disk and chlorophyll, can lead to erroneous conclusions and unnecessary costly management controls. Secchi disk measurements may be more useful as a predictor of ambient lake total phosphorus concentrations than of chlorophyll (Lambou et al., 1982).

2.5 PHOTOSYNTHESIS IN STANDING WATER

Photosynthesis is the assimilation of carbon dioxide, water, and nutrients into plant mass using light as an energy source and chlorophyll-a as en enzymatic catalyst. By the photosynthetic process the algal growth associated with eutrophication is created. Photosynthesis can be described by the equation:

$$6\,H_2O + 6\,CO_2 \rightarrow C_6H_{12}O_6 + 6\,O_2 \tag{2.1}$$

Photosynthesis depends on the availability of light, and lack of light is often a limiting factor. Descriptions of the photosynthetic process are given in numerous references e.g. by Goldman (1966), Fee (1973), and Megard (1984).

The specific rate of photosynthesis of algae depends on irradiance as shown in Figure 2.4. This relationship can be established either in the field using light and dark bottles or in incubators. Productivity can be measured by ^{14}C uptake or rate of oxygen production. The relationship can be expressed by a Haldane equation (Megard et al., 1984) of the form:

$$p = \frac{\mu_m I}{I + K_1 + I^2/K_2} \tag{2.2}$$

where K_1 and K_2 are species-specific coefficients and represent forward and backward reactions of the enzymatic process. The inhibition effects due to interaction of photons and pigments are included in this description. If it is known a priori that inhibition will not occur or if K_2 is very large, the term containing K_2 can be dropped and the expression becomes a Monod type equation, where K_1

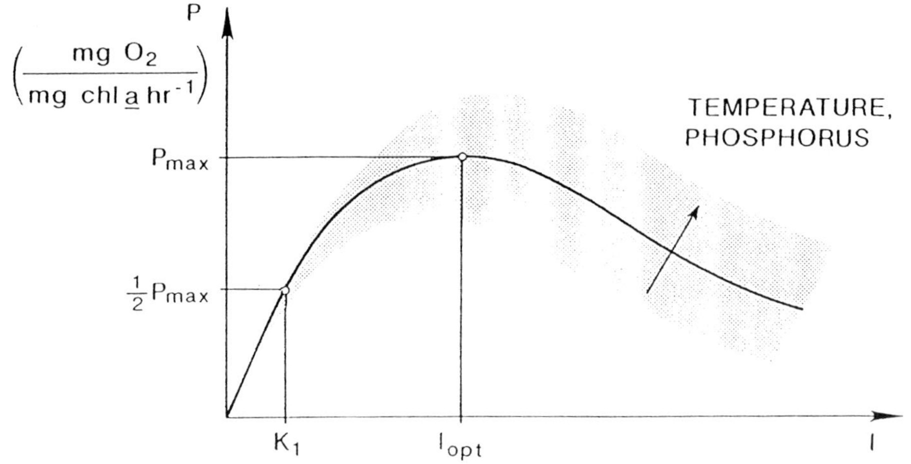

Figure 2.4. Photosynthesis response of phytoplankton to light (irradiance I)

thus is the Michaelis-Menten or half-saturation coefficient for light.

The maximum rate of photosynthesis is:

$$p_{\max} = \mu_m (1 + 2\sqrt{K_1/K_2}) \tag{2.3}$$

The $p(I)$ relationship (Equation 2.2) has also been given in other forms, for example, by Steele, Talling, and others. A review and details have been given by Jassby & Platt (1976) by Goldmann (1966) and by Hino (1981).

The irradiance available below the water surface is a function of incident light and attenuation. It is described by the relationship:

$$I(z) = I(0)e^{-kz} \tag{2.4}$$

where z = depth below water surface, $I(0)$ = incident irradiance at the water surface (μEm^{-2}day^{-1}), k = attenuation coefficient (m^{-1}).

Attenuation of light in water changes with the concentration of dissolved and suspended materials. A composite, total attenuation coefficient can be given as:

$$k = k_w + k_x X + k_c\,[\text{Chla}] + k_{SS}\,[\text{SS}] \tag{2.5}$$

where k_w = attenuation coefficient of water, k_x, k_c, k_{SS} = specific attenuation coefficients, X = concentration of dissolved organic materials, Chla = chlorophyll-a concentration, SS = concentration of inorganic suspended materials.

A typical value for productive lakes is $k_w \approx 0.3$ to 0.7 m^{-1}, and $k_c \approx 0.01$ to 0.04 m^{-1} (mg Chla m^{-3})$^{-1}$, k_x depends on color and k_{SS} on the nature of the suspended material (see Stefan et al., 1983). Absorption and scatter of light contribute to specific attenuation which has been related to the inverse of particle size (Hino & Utahara, 1978). The foregoing equation includes the process of selfshading by the algae.

The volumetric rate of photosynthesis of an algal population is:

$$p_v(z) = p \cdot C(z) \tag{2.6}$$

where $p_v(z)$ = volumetric rate of photosynthesis (μg Chla $l^{-1}hr^{-1}$), $C(z)$ = concentration of Chla (μg/l) as a function of depth z, p = photosynthetic productivity rate coefficient (hr^{-1}).

Because of selfshading the rate $p_v(z)$ varies with concentration of algae as shown in Figure 2.5.

The net growth of the algal concentration $C(z)$ can be found from the relationship

$$\frac{dC(z)}{dt} = (p - r)\, C(z) \tag{2.7}$$

where r = respiration (loss) rate (hr^{-1}).

If vertical mixing by wind is introduced (see Stefan et al., 1976), a value can be derived for the standing phytoplankton crop as a function of the mixed layer depth produced by the wind. The derivation uses the same basic relationships as above. It is, however, assumed that the algal cells are exposed to an average light intensity as they are moved around vertically. The residence time at each depth level is assumed to be the same. The resulting final equation is:

$$C_E = \frac{I(0)}{K_1 h\, k_c r}(p - r) - \frac{k_w}{k_c} \tag{2.8}$$

where the symbols are as previously defined and p is calculated by a Monod expression (K_1 = Michaelis-Menten coefficient and $K_2 = \infty$, in Equation 2.2), h = mixed layer depth. It can be seen that the algal concentration C varies inversely with mixed layer depth h and goes toward a constant $C' = -k_w/k_c$ when h becomes infinitely large. Increasing the mixed layer depth reduces standing algal crop.

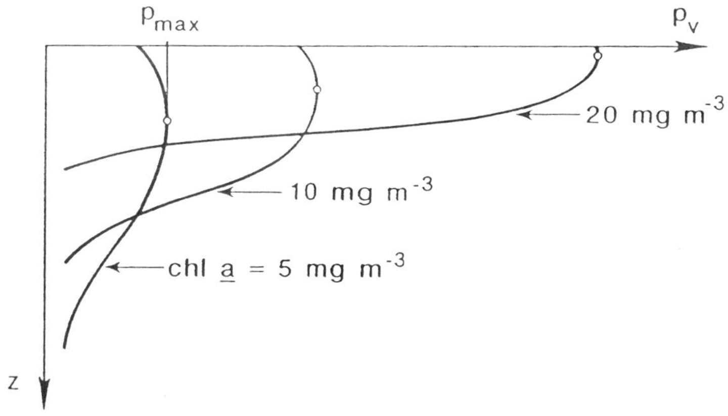

Figure 2.5. Volumetric rate of photosynthesis versus depth

2.6 PHOSPHORUS CYCLING AND PHYTOPLANKTON GROWTH

2.6.1 *Phosphorus uptake and cycling*

Phytoplankton growth in the epilimnion of a lake is largely determined by the available pool of limiting nutrients. Limiting nutrients can be phosphorus, nitrogen, silica or other elements. In many cases phosphorus is limiting, especially when the concentrations of nitrogen to phosphorus ratio (N/P) exceeds a critical value which lies in the vicinity of 15. When the N/P ratio falls below 10, nitrogen becomes limiting and control measures and modeling have to be done for nitrogen as a nutrient (see e.g. Sakamoto, 1966 or Baker et al., 1981).

Although input/output of phosphorus to predict lake trophic status is based on total phosphorus, there is good evidence that much of the phosphorus that enters lakes is unavailable for algal growth (Sonzogni et al., 1976; Tarapchak et al., 1986). The cycling of phosphorus among different forms is schematically shown in Figure 2.6.

Uptake of available phosphorus by algae, as for other nutrients, follows a Monod-type equation

$$\frac{\mathrm{dChla}}{\mathrm{d}t} = \mu_{\mathrm{max}} \left(\frac{I}{K_1 + I}\right) \left(\frac{P_a}{K_p + P_a}\right) \mathrm{Chla} \qquad (2.9)$$

$$\frac{\mathrm{d}P_a}{\mathrm{d}t} = -\mu_{\mathrm{max}} \left(\frac{I}{K_1 + I}\right) \left(\frac{P_a}{K_p + P_a}\right) \frac{\mathrm{Chla}}{Y} \qquad (2.10)$$

where μ_{max} = maximum growth rate (d^{-1}), Chla = chlorophyll-a concentration (mg/m^3), P_a = available phosphorus concentration, I = light intensity, K_1 = half-saturation constant for light, K_p = half-saturation constant for P, Y = yield coefficient (mg Chla/mg P).

In recent years several modifications have been proposed to model algae uptake. First, Equation 2.10 assumes that the yield coefficient, Y, is constant. In fact, Y is not constant but changes, depending on nutrient status. Droop (1973) and others have recognized that nutrient-depleted algae are capable of assimilating inorganic phosphorus very rapidly (much faster than growth rates) and storing it. Thus, Droop (1973) proposed a variable internal cell storage model which describes uptake and growth separately. Some models in current use utilize species-specific constants for nutrient uptake and growth. The advantage of this approach is that the growth of individual species can be modelled. The model may show, for example, a spring bloom of diatoms and a summer blue-green algae bloom. Vertical migration of blue-green algae is also a factor (Okada et al., 1983). The appropriate constants for individual species are available (Lorenzen, 1979; Baker, 1984; Tilman et al., 1982). In developing a model for a particular lake, it is suggested that only the dominant species (or groups) be modelled. Most

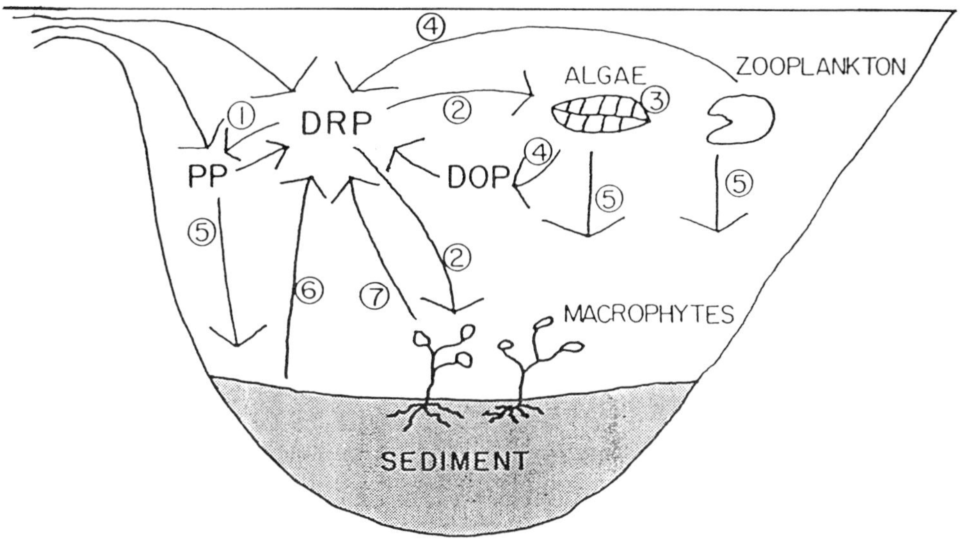

Figure 2.6. Phosphorus cycle in lakes. Processes: (1) adsorption/desorption, (2) assimilation, (3) grazing by zooplankton, (4) regeneration, (5) sedimentation, (6) sediment release, (7) regeneration. DRP = dissolved reactive phosphorus, DOC = dissolved organic phosphorus, PP = particulate phosphorus (after Baker, 1984)

modellers have restricted their model to one to four phytoplankton groups.

Epilimnetic phosphorus regeneration can be modelled using the equation:

$$\frac{dP_a}{dt} = k_r r \frac{\text{Chla}}{Y} \tag{2.11}$$

where k_r = losses by respiration/decay (d^{-1}), r = fraction of P immediately available for conversion to available phosphorus (P_a).

The use of a single term for excretion/mineralization is not strictly correct, and Scavia (1980) uses separate terms to represent excretion of P (by living algae) and decay of dead cells. It is not clear, however, whether this distinction is important, and most models use a single decay/respiration constant for the epilimnion. Values of k_r for several processes, as well as k_r values used in several models range from 0.01 to 0.1 day^{-1} at 20°C (Baker, 1984).

A fundamental, correct approach to modelling temperature effects is to use a modified van't Hoff-Arrhenius equation:

$$\ln\frac{k_1}{k_2} = \frac{[E(T_1 - T_2)]}{RT_1 T_2} \tag{2.12}$$

where k_1, k_2 = reaction rates at temperatures T_1 and T_2 respectively (degrees Kelvin), R = universal gas constant, E = activity energy (kcal/mole).

For a relatively narrow range of temperatures, the term $E/RT_1T_2 \approx$ constant and Equation 2.12 becomes:

$$k_1 = k_2\,\theta^{(T_1 - T_2)} \qquad (2.13)$$

Values for θ typically range from 1.0 to 1.12.

Growth rate is temperature-dependent and individual species have unique temperature optima and maxima. This can be significant in small, eutrophic lakes where summer temperatures may exceed the optimum temperature of some algae, particularly diatoms. Values of lethal and optimal temperatures for several groups of algae are given by Baker (1984), Scavia (1980), and Thomann et al. (1987).

Finally, Equation 2.11 includes the fraction of phosphorus that is convertible to available phosphorus. This accurately reflects findings by Foree & McCarty (1970) and others that a sizeable fraction of the phosphorus in algal cells is never released during decomposition.

2.6.2 *Phosphorus regeneration from lake sediments*

Lake sediments generally are a net sink for phosphorus; whole-lake mass balances show that a large fraction, typically 50 percent or more of the phosphorus load to lakes, is transferred to the sediments mostly via algal uptake and settling. However, in many eutrophic lakes, regeneration of P may be a significant internal source of phosphorus. This is particularly true for lakes in which P inputs have been suddenly reduced and for shallow lakes in which wind mixing periodically mixes the water near the sediments with the overlying water. Similarly, nitrogen release from the sediments has also been considered important (Hosomi et al., 1987).

Exchange of phosphorus across the sediment-water interface is complicated involving several processes including (1) sedimentation of organic and inorganic particulate phosphorus, (2) regeneration of inorganic phosphorus form organic phosphorus during decomposition, (3) adsorption and desorption of inorganic phosphorus on Fe and Al hydroxide surfaces, (4) chemical precipitation of phosphorus containing minerals, particularly apatite, and (5) diffusion.

Phosphorus release rates are largely controlled by adsorption of inorganic phosphorus onto Fe and Al hydroxides. Therefore, phosphorus release rates under anaerobic conditions are nearly always higher than phosphorus release under aerobic conditions. Reported measured values are from 0.2 to 48 mg $m^{-2}d^{-1}$ (Baker, 1984). Local measurements are highly recommended.

Aerobic release rates are usually only 10 to 50 percent of anaerobic release rates for the same sediment. Whole lake studies confirm that increased phosphorus concentrations in the hypolimnion generally coincide with the onset of anaerobic conditions (Kamp-Nielsen, 1974; Larsen et al., 1981; and others).

Several other factors affect phosphorus release rates. These include temperature, pH, bioturbation, and macrophyte growth.

Model calibration can be used to compute release rates from lake phosphorus concentrations and known input rates. This is commonly done for stratified lakes when inputs to the hypolimnion (settling of algae and inorganic particulate phosphorus) are known. The problem with model calibration in this manner is that it limits the predictive capability of the model.

Complex models of phosphorus release rates have been formulated (Kamp-Nielson, 1976; Lund et al., 1976), but these are both difficult to use and fail to recognize individual physical-chemical processes. More process oriented models are under development and include decomposition rates, adsorption of inorganic phosphorus, release of phosphorus upon onset of low redox (or low O_2) conditions, resuspension, and diffusion rates to or from the sediments.

2.7 PHOSPHORUS 'LOADING' MODELS FOR EUTROPHICATION

Dillon and Rigler (1974b) observed in northern dimictic lakes that summer Chla, in terms of average or maximum, relates to spring phosphorus or total annual phosphorus input. They give the relationship as

$$\text{Chla} = 0.073\, C_p^{1.45} \tag{2.14}$$

where Chla = summer chlorophyll-a (μg/l), C_p = total spring phosphorus (μg/l).

From continuous flow reactor theory the mean annual equilibrium phosphorus concentration can be calculated as

$$C_E = \frac{L}{Q_o + A_s V_s} \tag{2.15}$$

where L = loading = amount of total phosphorus added per year (kg/yr). Q_o = total annual outflow (m^3/yr), A_s = surface area (m^2), V_s = settling velocity (m/yr).

Apparent settling velocities V_s are on the order of 10 to 28 m/yr (Chapra, 1977; Vollenweider, 1975; Dillon, 1975; Edmondson, 1970). Chapra et al., (1976) added the empirical relationship

$$C_E = 0.9\, C_p \tag{2.16}$$

Combining Equations 2.14, 2.15, and 2.16 gives:

$$\text{Chla} = 1866 \left(\frac{L}{Q_o + AV_s} \right)^{1.45} \tag{2.17}$$

where L is in g/yr, Q_o is in m^3/yr, V_sA is in m^3/yr, Chla is in μg/l.

This can be inverted to

$$L = 0.005\, (\text{Chla})^{0.69}\, (Q_o + AV_s) \tag{2.18}$$

The acceptable phosphorus loading for an acceptable chlorophyll level can be

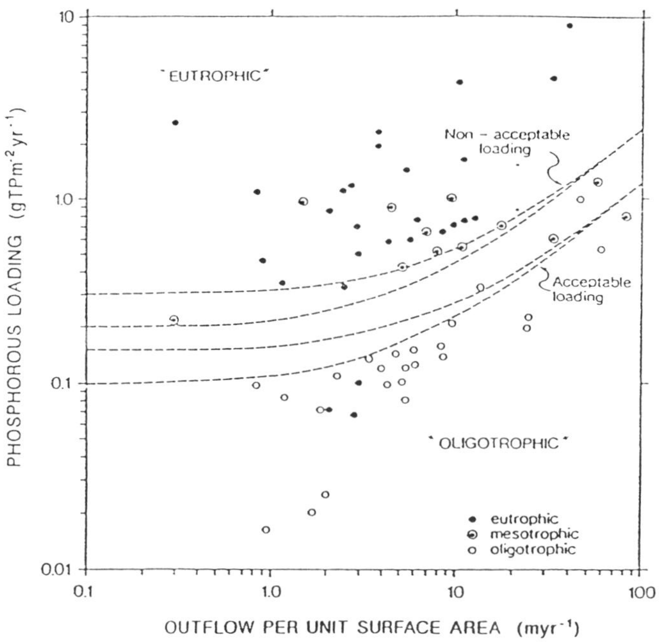

Figure 2.7. Lake loading tolerance for phosphorus (after Vollenweider, 1968)

specified from this equation plotted in Figure 2.7. Other similar models exist.

The best-known of these 'loading models' or input/output models are the Vollenweider (1968, 1969, 1975, 1976) model, the Dillon & Rigler (1974a, 1975) model, and the Jones & Bachman (1976) model. Chapra & Tarapchak (1976) show how these models can be developed. A recent summary was given by Walker (1985). An extension, including zooplankton, was prepared by Hino (1983).

2.8 DYNAMIC EUTROPHICATION AND WATER QUALITY SIMULATION MODELS

By combining hydrodynamic advection/diffusion transport relationships with chemical and biological kinetics, process-oriented, time-variable models have been developed (see review by Stefan et al., 1988).

2.8.1 *Zero-dimensional models of reservoirs, ponds, and lakes*

Zero-dimensional (0-D) water quality models of lakes, reservoirs, and ponds contain no information on hydrodynamics other than the hypothesis that the water body is well mixed. They are also referred to as input-output models, box models, and loading models (see preceding section). The input I_v of a material (nutrient, toxic, etc.) in units of kg $m^{-3}yr^{-1}$ and the concentration $C(t)$ of the material in units of kg m^{-3} in the well-mixed water body are then related by the conservation equation:

$$\frac{dC}{dt} = I_v - \frac{C}{\theta} \tag{2.19}$$

The term I_v is the sum of material inflows (a) from the watershed through tributaries, drains, sewers, and overland flow, (b) from the atmosphere, (c) from groundwater, and (d) from internal sources, e.g., the sediments. The input I_v is related to the loading L in kg yr^{-1} and the lake volume V in m^3 by $I_v = L/V$. A distinction between nonpoint sources and point sources for loading is common.

The parameter θ is a time constant (e.g. yr). It sums the rate coefficients of all first order processes by which the material is removed from the water, for example, by (1) an outflow, (2) a settling process, (3) a chemical transformation (reaction), or (4) a biological transformation all in units of kg $m^{-3}yr^{-1}$. The rate coefficients C_o/V for outflow, AV_s/V for settling loss, K_1 for biological transformations are related to the time constant θ:

$$\frac{1}{\theta} = \frac{Q_o}{V} + \frac{V_s A}{V} + K_1 + K_2 = \frac{1}{\theta_h} + \frac{1}{\theta_s} + \frac{1}{\theta_1} + \frac{1}{\theta_2} \tag{2.20}$$

It is evident from Equation 2.20 that different residence times exist for water and materials in lakes/reservoirs. Failure to recognize important processes in O-D models, such as material release from the sediments, has led to mispredictions of phosphorus in eutrophic lakes.

0-D models also can be used to simulate series of lakes or lake networks by using the output $C(t)$ of the most upstream lake as input to the following one, and so on. An application to the North American Great Lakes was given by Chapra (1977). It was used to evaluate and predict the effectiveness of several phosphorus management alternatives including P-ban in detergents, P-removal from municipal sewage by tertiary treatment, changes in agricultural practices, etc. A summary of the model and its application is given in Chapra & Reckhow (1983). Lake Biwa, Japan, was also simulated sucessfully with a 0-D model (Ikeda et al., 1978).

The idea has been further extended to interconnected bays and channel systems such as found in coastal plains or in shallow impoundments (Stefan & Anderson, 1980; Demetracopoulos & Stefan, 1983). In that case, gravity and wind-driven circulation could be simulated on a daily or even hourly time scale. Shortening the time scale is essential when flows are highly time-variable in response to weather. 0-D models are also applied to bays connected to larger water bodies (see for example, Chapra, 1979). In that case the exchange flow rates (in and out) across the mouth of the bay must be specified based on either tracer observations or hydrodynamic analysis. Exchange across the mouth of the embayment can be larger than inflow.

2.8.2 *Dynamic one-dimensional models of reservoirs, ponds, and lakes*

If water quality variations are to be predicted throughout a season, on a weekly, or even daily scale, internal hydrodynamics of a lake or reservoir cannot be ignored. Temperature stratification occurs in most 'standing waters' in temperate regions, and the associated density stabilities create vertical gradients in water quality during the summer and winter seasons. One-dimensional water quality models that describe water temperature and concentrations ($C(z, t)$) as a function of depth z and time t are therefore needed. In such models the lake or reservoir is described by a system of horizontal layers, each of which is well mixed. Vertical transport of heat or material compounds between layers is described by a diffusion equation in which a bulk vertical dispersion/diffusion coefficient $K_z(z)$ is incorporated.

The transport and transformation of dissolved substances in a 1-D vertically stratified water quality model is described by

$$V\frac{\partial C}{\partial t} = -Q_v\frac{\partial C}{\partial z}\Delta z + \frac{\partial}{\partial z}\left(K_z A\frac{\partial C}{\partial z}\right)\Delta z + Q_i C_i - Q_o C \pm S \qquad (2.21)$$

where Q_v is the vertical advective flowrate through a stationary layer of thickness Δz due to inflow to the reservoir or outflow from it (Orlob & Selna, 1970; Huber et al., 1972). The sink/source term 'S' includes chemical and biological transformations. One of the simplest is a single first order reaction, $S = K_1 CV$, where K_1 is a rate coefficient. For suspended materials a settling term, $-[\partial(V_s CA)/\partial z]]\,\Delta z$, must be added to the right-hand side of the equation, where V_s = settling velocity of the particles.

The vertical turbulent diffusion coefficient is usually introduced as an empirical function $K_z(\varepsilon)$ of a density stability parameter ε or N^2 where:

$$\varepsilon = \frac{1}{\rho}\frac{d\rho}{dz} \quad \text{and} \quad N^2 = \varepsilon g \qquad (2.22)$$

Uncertainty in the K_z values is often large. Attempts to relate K_z not only to stability, which hinders the development of vertical turbulence, but also to the forcing elements, particularly wind and inflows, have only been partially successful (Ford & Johnson, 1986; McCormick & Scavia, 1981).

A large number of 1-D water quality models exist for lakes and reservoirs. A survey was given by Orlob (1983). Examples are CE-QUAL-R1 (Env. Lab., 1986a), the Hydrologic Engineering Center HEC5 models, the Huber et al. model (1972), Imberger and Patterson's DYRESM model (Fischer, 1981), the Stefan et al. RESQUAL II (1982), and the Riley et al. MINLAKE model (1988b). Other similar 1-D reservoir models are TVA's RESTEMP (Brown & Shiao, 1981) and WRMMS (TVA, 1976) models, and a USGS model (House, 1981). All of these basically 1-D hydrodynamic models simulate temperature stratification, but CE-QUAL-R1 and MINLAKE also handle a large array of water quality parameters such as dissolved oxygen, algae, nutrients, and conservative substances.

CE-QUAL-R1 is designed for reservoirs; MINLAKE is intended for lakes. A similar model for small rural impoundments named FARMPOND was developed by the USDA/ARS. An example for the application of a 1-D eutrophication model to a reservoir in Japan was given by Miyanaga (1986).

2.8.3 *Two-dimensional models of reservoirs, ponds, and lakes*

Two-dimensional water quality models were developed for long, deep reservoirs in which significant vertical WQ gradients are coupled with horizontal ones. For example, Lake Powell and Lake Mead, reservoirs on the Colorado, River require such an approach. There are a few 2-D models available. One is the LARM model (Laterally Averaged Reservoir Model) developed by Edinger & Buchak (1983) and Buchak & Edinger (1984a, 1984b); another is the COORS (Computation of Reservoir Stratification) model by Waldrop et al. (1980), Harper et al. (1980a, 1980b) and TVA (1986). Both models solve two-dimensional advection/ diffusion equations in a vertical longitudinal plane through a reservoir. The models are width integrated. These models can predict the 2-D temperature structure of deep reservoirs throughout the annual stratification cycle and compute temporal and spatial hydrodynamics of reservoirs to provide advective components for water quality models. A later version of LARM referred to as CE-QUAL-W2 (Env. Lab., 1986b) has been modified by the Corps of Engineers to include 20 water quality constituents.

Selection of a width-integrated x, z-coordinate system in a vertical plane is appropriate for 2-D models of long and deep reservoirs. Depth-integrated models using x, y-coordinates are appropriate for shallow and wide lakes. Ligget, among others, has developed hydrodynamic depth integrated models for this purpose. Bennett et al. (1983) describe a typical 2-D lake circulation model. These models are typically driven by windshear, and include no stratification effects on vertical turbulence since they are depth integrated. This approach was used, for example, to model Lake Balaton in Hungary (van Straten & Somlyodi, 1980). Other examples are given by Shanahan & Harleman (1982).

2.8.4 *Three-dimensional models of reservoirs, ponds, and lakes*

Three-dimensional water quality models have not been used extensively because of the computational expense, as well as for other reasons. Three-dimensional flow simulations of lakes and reservoirs have been made by Simons (1973, 1975, 1976) and Leendertse et al. (1975), and verified in both small and large lakes. A special version was produced by Kielmann and used for pollutant transport in the Baltic Sea (Funkquist et al., 1984). Some results obtained for the North American Great Lakes have been used, for example, in a 3-D WQ model for Lake Ontario (Thomann et al., 1975, 1979). The WQ models are of the multiple box type, i.e., multi-layered in the vertical and each layer subdivided into very large elements representative of the coastal regions and the pelagic water. Advective and dispersive flows between boxes are determined by flow budgets based on long-term observed hydrology.

A 3-D model is being developed to study water quality problems in Chesapeake Bay, USA.

2.8.5 *Availability of models*

Numerous simulation models of phytoplankton, detritus, and nutrients have been applied to various water bodies. Of these, only a few have been maintained for general application. To be useful for eutrophication, nutrient and organic waste simulations, a model should allow for unsteady transport and loading, phytoplankton kinetics, nitrogen and phosphorus cycles, and carbon-dissolved oxygen balances. Furthermore, written documentation and a user manual must be available with the software.

Lake/reservoir modeling frameworks that meet these criteria are WASP4-(EUTRO4), CE-QUAL-R1, and CE-QUAL-W2. Such models as DEM, EXPLORE, RECEIVII, WRECEV, MINLAKE, and AUTO-QD meet these criteria but are not actively maintained and supported by a central, public distribution center. Different versions of these and other models may be maintained privately by consultants and local government agencies, and may be useful for those water bodies where set-up and calibration have been accomplished previously.

Eutrophication WASP (EUTRO4; Ambrose et al., 1987) combines a kinetic structure adapted from the Potomac Eutrophication Model with the WASP transport structure. EUTRO4 predicts dissolved oxygen, carbonaceous biochemical oxygen demand, phytoplankton, carbon, and chlorophyll-a, ammonia, nitrate, organic nitrogen, and orthophosphate in the bed and overlying waters.

WASP4 is linked to DYNHYD4 (Ambrose et al., 1987), a simple link-node hydrodynamic model capable of handling variable tidal cycles, wind, and unsteady inflows. It produces an output file that can be lined with WASP4 to supply the flows and volumes to the water quality model.

The Corps of Engineers' vertically one-dimensional model CE-QUAL-R1 (Environmental Laboratory, 1986a) simulates temperature plus as many as 34 other water quality parameters. Primary physical processes included are surface heat transfer, shortwave and longwave radiation and penetration, convective mixing, wind and flow-induced mixing, entrainment of ambient water by pumped-storage inflows, inflow density current placement, selective withdrawal, and density stratification as impacted by temperature and dissolved and suspended solids. Major chemical and biological processes in CE-QUAL-R1 include: the effects on DO of atmospheric exchange, photosynthesis, respiration, organic matter decomposition, nitrification, and chemical oxidation of reduced substances; uptake, excretion, and regeneration of phosphorus and nitrogen and nitrification-dentrification under aerobic and anaerobic conditions; carbon cycling and alkalinity-pH-CO_2 interactions; trophic relationships for phytoplankton and macrophytes; transfers through higher trophic levels (i.e., zooplankton and fish); accumulation and decomposition of detritus and organic sediment; coliform bacterial mortality; and accumulation and reoxidation of manganese, iron, and sulfide when anaerobic conditions prevail.

The Corps of Engineers has applied the one-dimensional water quality model CE-QUAL-R1 to a variety of existing and projected reservoirs, including Degray Lake, AK, (Kennedy & Nix, 1987) and Howard A. Hanson Reservoir, Wash., (Schneider & Price, 1988). The two-dimensional reservoir water quality model CE-QUAL-W2 is also available but has not been widely used.

The MINLAKE model is a dynamic, one-dimensional model simulating daily changes in the depth profiles of temperature, chlorophyll-a, biologically available phosphorus, cellular phosphorus, detritus and detrital phosphorus, dissolved oxygen, and suspended and dissolved solids with the optional simulation of nitrate-nitrite nitrogen and ammonium nitrogen. Chlorophyll-a can be simulated using up to three algal groups. One vertical migrating zooplankton population is also incorporated to simulate grazing effects. Zooplankton is the highest trophic level explicitly simulated in the model. It was decided that simulating higher trophic levels would significantly increase complexity and decrease reliability because of the lack of information on fish population dynamics in most lakes.

The MINLAKE model (Riley et al., 1987, 1988a, 1988b) was developed around six major criteria: (1) dynamic, process-oriented, simulating all dependent variables on a one-day time step or less; (2) general enough to handle a wide range of lakes with a minimum of alterations; (3) flexible to incorporate the unique qualities of a specific lake; (4) incorporate the dominant physical, chemical, and biological processes affected by lake treatment methods; (5) physical, chemical, and biological submodels of similar orders of complexity; and (6) economical to operate.

The lake is analyzed by subdividing the depth into a maximum of 40 layers with the layer thickness varying between upper and lower levels as set by the user.

The model is structured around four major subsections (Fig. 2.8). The first is model input, which determines how and what will be simulated in the current simulation. Second is the hydrodynamic or physical processes section. Central to this section are heat budget and wind mixing and the resulting density stratification of the water body and the simulation of inflow as either an epilimnion inflow or plunging density current. Third is the biological-chemical processes section that simulates the concentrations of dissolved and suspended substances in the lake. Finally, the model output routes the results to tabular and graphical output files. Outflow volumes and concentrations may be routed to an outflow data file in a format that allows the outflow to be used as the inflow in the simulation for a downstream lake.

The MINLAKE model treats phosphorus as the principal chemical affecting

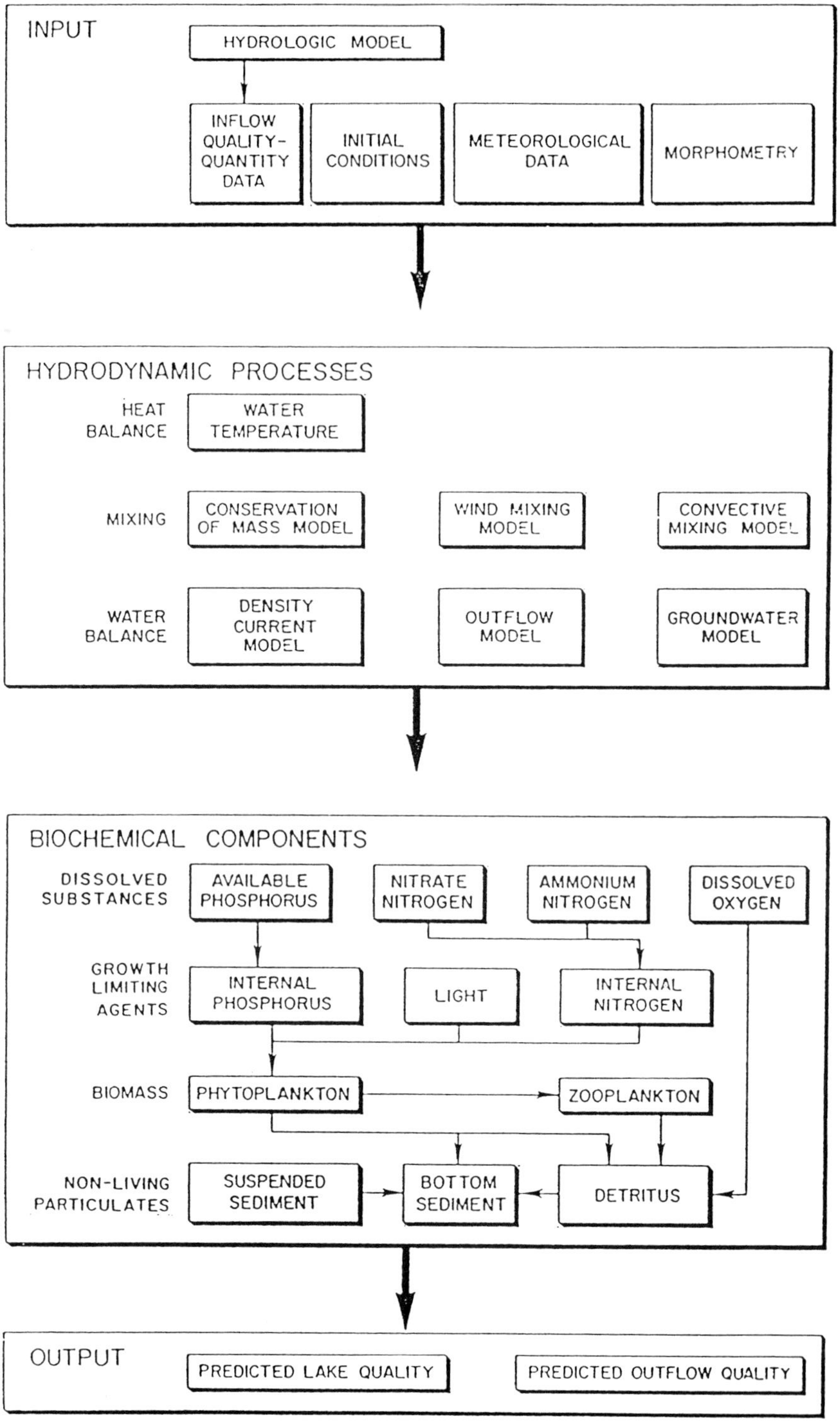

Figure 2.8. Flow chart of major model sections and submodels in the MINLAKE program

the chlorophyll-a concentration. Phosphorus is modeled in two pools: detrital phosphorus and dissolved reactive phosphorus. Often, only total phosphorus is measured in the field, in which case the dissolved reactive and detrital phosphorus are combined with algal phosphorus to represent total phosphorus.

Nitrogen simulation in the model is included in the third chlorophyll-a model level to aid in simulating shifts in the algal populations. The intent is to use the nitrogen simulation submodel only in the simulation of multiple algal groups. Algal preference for the uptake of ammonium nitrogen over nitrate-nitrite nitrogen is expressed through an ammonium preference coefficient included in the uptake computations (Scavia, 1980). Nitrification is included as a first order reaction on the concentration of ammonium.

Details on the model formulation can be found elsewhere (Riley et al., 1987, 1988b).

2.9 PROTECTION AND RESTORATION

2.9.1 *Background information*

Protection of lakes and reservoirs from eutrophication has been of international concern for a number of years. Beginning with a major review of the problem in 1967 (NAS, 1969), the literature on eutrophication and its control has become most prolific. Summaries and milestones that provide overviews and further references have been given by the WHO (1981), OECD (1982), ASCE (1981, 1983), Chapra & Reckhow (1983), Cooke et al. (1986), USEPA (1975, 1979, 1980b, 1984, 1985, 1988), Henderson-Sellers (1984), TVA (1979), and numerous publications by NALMS (North American Lake Management Society, founded in 1980). Simulation (modeling) of eutrophication is discussed in books by the USEPA (1976, 1980a), Hydroscience (1976), Thomann & Mueller (1987), Middlebrooks et al. (1974), and Canale (1976). There are numerous excellent publications on nutrient management, such as by USEPA (1983, 1985, 1987), NALMS (1985, 1986, 1987), Sas (1989), Ryoling & Rast (1989), Jørgensen & Löffler (1990), and the National Research Council (1992).

Common lake and reservoir problems are excessive algal growth, deficient fishery, poor potable water quality, and excessive macrophyte growth. Water quality in terms of transparency, color, and odor is often related to the amount of algae present.

Eutrophication is responsible for many water supply problems (WHO, 1981).

2.9.2 *Methods of lake and reservoir protection and restoration and their simulation and evaluation in models*

Lake treatment processes can be divided into watershed practice methods, inflow-outflow methods, and in-lake treatment methods. Guidance manuals for lake and reservoir restoration projects have been provided by the Wisconsin DNR (1974), USEPA (1988), National Water and Soil Conservation Authority (Vant, 1987), Ryoling & Rast (1989), and Sas (1989), among others.

Any lake management strategy may be of limited effectiveness, and if effective, may have secondary and undesirable effects on other lake processes. For example, removing algae increases light penetration and may foster undesirable macrophyte growth.

Cultural eutrophication is a central concern in lake management. Although many methods have been developed to rectify the adverse impacts of cultural eutrophication, their effectiveness has varied widely among different lakes. This is in part due to a lack of detailed, quantitative methods to assess the relative importance of processes affecting eutrophication and the potential success of different lake treatment strategies.

Mathematical models are developed to address this need. Some eutrophication models are simplistic, addressing only the impact of one process and ignoring the interplay of physical, chemical, and biological processes (Jones & Bachman, 1976; Dillon & Rigler, 1974b; Vollenweider, 1975; Walker, 1985). Advanced mathematical models become large, cumbersome, and expensive to operate, and highly specific to one lake (Scavia, 1980; Bierman, 1981; Matsuoka et al., 1986; Cochrane et al., 1987). Models that bridge the gap between simplistic and complex models, and are versatile and management-oriented, are being developed.

Lakes are in a continuous state of change. Some changes occur over very short periods of time, on the order of hours; other changes mark long-term trends in the biological and physical conditions of a lake. Among the long-term trends in many lakes are a decrease in depth and an increasing productivity.

The uniqueness and complexity of each lake and lake system require an individualized assessment for lake management and restoration techniques.

The lake manager must acquire a sufficient data base to determine what dominant lake processes cause a lake problem, decide which management techniques affect lake processes sufficiently to reduce the problem, and evaluate cost and benefits of management techniques. Modeling is a tool that can provide the information necessary for this decision-making process.

Ideally, the modeling of lake water quality simulates lake processes and their interconnected and interdependent relationships. Modeling offers the lake manager a sophisticated evaluation tool at a comparatively low cost to analyze expensive lake management techniques and to optimize the money spent on a lake management program. A primary limitation of model prediction is the data base with which the model is verified.

Almost all lake improvement techniques are designed to change one or more of the following (Fig. 2.9):

1. The inflow of water, nutrients, and sediment into a lake;
2. The loss of algal population by settling or grazing or outflow;

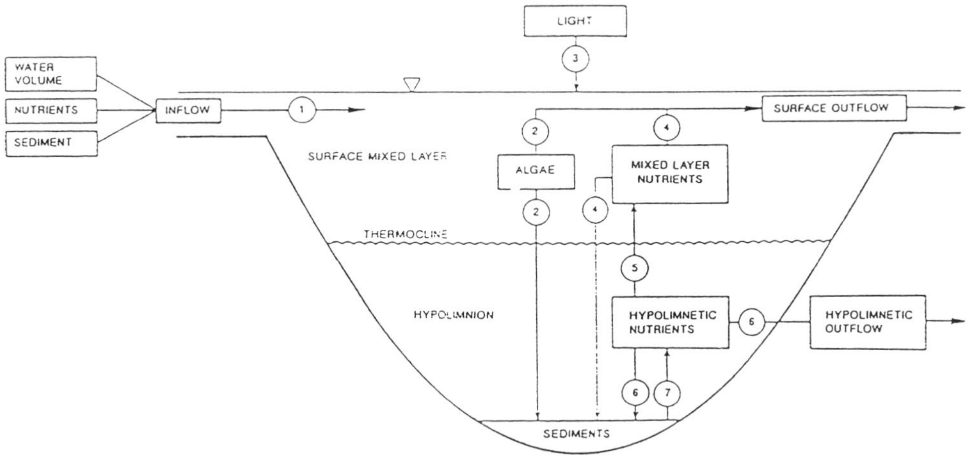

Figure 2.9. Control points (1 through 7) in eutrophication control techniques

3. The availability of light to an algal population within a lake;
4. The nutrients in the photogenic or well-mixed surface zone of a lake;
5. The transfer of nutrients from a thermally isolated hypolimnion to a surface layer where the nutrients can become available for algal growth;
6. The quantity of nutrients in the hypolimnion;
7. The sedimentary release of nutrients.

Inflow control methods affect the quantity and quality of the inflow. If the inflow is important, a hydrologic runoff quality model can be used to simulate inflow quantity and quality. Submodels can be added to the hydrologic runoff quality model to simulate individual inflow control alternatives.

Effects of inflow diversion or wastewater treatment can be specified in both quantity and quality. These types of point source inflow control methods are easier to simulate than non-point source methods.

Lake deepening methods are usually intended to reduce resuspension of bottom sediments and to reduce the release of sedimentary nutrients. These methods include in situ consolidation of sediment by physical, chemical, and biological methods. Lake deepening methods can be modeled by changing the depth area curves in the model to reflect the new morphology of the lake. The bottom release simulation may have to be altered if lake deepening is accomplished to either a nutrient-rich or a nutrient-poor layer of sediment.

Sediment nutrient inactivation is intended to reduce hypolimnetic nutrients by either precipitating them during treatment and/or preventing further release of nutrients from the sediments.

Water column nutrient inactivation has two objectives: to reduce the nutrients in the photogenic surface waters and to reduce hypolimnetic nutrients. The water column nutrient inactivation methods precipitate nutrients from the entire water column. The simplest method of simulating water column nutrient inactivation is by removing a set quantity of the nutrient from each layer as determined from laboratory analysis for the effectiveness of the inactivation techniques.

Water column mixing methods are designed (1) to reduce the light availability to algae by increasing the mixed-layer depth and (2) to maintain aerobic conditions in the bottom waters to prevent or reduce sediment nutrient release. A USEPA manual (Lorenzen & Fast, 1977), among other similar publications, provides some guidance to selection and design.

Mixing techniques have been tried on a large number of lakes with varied results. If compressed air is used, it is difficult to predict (without analysis) what the resulting mixing in the lake will be.

Hypolimnetic oxygenation is designed to improve fish habitat and/or to reduce sedimentary phosphorus release. Based on the increased concentration of oxygen in the hypolimnion, the rate of exchange of nutrients between the sediment and the water column have to be simulated.

Nutrient outflow acceleration by selective discharge affects (1) the algal population by removing some of that population, (2) nutrient transfer either by removing nutrient-laden layers or by decreasing the size of nutrient-laden layers. Flushing and dilution methods have the same effect but also alter the inflow input.

Herbicide treatment methods are designed to directly reduce algal populations within the lake. *Macrophyte control methods* affect the population of macrophytes and are usually designed from a water quality perspective to reduce the nutrient recycling from the macrophytes. *Fish control methods* that are used to improve water quality are usually designed to reduce nutrients in the surface layers by reducing the fish population, especially bottom-feeding fish, that causes the recycling of nutrients from the sediment.

2.10 DETERMINATION AND SIMULATION OF WATERSHED INPUTS TO LAKES AND RESERVOIRS

Lakes and reservoirs receive water, sediments, and nutrients from a watershed. To apply loading models or dynamic simulation models, it is necessary to specify that input from the watershed.

It is customary to distinguish between point sources and nonpoint (diffuse) sources of nutrients. Point sources are relatively easily identified and control technology is available. Details can be found in the USEPA *Design Manual on Phosphorus Removal* (1987).

In highly developed areas surface water loadings of nitrogen and phosphorus from human excrement and other sources are on average 12 g N person^{-1}day^{-1} and 2.25 g P person^{-1}day^{-1}. Other values apply in lesser developed regions.

Input from nonpoint sources is usually related to agricultural practices, in particular feedlot operations, fertilizer applications and tillage methods. Nonpoint source loading rates are much more difficult to obtain and highly variable from site to site. Many field measurements of phosphorus loading have been made, but there is still a large uncertainty in the actual values. Detailed rates of loading from urban watersheds were summarized by USEPA (1983) and Jennings (1988b) after EPA research.

Nutrient input from the watershed can be determined either by direct field measurements or with the aid of watershed models. A major effect may be necessary to obtain watershed input information. It took several years, for example, to generate the phosphorus loading information for the North American Great Lakes loading model.

Since the desired water quality varies among countries and geographical areas, the same specific phosphorus load may be interpreted differently in terms of water quality objectives in these areas. For example, the natural phoshorus load in a lake surrounded by high natural soil fertility may make the reservoir eutrophication prevention objective unrealistic. In heavily populated, intensely cultivated areas, determination of the 'natural' phosphorus load may be difficult. In agricultural and urban watersheds, nutrient runoff is so large that shallow lakes and reservoirs can hardly be protected from eutrophication.

In dynamic lake and reservoir models, the watershed input must be specified at time steps of a day, week, or month rather than a whole year as is common for loading models. Watershed runoff models are usualy available for event simulation, but not for continuous flow simulation. Watershed models that simulate water quality have been developed for urban watersheds (for example SWMM) and for agricultural watersheds. A recent review of the latter can be found in the proceedings of the International Symposium on Water Quality Modeling of Agricultural Nonpoint Sources (DeCoursey, 1988). Nonpoint source loading models in use by the USGS have been reviewed by Jennings (1988a, b). At present most of the available watershed models do not produce the kind of output needed as input to dynamic lake and reservoir water quality models. The applicability of different levels of watershed models to dynamic lake and reservoir water quality modeling has been reviewed by Hendrickson et al. (1985), with the conclusion that at present simpler models based on, for example, the universal soil loss equation or curve number procedures may be as reliable to generate watershed nutrient input as more sophisticated ones. The top-of-the-line model in this category may be the Hydrological Simulation Program – FORTRAN (HSPF) (Johanson et al., 1984) which is a comprehensive package for simulation of watershed hydrology and water quality for both conventional and toxic organic pollutants. HSPF incorporates the watershed-scale ARM (Agricultural Runoff

Model) and NPS (Nonpoint Source) models into a basin-scale analysis framework that includes pollutant transport and transformation in stream channels.

The uncertainty of phosphorus loading by runoff from watersheds was reviewed by Reckhow et al. (1980) for the USEPA. Reviews by Uttormark et al. (1974) provide loading functions based on best estimates. Loading functions depend on land use that varies from country to country, and values given e.g. for US sources should not be applied without field verification to other countries.

From a mechanistic point of view, nutrient loading is the result of many parameters and processes interactions. They include topography, soil characteristics, rainfall intensity, fertilizer application, etc. The reader interested in these processes will find information in the agricultural literature, e.g., on Modeling of Chemical Transfer (Ahuja, 1986), or on the Impact of Fertilizer Phosphorus (Sharpley & Menzel, 1987).

REFERENCES

Ahuja, L.R. 1986. Characterization and modeling of chemical transfer to runoff. In B.A. Stewart (ed.), *Advances in Soil Science*, 4: 149-188. New York: Springer.

Ambrose, R.B., Jr. 1987. WASP4, A general water quality model for toxic and conventional pollutants. US Environmental Protection Agency, Athens, GA.

ASCE 1981. Surface water impoundments. H.G. Stefan (ed.), *Proceedings of a Conference.*

ASCE 1983. Accomplishments and impacts of reservoirs. In G.G. Green & E.E. Eiker (eds), *Proceedings of a symposium, Houston, Texas, Oct.* 20-21, 1983.

Baker, L.A. 1984. A model for phosphorus cycling in lakes. Internal memo No. 100, St. Anthony Falls Hydraulic Laboratory, Univ. of Minnesota, Minneapolis, MN, 31 pp.

Baker, L.A., P.L. Brezonik & C.R. Kratzer 1981. Nutrient loading – Trophic state relationships in Florida Lakes, Publ. No 56. Water Resources Research Center, University of Florida, 126 pp.

Bennett, J.R., A.H. Clites & D.J. Schwab 1983. A two-dimensional lake circulation modeling system: Programs to compute particle trajectories and the motion of dissolved substances. NOAA Tech. Memorandum ERL GLERL-46, Ann Arbor, MI, 51 pp.

Bierman, V.J., Jr. 1981. Modelling phytoplankton-nutrient dynamics in Saginaw Bay. *Jour. of Great Lake Research*, 7(4).

Brown, R.T. & M.C. Shiao 1981. Reservoir temperature modeling: Case study of Normany Reservoir. TVA Water Systems Development Branch, Rep. WR28-1-86-101.

Buchak, E.M. & J.E. Edinger 1984a. Generalized, longitudinal-vertical hydrodynamics and transport: Development, programming and applications. Document No. 84-18-R, US Army Corps of Engineers, WES, Vicksburg, Mississippi, June 1984.

Buchak, E.M. & J.E. Edinger 1984b. Simulation of a density underflow into Wellington Reservoir using longitudinal-vertical numerical hydrodynamics. Document No. 84-18-R, US Army Corps of Engineers, WES, Vicksburg, MS, March.

Canale, R.P. (ed.) 1976. *Modeling Biochemical Processes in Aquatic Ecosystems*. Ann Arbor Science, 389 pp.

Carlson, R.E . 1977. A trophic state index for lakes. *Limnol. Oceanogr.*, 22: 361-369.

Chapra, S.C. 1977. Total phosphorus model for the Great Lakes, *Jour. of Environmental Engineering, ASCE*, 103:147-161.

Chapra, S.C. 1979. Applying phosphorus loading models to embayments. *Limnology and Oceanography*, 24(1): 163-168.

Chapra, S.C. & K.H. Reckhow 1983. *Engineering approaches for lake management, 2: Mechanistic modeling*. Woburn, MA: Butterworth Publishers.

Chapra, S.C. & S.J. Tarapchak 1976. A chlorophyll-a model and its relationship to phosphorus loading plots for lakes. *Water Resources Research*, 12(6): 1260-1264.

Cochrane, K.L. et al. 1987. An ecosystem model of phosphorus cycling in a warm monomictic, hypertrophic impoundment. *Ecological Modelling*, 37: 207-233.

Cooke, D.G., E.B. Welch, S.A. Peterson & P.R. Newroth 1986. *Lake and reservoir restoration*. Boston: Butterworth.

DeCoursey, D. (ed.) 1988. Water quality modelling of agricultural nonpoint sources. *Ag. Res. Service, USDA* (in print).

Demetracopoulos, A.C. & H. Stefan 1983. Model of Mississippi River pool. *ASCE, Jour. Hydraulics Division*, 1006-1029.

Denny, P. 1972. Sites of nutrient limitation in algae. *Jour. Phycol*, 9: 264-272.

Department of Natural Resources 1974. Survey of lake rehabilitation techniques and experiences. Tech. Bull. No. 75, Madison, Wisconsin, 179 pp.

Dillon, P.J. 1975. The phosphorus budget of Cameron Lake, Ontario: The importance of flushing rate to the degree of eutrophy of lakes. *Limnol. Oceanogr.*, 20: 28-39.

Dillon, P.J. & F.H. Rigler 1974a. A test of a simple nutrient budget model for predicting the phosphorus concentration in lake water. *J. Fish. Res. Bd. Can.*, 31: 1771-1778.

Dillon, P.J. & F.H. Rigler 1974b. The phosphorus-chlorophyll relationship in lakes. *Limnology and Oceanography*, 19(4): 767-773.

Dillon, P.J. & F.H. Rigler 1975. A simple method for predicting the capacity of a lake for development based on lake trophic status. *J. Fish Res. Board*, Canada, 31(9): 1519-1531.

Droop, M.R. 1973. Some thoughts on nutrient limitation in algae. *Jour. Phycol.*, 9: 264-272.

Edinger, J.E. & E.M. Buchak 1983. Developments in LARM2: A longitudinal-vertical, time-varying hydrodynamic reservoir model. Technical Report E-83-1, USA Waterways Experiment Station, Vicksburg, MS.

Edmondson, W.T. 1970. Phosphorus, nitrogen and algae in Lake Washington after diversion of sewage. *Science*, 169: 690-691.

Environmental Laboratory 1986a. CEQUAL-R1: A numerical one-dimensional model of reservoir water quality. User's Manual, Instruction Report E-82-1, Revised Edition, US Army Engineer Waterways Experiment Station, Vicksburg, MS.

Environmental Laboratory 1986b. CE-QUAL-W2: A numerical two-dimensional, laterally averaged model for hydrodynamics and water quality. User's Manual, Instruction Report E-86-5, US Army Engineering Waterways Experiment Station, Vicksburg, MS.

Fee, E.J. 1973. A numerical model for determining integral primary production and its application to Lake Michigan. *J. Fish. Res. Board Can.*, 30: 1447-1468.

Fischer, H.G. (ed.) 1981. Transport models for inland and coastal water. *Symposium Proceedings*, Academic Press.

Ford, D.E. & L.J. Johnson 1986. An assessment of reservoir mixing processes. No. E-86-7, Environmental Laboratory, US Army Engineers Waterways Experiment Station, Vicksburg, MS.

Foree, E.G. & P.L. McCarthy 1970. Anaerobic decomposition of algae. *Env. Science & Technology*, 4: 842-849.

Funkquist, L. & L. Gidhagen 1984. A model for pollution studies in the Baltic Sea. Swedish Meteorological and Hydrological Inst., Norrkoping, Report No. RHO 39.

Goldman, C.R. (ed.) 1966. *Primary productivity in aquatic environments*. Berkeley: University of California Press, 466 pp.

Harper, W.L. & W.R. Waldrop 1980a. A two-dimensional, laterally averaged hydrodynamic model with application to Cherokee Reservoir. *Proceedings of Symposium on Surface-Water Impoundments*, University of Minnesota, Minneapolis, MN.

Harper, W.L. & W.R. Waldrop1980b. Numerical hydrodynamics of reservoir stratification and density currents, IAHR, Trondheim, Norway. *Proceedings of the Second International Symposium on Stratified Flows*.

Henderson-Sellers, B. 1984. Reservoirs, London: The MacMillan Press Ltd. *Engineering Limnology*. Pitman Advanced Publishing Program, 356 pp.

Hendrickson, A.C.H. & H.G. Stefan 1985. Models for daily quantity and quality of runoff: Input to a dynamic lake water quality simulation model. External Memorandum No. 196, St. Anthony Falls, Hydraulic Laboratory, Univ. of Minnesota, Oct., 119 pp.

Hino, M. 1981. *Ecohydrodynamics, Advances in Hydroscience*, 12: 143-149. Academic Press.

Hino, M. 1983. Prediction of nutrient loading limit to prevent eutrophication of reservoirs by the Three Ranks Biological System Model. *Proc. 20th IAHR Congress, Moscow*, V: 510-517.

Hino, M. & H. Utahara 1978. Light attenuation by suspended particles and growth of phytoplankton. *Proc. 22nd Japanese Conf. on Hydraulic Research, JSCE*, 49-54.

Hosomi, M. & R. Sudo 1987. Nutrient concentrations in the interstitial water of the sediments in Lake Kasumigaura. *Jap. Journ. Limnology*, 48: 119-129.

House, L.B. 1981. One-dimensional reservoir-lake temperature and dissolved oxygen model. US Geological Survey, Water-Resources Investigations, 82-5: 114

Huber, W.C., D.R.F. Harleman & P.J. Ryan 1972. Temperature predictions in stratified reservoirs. *ASCE, Jour. Hydraulic Division*, 98(HY4).

Hutchinson, E.G. 1957. *A Treatise on Limnology*. New York: John Wiley.

Hydroscience, Inc. 1976. Assessment of the effects of nutrient loadings on Lake Ontario using a mathematical model of the phytoplankton. Windsor, Ont.

Ikeda, S. & N. Adachi 1978. A dynamic water quality of Lake Biwa – A simulation study of the lake eutrophication. *Ecol. Modelling*, 4: 151-172.

Jassby, A.D. & T. Platt 1976. Mathematical formulation of the relationship between photosynthesis and light for phytoplankton. *Limnology and Oceanography*, 21(4): 540-547.

Jennings, M.E. 1988a. Estimation of urban stormwater quality. *ASCE, Hydraulic Engineering, Proc. of the 1988 National Conf.*, 78-83.

Jennings, M.E. 1988b. Federal microcomputer software for urban hydrology. *ASCE, Hydraulic Engineering, Proc. of the 1988 National Conf.*, 616-621.

Johanson, R.C. et al. 1984. Hydrological simulation program-Fortran (HSPF): Users manual for research 8.0. US Environmental Protection Agency, Athens, GA, EPA-600: 3-84-006.

Jones, J.R. & R.W. Bachman 1976. Prediction of phosphorus and chlorophyll levels in lakes. *Jour. Water Pollut. Control Fed.*, 48(a): 2176-2182.

Jørgensen, S.E. & H. Löffler 1990. *Guidelines of lake management*. International Lake Environment Committee Foundation, 3:174 pp. Shiga-Kaikan Bld., Shiga, Japan.

Kamp-Nielson, L. 1974. Mud-water exchange of phosphate in undisturbed cores and factors affecting exchange rates. *Arch. Hydrobiol.*, 73: 218.

Kamp-Nielson, L. 1976. Modelling the temporal variation in sediment phosphorus fractions in interactions between sediments and freshwater. *Proceedings, international Symposium, Amsterdam, Sept. 1976.*

Kennedy, R.H. & J. Nix (eds) 1987. *Proceedings of the Degray Lake Symposium*, Tech. Rep. No. E1-87-4, Env. & Water Quality Operational Studies, Waterways Experiment Station, Army Corps of Engineers, Vicksburg, MS, 633 pp.

Lambou, V.W., S.C. Hern, W.D. Taylor & L.R. Williams 1982. Chlorophyll, phosphorus, secchi disk, and trophic state. *Water Resources Bulletin, American Water Resources Assoc.*, 18(5): 807-813.

Larsen, D.P., D.W. Schults & K.W. Malueg 1981. Summer internal phosphorus supply in Shagawa Lake, Minnesota. *Limnology and Oceanography*, 26: 740-753.

Leendertse, J.J. & S.K. Liu 1975. A three-dimensional model for estuaries and coastal seas: Aspects of computation. US Dept. of Interior, Washington, D.C., R-1764-OWRT, 123 p.

Likens, G.C. (ed.) 1972. Nutrients and eutrophication – The limiting nutrient controversy. *Special Symposia*, Vol. I, American Society of Limnology and Oceanography. Lawrence, Kansas: Allen Press.

Likens, G.E., W. Rodke & C. Serruya (eds) 1979. Lake metabolism and management. *Ergebnisse der Limnologie*, 13: 349. Schweizerbartsche Verlagsbuchhandlung, Stuttgart.

Lorenzen, W.M. (ed.) 1979. Phytoplankton – Environmental interaction in reservoirs. *Proceedings of a Workshop. Waterways Experiment Station, US Army Corps of Engineers*, 377 pp.

Lorenzen, M. & A. Fast 1977. A guide to aeration – Circulation techniques for lake management. US EPA, EPA-600: 3-77-004.

Lung, W.S., R.P. Canale & P.L. Freedman 1976. Phosphorus models for eutrophic lakes. *Water Research*, 10: 1101-1114.

Matsuoka, Y., T. Goda & M. Naito 1986. A eutrophication model of Lake Kasumigaura. *Ecological Modelling*, 31: 201-219.

McCormick, M.J. & D. Scavia 1981. Calculation of vertical profiles of lake-averaged temperature and diffusivity in Lakes Ontario and Washington. *Water Resources Research*, 17(2): 305-310.

Megard, R.O., D.W. Tonkyn & W.H. Senft, II 1984. Kinetics of oxygen photosynthesis in planktonic algae. *Jour. of Plankton Research*, 6(2): 325-337.

Middlebrooks, E.J., D.H. Falkenborg & T.E. Maloney (eds) 1974. In *Modeling the Eutrophication Process*, Ann Arbor Science.

Miyanaga, M. 1986. Modelling of stratified flow and eutrophication in reservoirs. *Ecol. Modelling*, 31: 133-144.

NALMS 1985. Lake and reservoir management – Practical application. *Proceedings of the Fourth Annual Conference and International Symposium, No. 16-19, 1984.*

NALMS 1986. Lake and reservoir management. *Proceedings of the Fifth Annual Conference and International Symposium on Applied Lake & Watershed Management, No. 13-16, Lake Geneva, Wisconsin.*

NALMS 1987. Lake and reservoir management. *Proceedings of the Sixth Annual Conf.*

National Academy of Sciences 1969. Eutrophication: Causes, consequences, correctives. *Proceedings of the International Symposium, June, 11-15, 1967*, University of Wisconsin, Madison.

National Research Council 1992. Restoration of aquatic ecosystems/science, *technology and public policy*. National Academy Press, Washington D.C., 552 pp.

OECD 1982. Eutrophication of waters, monitoring, assessment and control. OECD Publication and Information Center, Washington D.C.

Okada, M. & S. Aiba 1983. Simulation of water-bloom in a eutrophic lake – III: Modeling the vertical migration and growth of Microcystis aeruginosa. *Water Res.*, 17(8): 883-893.

Orlob, G.T. (ed.) 1983. *Mathematical modeling of water quality of streams, lakes, and reservoirs*, 518 pp. Wiley & Sons.

Orlob, G.T . & L.G. Selna 1970. Temperature variation in deep reservoirs. *ASCE, Jour. Hydraulic Division*, 96(HY2): 391-410.

Reckhow, K.H., M.N. Beaulac & J.T. Simpson 1980. Modelilng phosphorus loading and lake response under uncertainty: A manual and vompilation of rxport voefficients. Proj. Rep. for USEPA, Clean Lakes Section, Washington D.C., June 1980, 240 pp.

Riley, M.J. & H.G. Stefan 1987. Dynamic lake water quality simulation model MINLAKE. St. Anthony Falls Hydraulic Laboratory, University of Minnesota, Minneapolis, MN, Rep. 263, 140 pp.

Riley, M.J. & H.G. Stefan 1988a. Development of the Minnesota Lake Water Quality management model 'MINLAKE'. *Lake and Reservoir Management, NALMS* 4(2): 73-83.

Riley, M.J. & H.G. Stefan 1988b. MINLAKE: A dynamic lake water quality simulation model. *Ecological Modelling*, 43: 155-182.

Ryoling, S.-O. & W. Rast 1989. The contral of eutrophication of lakes and reservoirs, *UNESCO Man and Biosphere Series*. The Parthenon Publishing Group, New Jersey, I:314 pp.

Sakamoto, M. 1966. Primary production by phytoplankton community in some Japanese lakes and its dependence on lake depth. *Arch Hydrobiol.* 62: 1-28.

Sas, H. 1989. *Lake restoration by reduction of nutrient loading/expectations, experiences, extrapolations*. Academia Verlag, Richarz GmbH, Sankt Augustin I, Germany, 497 pp.

Scavia, D. 1980. An ecological model of Lake Ontario. *Ecological Modelling*, 8: 49-78.

Schaffranek, R.W., R.A. Baltzer & D.E. Goldberg 1981. A model for simulation of flow in singular and interconnected channels. Techniques of Water Resources Investigations of the USGS, Chapt. C3.

Schneider, M.L. & R.E. Price 1988. Temperature analysis. Howard A. Hanson Reservoir, Washington. Tech. Rep. HL-88-22, Waterways Experiment Station, Vicksburg, MS, Sept. 1988, 63 pp.

Shanahan, P. & D.R.F. Harleman 1982. Linked hydrodynamic and biogeochemical models of water quality in shallow lakes. Report 268, Ralph M. Parson Laboratory, MIT, Cambridge, MA.

Shapiro, J., B. Forsberg, V. Lamarra, G. Lindmark, M. Lynch, E. Smeltzer & G. Zoto 1982. Experiments and experiences in biomanipulation. Limnol. Res. Center, Univ. of Minnesota, Minneapolis, MN and USEPA, Corvallis-ERL, EPA-600/3-82-096.

Sharpley, A.N. & R.G. Menzel 1987. The impact of soil and fertilizer phosphorus on the environment. *Advances in Agronomy*, 41: 297-324.

Simons, T.J. 1973. Development of three-dimensional numerical model of the Great Lakes. *Scientific Series*, 12. Canada Centre for Inland Waters, Burlington, CA.

Simons, T.J. 1975. Verification of numercial models of Lake Ontario – II. Stratified circulation and temperature changes. *Jour. Phys. Oceanogr.*, 5: 98-110.

Simons, T.J. 1976. Verification of numerical models of Lake Ontario – III. Longterm heat transports. *Jour. of Phys. Oceanogr.*, 6(3): 372-378.

Sonzogni, W.C. et al. 1976. A phosphorus residence time model: Theory and application. *Water Resources*, 10: 429-435. Great Britain: Pergamon Press.

Stefan, H. & K.J. Anderson 1980. Wind-driven flow in Mississippi River impoundment. *ASCE, Jour. Hydraulic Division*, 1503-1520.

Stefan, H.G. (ed.) 1981. *Surface water impoundments*. American Society of Civil Engineers, New York, 1682 pp. (2 Volumes).

Stefan, H.G., R. Ambrose & M. Dortch 1988. Surface water quality models: Modeler's prospective. *Proceedings, International symp. on water quality model of agricultural non-point sources, Utah State Univ.*

Stefan, H.G., J.J. Cardoni & A.Y. Fu 1982. RESQUAL II: A dynamic water quality simulation program for a stratified shallow lake or reservoir: Application to Lake Chicot, Arkansas. Report No. 209, St. Anthony Falls Hydraulic Laboratory, University of Minnesota, Minneapolis, MN 55414.

Stefan, H.G., J. Cardoni, F. Schiebe & C. Cooper 1983. Model of light penetration in a turbid lake. *Water Resources Research*, 19(1): 109-120.

Stefan, H., T. Skoglund & R. Megard 1976. Wind control of algae growth in eutrophic lakes. *Jour. Environmental Engineering Division, ASCE*, 102(EE6): 1201-1213. Proc. Paper 12619.

Tarapchak, S.J. & C. Nalewajko 1986. A review: Phosphorus – Plankton dynamics and phosphorus cycling in aquatic systems. NOAA TM ERL GLERL, 60: 51 pp.

Thomann, R.V., K.M. DiToro, R.P. Winfield & D.J. O'Connor 1975. Mathematical modeling of phytoplankton in Lake Ontario. 1. Model development and verification. Report No. EPA-660/3-75-005, USEPA, August.

Thomann, R.V. & J.A. Mueller 1987. *Principles of surface water quality modeling and control*, 608 pp. Harper and Row

Thomann, R.V., R.P. Winfield & J.J. Segna 1979. Verificationi analysis of Lake Ontario and Rochester Embayment three dimensional eutrophication models. Report No. EPA-600/3-79-094, USEPA, August.

Thornton, K.W., B.L. Kimmel & F.E. Payne 1990. *Reservoir limnology/ecological perspectives*. J.Wiley & Sons Inc., New Jersey, 246 pp.

Tilman, D., S.S. Kilham & P. Kilham 1982. Phytoplankton community ecology: The role of limiting nutrients. *Annual Review Ecol. Syst.*, 13: 349-377.

TVA 1976. Water temperature prediction model for deep reservoirs (WRMMS). Report A-2.

TVA 1978. Environmental effects of hydraulic engineering works. *Proceedings of International Symposium., Sept.*, 474 pp.

TVA 1979. Environmental effects of hydraulic engineering works. In E.E. Driver & W.O. Wunderlich (eds). *Proceedings of an International Symposium Held at Knoxville, Tennessee, Sept. 12-14, 1978*

TVA 1986. Hydrodynamic and water quality models of the TVA Engineering Laboratory, Norris, TN, Issue No. 3, Sept.

US Environmental Protection Agency 1975. National Eutrophication Survey. Environmental Research Laboratory, Las Vegas, NV.

US Environmental Protection Agency 1976. *Proceedings, Conference on Environmental Modelling and Simulation, April 19-22, 1976, Cincinnati, OH.*

US Environmental Protection Agency 1979. Lake restoration. *Proceedings of National Conf., March*, 254 pp.

US Environmental Protection Agency 1980a. Environmental Research Laboratory. Workshop on verification of water quality models. Report EPA-600/9-80-016.

US Environmental Protection Agency 1980b. Restoration of lakes and inland waters. Portland, Maine, EPA 4005-81-010, 552 pp.

US Environmental Protection Agency 1983. Results of the nationwide urban runoff program, Vol. 1 – Final Report, Water Planning Division, Washington, D.C.

US Environmental Protection Agency 1984. Lake and reservoir management. *Proceedings, Third Annual Conf., North American Lake Management Society*, EPA 440/5/84-001, 604 pp.

US Environmental Protection Agency 1985. Perspectives on nonpoint source pollution. *Proceedings of a National Conference, Kansas City, MO, May 19-22.*

US Environmental Protection Agency 1987. *Design manual: Phosphorus removal.* EPA/625/1-87 /001. 116 pp.

US Environmental Protection Agency 1988. The lake and reservoir restoration guidance manual. EPA/440/5-88-002.

Uttormark, P.D., J.D. Chapin & K.M. Green 1974. Estimating nutrient loadings of lakes form non-point sources. EPA-66013-74-020.

van Straten, G. & L. Somlyodi 1980. Lake Balaton eutrophication study: Present status and future program. Paper WP-80-187, Int'l. Institute for Applied Systems Analysis, Laxenburg, Austria, Dec.

Vant, W.N. (ed.) 1987. *Lake Managers Handbook.* Water and Soil Miscellaneous Publ. No. 103, Water Quality Centre, Ministry of Works and Development, Hamilton, New Zealand, 230 pp.

Vollenweider, R.A. 1968. Scientific fundamentations of the eutrophication of lakes and flow waters, with particular reference to nitrogen and phosphorus as factors in eutrophication. *Water Management Research.* DAS/CSI.68.27. OECD, Paris, 183 pp.

Vollenweider, R.A. 1969. Moglichkeiten und Grenzen elementarer Modelle der Stoffbilanz von Seen. *Arch. Hydrobiol.*, 66: 1-36.

Vollenweider, R.A. 1975. Input-output models with special reference to the phosphorus loading concept in limnology. *Schweiz. Z. Hydrol.*, 37: 58-83.

Vollenweider, R.A. 1976. Advances in defining critical loading levels for phosphorus in lake eutrophication. *Mem. Ist. Ital. Idrobiol.*, 33: 58-83.

Waldrop, W.R., C.D. Ungate & W.L. Harper 1980. Computer simulation of hydrodynamics and temperatures of Tellico Reservoir. Report No. WR28-1-65-100, TVA Water Systems Development Branch, Norris, TN.

Walker, W.W. Jr. 1985. Models and software for reservoir eutrophication assessment. *Lake and Reservoir Management. Proceedings, 4th Annu. Int. Symp. N. Am. Lakek Manage. Soc., McAfee, NJ.*

Wetzel, R. 1975. *Limnology*. Saunders Publishing Co., Philadelphia, Pa.

World Health Organization 1981. Eutrophication: A global problem – Parts 1 and 2. *Water Quality Bulletin*, 6(3 & 4).

Control of water quality in lakes and reservoirs 3

H. BERNHARDT
Wahnbachtalsperrenverband, Siegburg, Germany

3.1 INTRODUCTION – EUTROPHICATION AS THE MAIN WATER QUALITY PROBLEM

Measures taken to control the quality of the water in natural and man-made lakes primarily serve to prevent eutrophication. They are applied above all in lakes which are used for the supply of drinking water.

3.2 PHOSPHORUS AS A GROWTH-LIMITING FACTOR

In most lakes, the extent of the algal development, and thus the eutrophication, is dependent on the content of phosphorus compounds; i.e. phosphorus is the limiting factor. The results of the OECD monitoring programme (Vollenweider and Kerekes, 1982) confirmed this fact in that they showed a statistically significant correlation between the annual mean value of the chlorophyll-a content in the euphotic zone and the average annual total P-concentration in the lakes and reservoirs investigated (Clasen & Bernhardt, 1980) (Fig. 3.1).

If the average annual input from all the tributaries to a lake exceeds the tolerable value for that lake, the probability is high that the amount of phytoplankton production will increase.

Sediment deposited on the bottom of lakes also plays a decisive role as a nutrient source because it can continually release nutrients into the water by diffusion, even if aerobic conditions are present at the sediment-water interface (0.1-1 mg/m^2d) (Boström et al., 1983; Golterman, 1977; Stumm & Leckie, 1971; Hoyer et al., 1982). The nutrients allow larger algal developments to occur in the shallow areas, and these can then expand into the rest of the impoundment. An input of nitrogen compounds does not affect primary production if phosphorus is the controlling factor (Bernhardt, 1983a). Nevertheless, inputs of nitrate, nitrite and ammonium are important because they can disturb the water treatment process and impair the quality of drinking water.

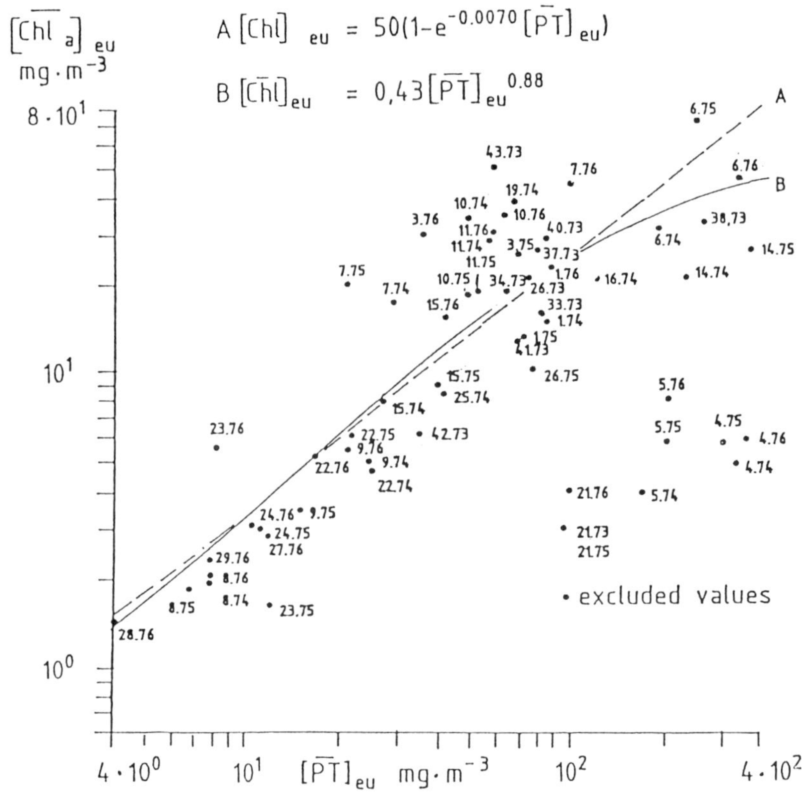

Figure 3.1. Annual average chlorophyll-a concentration versus annual average concentration of total phosphorus in the euphotic zone. Each point represents a lake studied in the OECD cooperative programme for inland waters (From Clasen & Bernhardt, 1980)

3.3 DISADVANTAGEOUS EFFECTS OF EUTROPHICATION ON DRINKING WATER

A lake planned to provide drinking water should contain water of such a quality that drinking water can be obtained from it through normal treatment processes that are not too complicated. This condition is satisfied if the catchment area is used mainly for forestry and is sparsely populated. Larger population densities and a predominance of agricultural land use can cause lakes to become loaded with pollutants. Experience obtained with the Wahnbach Reservoir (Germany) that became eutrophic within 10 years of its existence has shown that mass developments of algae and the associated algogenic matter impair the water purification process (Bernhardt, 1984).

The most unfavourable effects are:

1. Formation of large amounts of particulate organic matter (microorganisms) that causes premature filter clogging or if they pass through the filter, they are

responsible for renewed bacterial growth in the distribution system.

2. Some kind of organic compounds of algal origin that produce odour and taste even if they are present only in very small quantities (ng/l = 10^{-9}g/l).

3. Release of algogenic organic compounds that inhibit flocculation and impair treatment with chlorine or chlorine dioxid.

4. Formation of humin-like substances that are biologically poorly degradable and result from the decomposition of organisms. These substances are precursors of trihalogen methanes if chlorine is used as an oxidizing agent.

5. Development of reducing (anaerobic) conditions at the water-sediment interface on the bottom of lakes that increase the concentrations of Fe- and Mn-compounds in raw water and necessitates the removal of manganese by chemical or biological treatment processes.

6. Reduction of sulphate to hydrogen sulphide. Interference with phosphate fixation at the sediment that accelerates the eutrophication process.

7. Increased concentrations of NH_4^+ ions in the water which disturb the process of disinfection with chlorine. They can be the cause of the formation of nitrites in the distribution system.

8. Development of larger organisms in the distribution system, e.g. nematodes, freshwater sponges, bryozoans, hydrozoans, insect larvae etc., as the result is increased amounts of organic compounds in the drinking water and the growth of bacterial films on the pipewalls.

9. Clogging of water meters, sieves and water taps by zooplankton and macroorganisms that either survive the water treatment process and pass into the drinking water or develop in the distribution system and in the pressure tanks. The detrimental effect of the eutrophication of standing waters on the production of drinking water has been described elsewhere (Bernhardt, 1984).

3.4 MEASURES FOR CONTROLLING THE WATER QUALITY OF LAKES

If the water quality of a lake is impaired mainly by eutrophication processes, measures have to be taken to control the eutrophication. Such a measure is to change one of the conditons needed for growth of the algae so as to limit that growth. Because in most cases phosphorus is the limiting factor, practical measures are aimed at reducing significantly the phosphorus loading of the lake. Either chemical or biological methods can be used. Some methods reduce the content of organic particulate and soluble substances.

In many cases, particularly with enclosed reservoirs, the nutrient loading is so high that algal growth cannot be restricted by limiting nutrients. In these cases, a restriction of algal growth can often be achieved by applying measures that influence other growth factors. For example, the phytoassimilation rate can be reduced by preventing light from entering the water, and a shortening of the residence time of the water in the reservoir can also limit bioproduction.

Biomanipulation methods can also be used and are particularly successful in shallow lakes with deep sediments in which an intensive exchange of substances occurs between the water body and the sediment. A summary is given in Table 3.1.

Table 3.1. Measures for controlling eutrophication

Limitation of the phosphate loading in the catchment area

Point sources:
Wastewater measures
- advanced wastewater treatment with precipitation and sedimentation, when necessary flocculation and filtration;
- wastewater collection from as many sources as possible; ring collectors and diversion channels.

Legislative measures
- reduction of the amount of phosphates in detergents and cleansing agents.

Non-point sources:
Measures taken in agricultural practice
- prevention of erosion and run-off with special agricultural techniques (spreading of animal manure in a way that is compatible with plant growth and lake protection).
- evergreen strips of land along the tributaries
- compatibility of crops with ground slope, the condition of the soil, distance from impounded water.

Treatment of inflowing streams
- seepage trenches
- seepage basins with artificial soil filters
- biological measures in pre-reservoirs and bioreactors

Limitation of the phosphate content in the lake:
Measures in the lake
- phosphate precipitation in the lake with A1- or Fe-salts
- hypolimnetic aeration (increasing the respiration capacity, increasing the P fixation potential)

Limitation of the internal loading:
Measures taken for the sediment on lake bottom
- dredging;
- covering;
- injection of nitrate into the sediment as an oxygen carrier (Riplox-method);
- removal of hypolimnetic water.

Limitation of the algal development in the lake without changing the P budget:
- artificial mixing (light limitation);
- manipulation of the pelagic food web (Biomanipulation);
- reduction in the residence time of the water in the lake;
- use of non-toxic natural products to control algal growth.

Before applying any of these measures, one must investigate whether it will achieve the desired aim. All factors that determine the extent of eutrophication have to be considered in order to estimate their relative importance. With the help of study programmes, which have to be suited to the respective properties and conditions of the catchment area, the substantial causes of eutrophication must be determined. Normally, these study programmes should run over a period of at least one year, better two years, and should be so organized that not only the nutrient concentration but also the nutrient loading can be determined with sufficient accuracy. In addition to the nutrients, other factors that influence algal growth also have to be considered; these include light, climate, temperature, residence time of the water in the lake, presence and availabililty of trace elements, current and stratification conditions.

The results of the study programmes serve as a guide in selecting the most effective method for reducing algal development in each particular case (Table 3.1). The following aspects are particularly important:

1. Effectiveness with respect to the desired water condition (trophic grade);
2. Effectiveness with respect to the desired secondary effects (e.g. elimination of dissolved and undissolved organic and inorganic matter that has no effect on eutrophication)
3. Technical practicability for the conditions present in each case;
4. Initial cost (i.e. investment) and running costs.

3.5 CASE STUDIES OF THE APPLICATION OF THESE MEASURES*

The water quality in natural lakes and reservoirs can be restored by carrying out the measures

- in the catchment area;
- at the inlet;
- in the lake itself.

Measures carried out in the catchment area and at the inlet are solely for the purpose of reducing the nutrient influx. In contrast, measures taken in the lake, serve mainly to limit algal development, either directly or indirectly, by limiting other growth factors.

3.5.1 *Measures in the catchment area*

The measures to be applied in the catchment area depend upon whether the nutrients arise from point sources or non-point sources.

*This compilation is a survey of technically usable methods with examples.

1. *Point sources*

Point sources of nutrients come from specific points in communal and industrial sewage systems and from urban runoff. The nutrients can be collected in sewers, pipes and canals. The concentration of P in communal waste waters amounts to 5-10 (mg/l) P (Hamm, 1989); it depends on the amount of water discharged, on the proportion of industrial waste water, on the shape of phosphate in the detergents, on the degree of dilution (water from other sources) and on the type of waste water conveyance, i.e. rain water and sewage mixed or separate. The amount of nitrogen excreted per person per day is about 11 g and the average nitrogen concentration in communal waste water is 20-40 (mg/l) N (Pöpping-haus, 1975).

Measures that serve to eliminate nutrients can be coupled with those already in use in existing wastewater treatment plants. A reduction of phosphates can be achieved by

– Chemical precipitation processes as part of waste-water treatment followed by sedimentation or filtration; the processes to be used are different for the pre-precipitation, simultaneous precipitation and post-precipitation.

– Use of diversion channels to prevent sewage from entering the reservoir.

Advanced wastewater treatment. The extent to which advanced wastewater treatment is needed in order to reduce the phosphorus load depends on the conditions in the catchment area and on the contribution of the P load from point sources in comparison to the total P loading of the lake. A general stipulation of a maximum P concentration in the outflow of a waste water treatament plant, e.g. 1 (mg/l)P, is not recommended and can have little effect if the principle P loading of a lake arises from point sources (wastewater treatment plants). Therefore, in Switzerland, for example, efforts are being made to ensure that effluents from wastewater treatment plants that flow directly or indirectly into lakes should not contain more than 0.2 (mg/l)P and must not contain more than 0.8 (mg/l)P (Boller, 1985). Such a low residual content of total P compounds (P_{tot}) can, however, be achieved only with chemical precipitation (usually simultaneous precipitation) followed by post-precipitation and filtration through multi-media filters (Boller, 1985) or by employing a combination of chemical precipitation and biological P elimination according to the aerobe/anaerobic method.

Many other methods are not dealt with here in; more comprehensive information is contained in the reviews that have been published during the last few years (Harremöes et al., 1985; Böhnke, 1980). The efficiency of the different technical methods of advanced wastewater treatment with respect to P removal and reduction of BOD and COD are presented in Table 3.2.

Sewerage Conveyance Network Systems. Good results are obtained in the controlling of eutrophication if a sewerage network is present that receives waste water from point sources that would otherwise be released directly into the lake. Depending upon the local conditions, these results can be realized with diversion pipes and ring collectors. Typical examples of this approach are those for the lakes

Table 3.2. Average values of removal in % of P_{tot}, BOD_5 and COD with methods of advanced waste water treatment, summarized and extended from Klute (1984)

Method	BOD_5	COD	P_{tot}
Pre-precipitation	94.4	85.1	80.0
Simultaneous-precipitation	95.8	88.0	87.6
Post-precipitation	95.6	88.8	90.0
Combination of different methods for precipitation/flocculation	92.7	83	88.4
Precipitation, sedimentation followed by flocculation and filtration (Boller et al., 1981)	97.2	88.9	98
Activated alumina filtration (Donnert & Eberle, 1982)	–	40*	~93

*Unfiltered sample sometimes containing bulking sludge

in Bavaria (Germany) (Hamm, 1976), Austria (Sampl, 1980), Lake Washington (USA) (Edmondson, 1985) and Lake Bracciano (Bruno, 1985; Martini, 1985). If the P loading of a lake can be sufficiently reduced in this way, oligotrophic conditions will develop.

The study on Lake Washington clearly produced such results with these measures, the average P concentration in the lake was reduced to below 20 μg/l; this value appears to be a condition for the occurence of Daphnia, which have been dominant in the lake since 1976. The decrease in phytoplankton has been achieved due to the low P availability and the grazing effect of the Daphnia (Edmondson, 1985).

Wastewater produced by communities near the lakes in Kärnten (Austria) (Sampl, 1980) is collected in peripheral pipe lines and, following treatment, is released at the outflow of the lake. A continual decrease in the P_{tot}-concentration has occurred, and the biomass production has decreased correspondingly. The transparency of the water increased from ~ 2 m to 5-8 m. The technical application, running experience on cost of ring collectors and piping lines are given by Flögel & Flögel (1985) and Martini (1985).

2. *Non-point sources*

Non-point sources are defined as nutrient sources that contribute to the loading but cannot be intercepted by technical means:

– waste water from populated areas not connected to a sewage system (septic tanks, domestic wastes);

– effluents from isolated farmhouses;
– precipitation;
– ground leaching;
– soil erosion;
– leaf litters and their leachates;
– surface runoff containing mineral fertilizers, and organic compounds;
– manure and dung loss from stock farms;

- watering places for cattle that contribute directly to reservoirs;
- aquaculture and fisheries;
- water birds
- open air bathing facilities.

Non-point sources can be the sole cause of eutrophication if agriculture and livestock farming as well as small farms are more prevalent in the catchment area than woodlands. When non-point sources dominate, methods based on the collection and treatament of wastewater cannot be employed. The methods that are effective in this instance are diverse and must be suited to the local conditions. For example, if the nutrients arise from the sewage of scattered villages, they can be disposed of harmlessly by land irrigation provided that the soil has a sufficient capacity for binding the phosphate and is able to absorb the water. If the nutrients arise as a result of agricultlural practice (erosion, waste water from farms, domestic animals or pasture farming), control measures aimed at protecting the water quality must take the form of changing or modifying practices that produce pollution. Maize planted on sloping fields is particularly disadvantageous because catastrophic erosion can take place during heavy rainfalls. Pasture farming alone can cause large amounts of nutrients in the form of animal excrements to enter tributaries to lakes, particularly if the tributaries and pastures are close together (Schulte-Wülwer-Leidig, 1985; Müller, 1984) or if the farmland is on slopes and the nutrients can be washed into the streams (Briese, 1984). The rates at which phosphorus is released from catchment areas of different structures are given in the following for guidance (Scharf et al., 1984).

P delivery from	kg P/km^2.y
Forest on rock substratum	<5
Forest in alpine regions	10 - 20
Meadows not used for grazing	20
Agricultural areas	40
Pastures and access ways	80

3.5.2 *Treatment of inflowing streams*

The measures that can be applied to the tributaries serve solely to reduce the influx of phosphorus into the lake. Such measures are:

- seepage of small tributaries, with the help of ponds and seepage trenches, into the ground whereby the water is purified on its way through the soil;
- seepages of small tributaries through artificial filters;
- biological P-elimination in pre-reservoirs or bio reactors;
- filtration of small tributaries through aluminium oxide filters;
- treatment of the water by means of precipitation, flocculation and filtration.

The first two methods given above can also be employed in catchment areas below individual point sources, e.g. scattered farms that are not connected to a sewerage system and that release their wastewater into small streams.

Measures that can be used for removing unwanted substances from the water at the inlet have been described recently and in detail in reference (Bernhardt, 1983). The technically usable processes are, therefore, only be briefly described herein.

1. *Seepage trenches*. Seepage trenches are used to be eliminate phosphorus compounds, organic substances and bacteria, and they are effective if the amounts of water to be treated are low and the runoff fluctuation is small. These conditions hold for small catchment areas in which storm water does not cause overflows and the normal influent rate does not exceed 100 l/s. The design concept of seepage trenches is given in Figure 3.2 (Grau, 1978). A seepage trench functions because phosphorus is removed from the water as it passes through the sub-soil; the process is effective in direct proportion to the amount of fine-grained sandy clay in the soil. Further details on this process as well as the basis for calculating the size of a seepage trench can be obtained from Grau (1977).

In spite of the relatively low elimination rates, phosphorus concentrations achieved are 7-12 μg/l o-PO_4-P in the water flowing into the reservoir (Table 3.3), and these concentrations are so low that eutrophication is unlikely. The elimination rate increases with increasing o-PO_4 concentrations in the seepage trenches. Over 99% of the P-content of a biologically treated waste water can be eliminated, i.e. from 3-6 mg/l o-PO_4-P to 0.007 mg/l o-PO_4-P. However, the soil

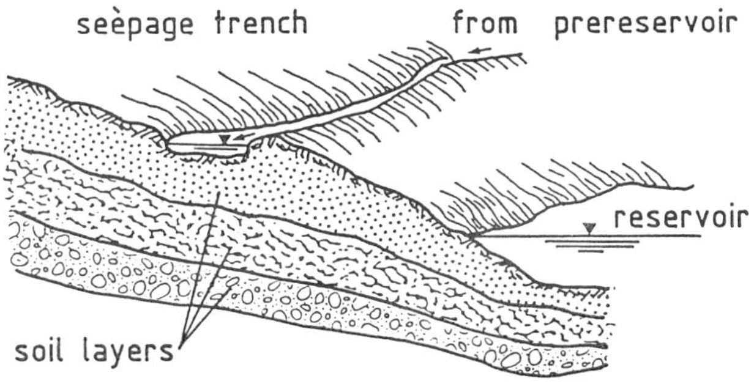

Figure 3.2. Scheme of a seepage trench (Grau, 1978)

Table 3.3. o-PO_4 elimination (E%) by seepage trenches. Mean values from a three-year study at the Kerspe Dam (Grau, 1977)

o-PO_4 concentration	Inflow pre-reservoir μg/l	Inflow seepage trench μg/l	E %	Below the seepage trench μg/l	E %
Creek A	51	31	39	20	35
Creek B	61	57	7	36	37
		76*	–	38*	50

*Values obtained during the summer months

must have a high adsorption capacity, i.e. be composed of sandy silt. Quartz sand has a comparatively low capacity for removing phosphate from water, and it is quickly exhausted.

2. *Seepage filter basins.* If the natural soil is not suitable for seepage trenches, seepage filter basins that contain a seepage layer can be built. They must be constructed in such a way that sufficient phosphate is adsorbed and that most of the incoming water is filtered. The filter must be filled with fine particulate materials (< 0.5 mm diameter) which contain a high proportion of Al- and Fe-oxides. These oxides guarantee a high capacity for phosphate adsorption. Moreover, these materials must be sufficiently permeable to ensure that the water flows through the filter and does not by-pass it. The permeability K_f-values of 10^{-4} m/s in the filter have proved to be practicable. A drainage system is incorporated in the seepage basin. In order to prevent algal development and microbial growth in the seepage basin, the seepage layer is covered by coarse material.

3. *Filtration through activated alumina filters.* Small inflowing streams and effluents from fish breeding farms and scattered communities can be treated by filtration through activated alumina filters. A suitable method, based on the work of Eberle et al. (1980), has been developed on a practical scale by the Wahnbach Reservoir Association (Bernhardt et al., 1981).

Activated alumina is a technical aluminium oxide that is manufactured by incinerating aluminium hydroxide. It selectively adsorbs phosphate under slightly acid conditions and has a specific capacity of approximately 110 mg P/kg and m^2 of interior surface (200-300 m^2/g) (Yee, 1966). The arrangement of the activated alumina filter at the Wahnbach reservoir is shown in Figure 3.3. Each filter, filled with activated alumina, has a surface area of 0.8 m^2 and a height of 1 m. The filtration speed is approximately 1 m/h. The filters are situated in a collection pond and must be submerged. The number of filters needed depends on the amount of water to be treated. The filters are backwashed every 4 weeks (velocity ~ 80 m/h). The backwash water is taken from the collection pond and fed back into the pond where the sludge quickly settles. It does not flow back into the filters because the velocity is too low. The activated alumina must be regenerated periodically in accordance with the water quality (o-PO_4 content). Regeneration is performed with NaOH. The plant at the Wahnbach Reservoir required regeneration after about 2 years of use with a throughput of 12 000 m^3 water (100-200 µg/l o-PO_4-P) per m^3 of activated alumina (γ = 0.8 kg/l) (2-3 g P/kg alumina) (Fig. 3.4). The activated allumina was regenerated and reinstalled in the filter on 1.4.1984; the regenerated activated alumina had the same efficiency as the new material (~ 80% P-elimination). The results of the plant operations (Table 3.4) show that not only P compounds, but also COD and NH were eliminated. The process functions without electrical energy. Even in winter when the collection pond is covered with ice, the p_{tot} content is reduced to < 50 µg/l.

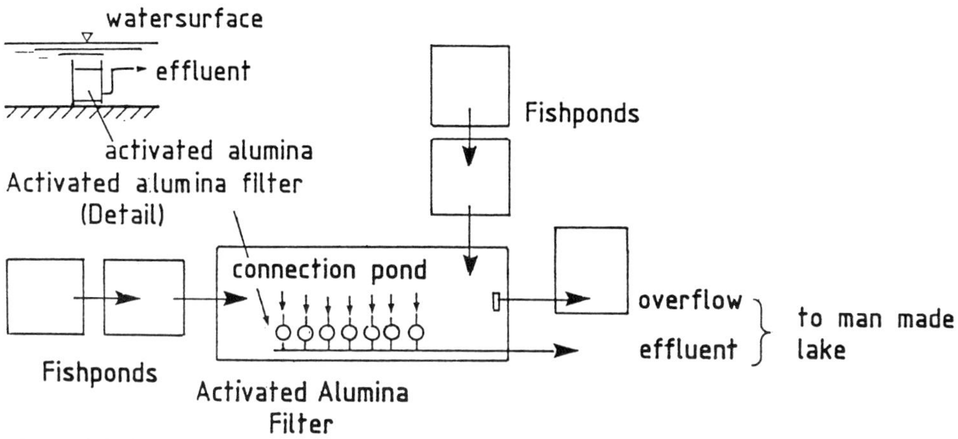

Figure 3.3. Schematic arrangement of an activated alumina filter in the collection pond for fish culture

Table 3.4. Mean values and elimination capacities (E) of the activated alumina filter plant for various parameters during the period June 1980 to October 1981 and, following regeneration, from April 1984 to May 1985

Parameter	June 1980 - October 1981			April 1984 - May 1985		
	Inflow (μg/l)	Outflow (μg/l)	(E) (%)	Inflow (μg/l)	Outflow (μg/l)	(E) (%)
o-PO_4^{3-}-P	222	39	82	88	7	84
Particulate P-compounds	87	34	61	50	14	72
Total P-compounds	341	85	75	158	30	81
DOC	3.3	2.2	39	2.8	1.6	43
NH_4^+	1785	846	52	1510	490	68
NO_2^-	1345	726	46			–
NO_3^-	42.2	43.0	–	43.3	44.4	–
Al^{3+}	23	(10-25)*	–	20	(10-35)*	–

*Minimum and maximum single values

4. *Pre-reservoirs as bioreactors.* The use of pre-reservoirs as bioreactors for phosphorus elimination is founded on investigations by Uhlmann & Benndorf (1980), Benndorf & Pütz (1984), Benndorf et al. (1976), and it is based on the growth of microorganisms that fix the phosphorus in their biomass. Following sedimentation of the biomass, the phosphorus remains in the sediment provided that the conditions are sufficiently aerobic. Anaerobic conditions at the bottom of the impoundment should be avoided. During the summer in middle Europe, such pre-reservoirs have been shown to eliminate 70-90% of the ortho phosphate. During the winter, the efficiency decreases to 0-30% due to light limitation and low water temperatures. Phosphorus elimination in winter is caused mainly by the sedimentation of particulate matter (Wilhelmus et al., 1967). The average value

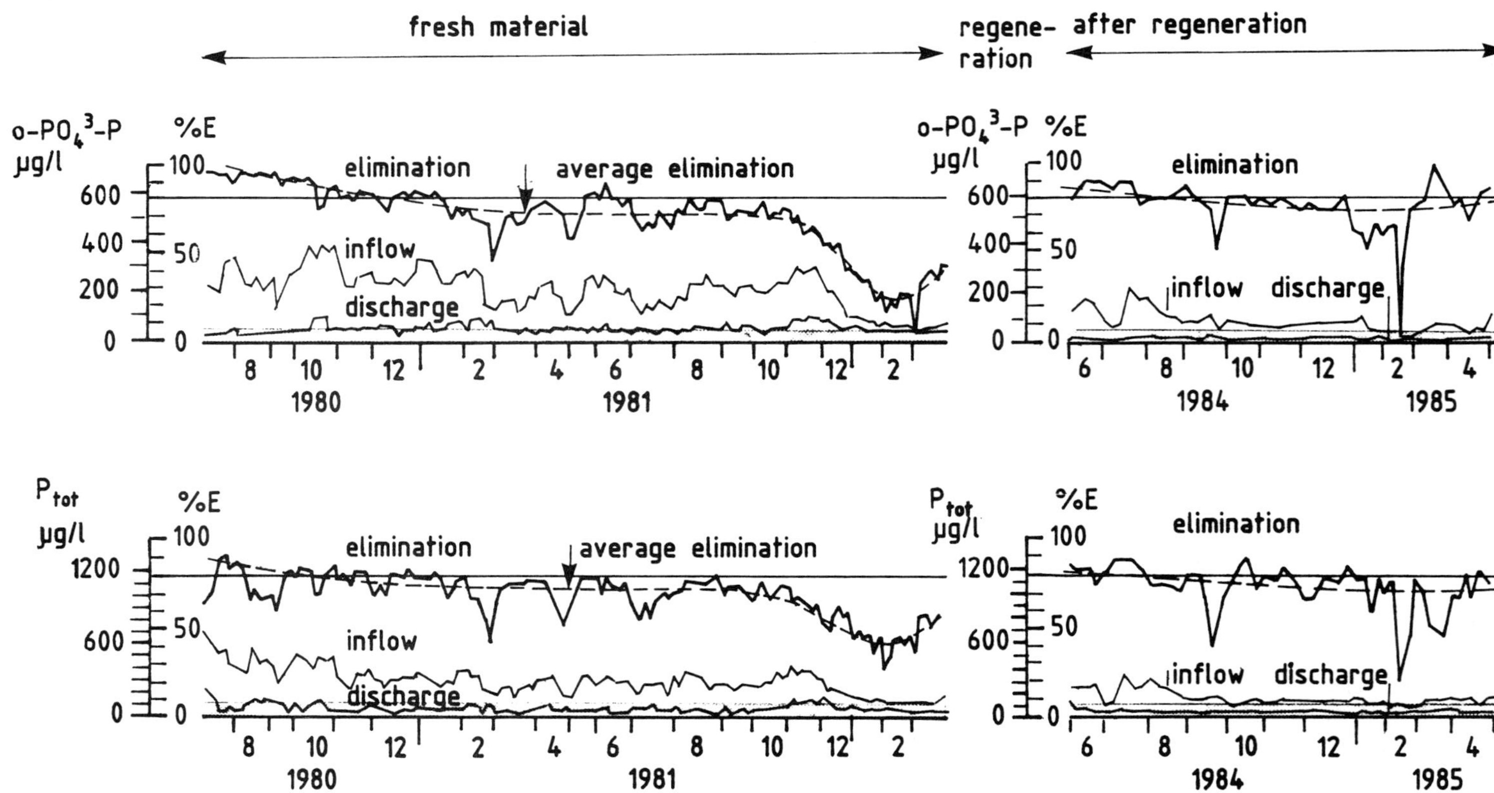

Figure 3.4. Results of P-elimination using activated alumina filters. P-concentrations of in- and outflow of the collection pond for the period June 1980 to March, 1982 with new filter material and from May, 1984 to 1985 with regenerated material

for the elimination of dissolved, reactive phosphate calculated for each complete year amounts to 50-60% P (Uhlmann & Klapper, 1985).

In order to ensure a reliable elimination of phosphorus, the retention time in the production zone should be optimal and not too long. In accordance with the light intensity needed for algal development, the production zone (upper water layer) can have a depth of up to 3 m. The impoundment should have a depth of not more than 5 m. The residence time of the water in the 3 m layer must be longer than the critical residence time for the development of the diatoms that are the preferred dominant plankton for the elimination of P. The residence time should, however, be shorter than the time needed for the development of blue-green algae that are not required for P-fixation. Daphnia are also undesirable (grazing effect), and they should be removed with controlled fish keeping.

The process of biological P elimination in pre-reservoirs has been developed in Germany to a technical scale (N.N., 1983). With the help of calculations of the probable average o-PO_4 removal, 4 pre-reservoirs were built in different areas according to these guidelines, and the expected P removal was largely achieved.

P elimination	Case 1	Case 2	Case 3	Case 4
Calculated (%)	66	87	79	58
Measured (%)	52	68	74	56

5. *Phosphorus elimination from tributaries using precipitation, flocculation and filtration (Wahnbach system).* The most efficient, but also, the most laborious and expensive method, is the treatment of a whole tributary with a process involving precipitation, flocculation and filtration. This procedure can reduce the total phosphorus compounds to residual amounts of 5 μg/l P_{tot}. It has been developed at the Wahnbach reservoir (43 Mio m^3 content, mean depth 20 m, max depth 43 m, area 2.15 Mio m^2, water residence time ~ 1 year). It has operated successfully since 1978. A survey of the process is given in (Bernhardt, 1983). The principles of operation are shown in Figure 3.5 (Bernhardt & Schell, 1979, 1982). The Wahnbach system is suitable for those lakes which have one or, at most, two tributaries. The phosphorus elimination plant (PEP) at the Wahnbach reservoir is designed for a maximum flow of 5 m^3/s because fluctuations in the amount of incoming water can produce flows well above the mean of 1 m^3/s delivered by the Wahnbach River.

The PEP fulfils the following requirements:

– According to water flow, operation for only a few hours with intervals of several days between them (in summertime), and frequent stopping and starting without decrease in quality of the filtrate; operation with a maximum throughput of 5 m^3/s for several weeks.

– No drop in efficiency at a water temperature of 0°C.

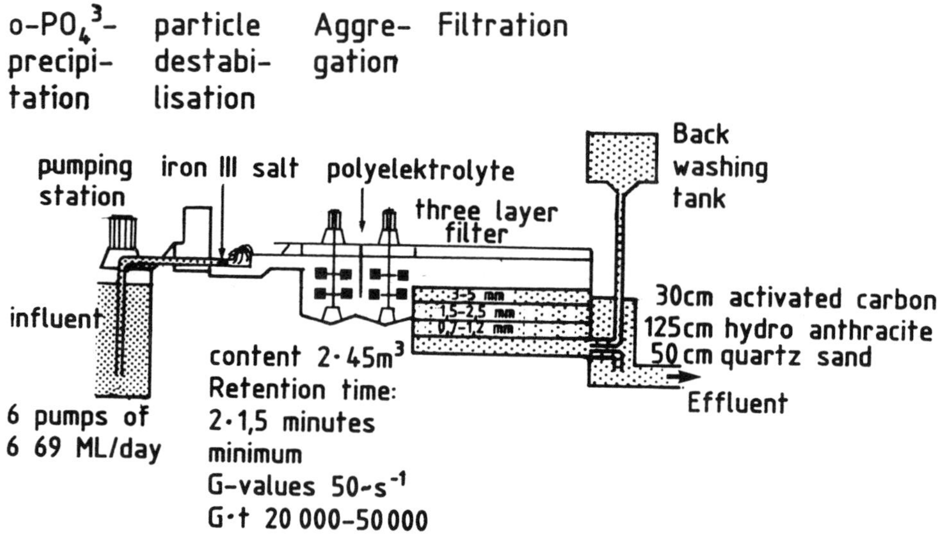

Figure 3.5. Scheme of the Phosphorus Elimination Plant (PEP) with P-precipitation, flocculation and filtration through a three layer filter according to the Wahnbach-system (energy input controlled direct filtration)

– Treatment of water with high turbidity (up to the solid concentration of 100 mg/l by dry mass processed at the drying temperature of 105°C, without shortening filter-run times below 10 h at a maximum filtration rate of 15 m/h.

– Decrease of total phosphorus content to values < 5 µg/l.

– Removal of > 99% of the plankton blooming during the summer months (max 400 000 cells/ml). Algal cells which break through the filter are the cause of high concentrations of undissolved and dissolved organically bound phosphorus compounds in the filtrate; they cause an excessive phosphorus load on the reservoir.

– Removal of > 99% of inorganic suspended material that is rich in phosphates because of the erosion of agricultural land.

– Flocculation with 4-12 mg/l Fe^{3+}. The total iron concentration in the effluent should not exceed 50 µg/l.

– Compact construction of the plant with 10 filter beds with a filter area of 1100 m^2 in order to be economically and technically feasible.

Figure 3.6 shows the amounts of phosphorus from different sources that were carried into the reservoir from 1977-1984. The black columns (1970-1977) represent the quantities of phosphorus that the Wahnbach River transported into the reservoir, before the PEP was in operation. The areas marked with diagonal lines represent the phosphorus load that was retained by the PEP from 1978 onwards. The black part below these areas represents the quantities of phosphorus that could not be treated by the PEP owing to pre-reservoir overflow. The dotted

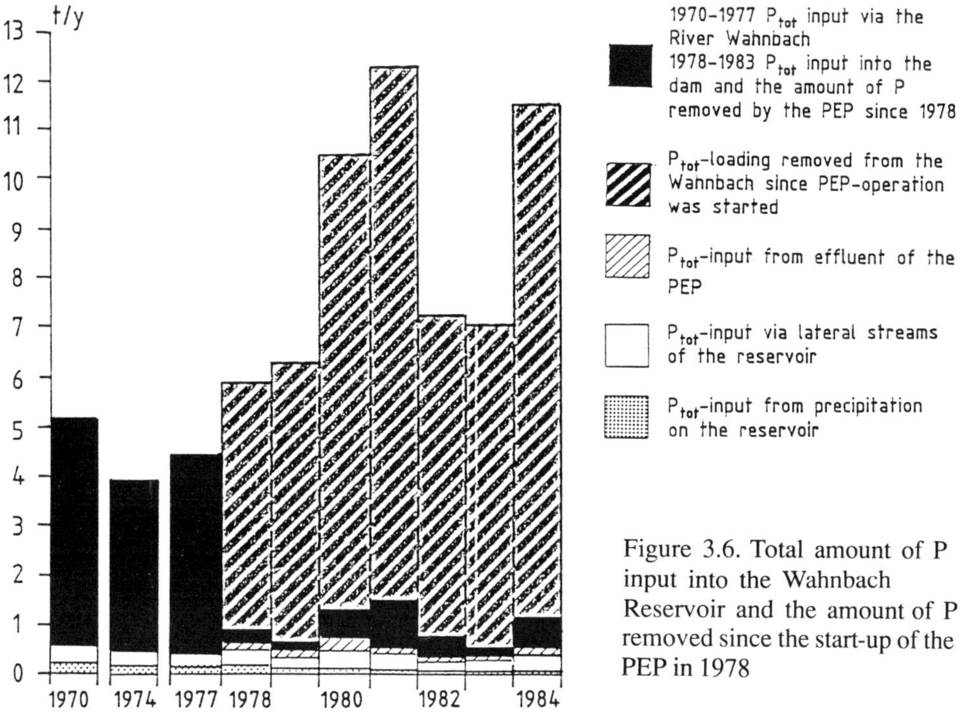

Figure 3.6. Total amount of P input into the Wahnbach Reservoir and the amount of P removed since the start-up of the PEP in 1978

parts of the columns are the small quantities of P that were transported into the main reservoir in the PEP effluent (< 5 μg/l P).

The quality of the filtrate and the average elimination are shown in Table 3.5. About 95-99% of the phosphorus compounds were eliminated and the total phosphate concentration in the filtrate amounted to 5 μg/l based on a seven year average. The percentage elimination of organic compounds varied between 60 and 70%. More than 99% of the algae were eliminated and more than 99% of the turbidity was removed.

The decrease in P-input has decreased considerably the P-concentration in the Wahnbach Reservoir. As a consequence, there is less algal growth, and blue-green algae, which had been a dominant part of the algal flora, have disappeared from the water (Bernhardt & Coworkers, 1985). The average Secchi depth has increased to 3-6 m. Moreover, mass developments of algae no longer disturb the drinking water treatment process to any appreciable extent. The biomass production in the reservoir, measured as chlorophyll, ranges from 5 μg/l to 8 μg/l with a long-term average of 6 μg/l, an amount that corresponds to a mesotrophic state (Vollenweider & Kerekes, 1982).

This full scale experiment could demonstrate the possibility of converting a lake with a capacity of about 40 million m^3 and a retention time of about 1 year

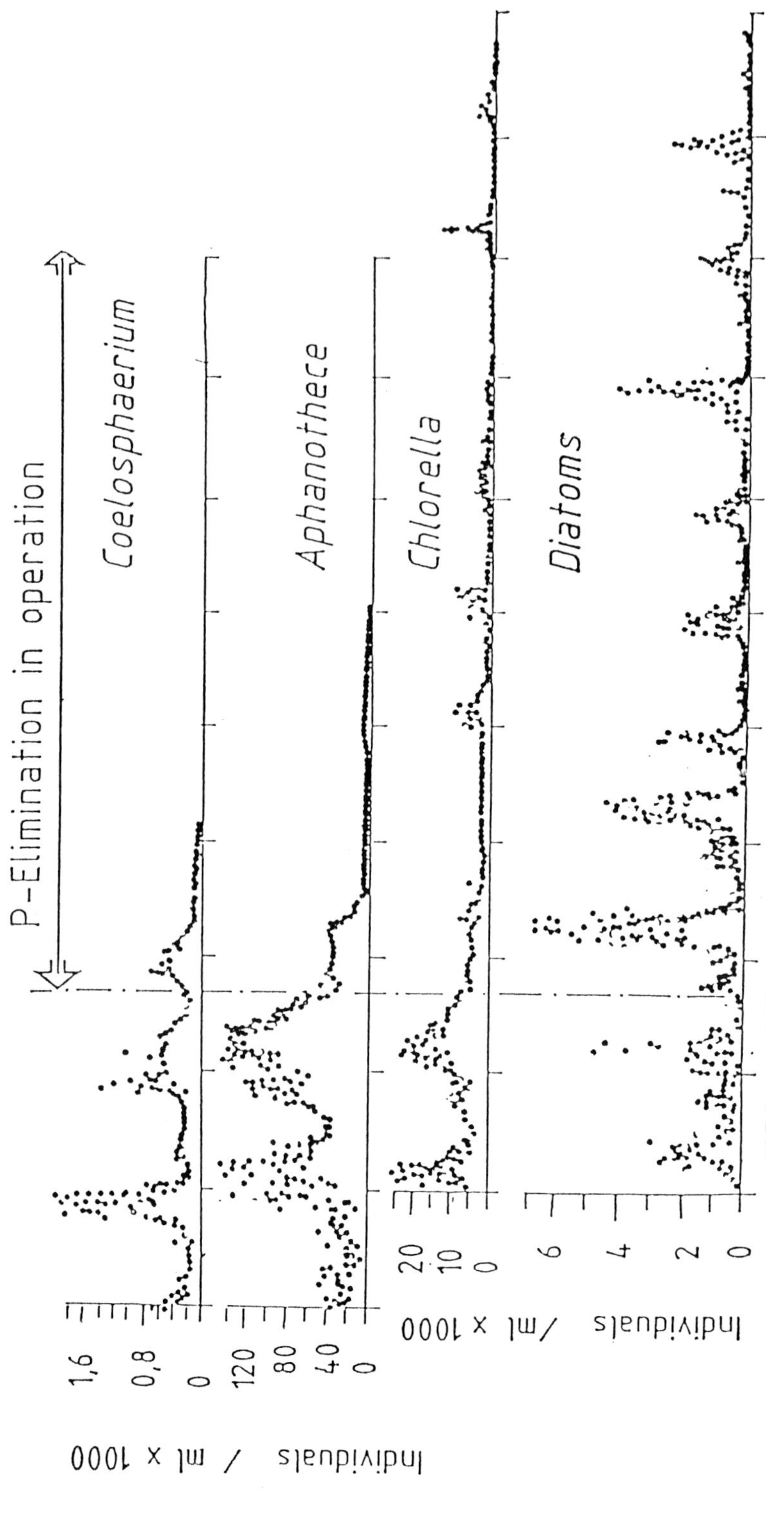

Figure 3.7. Effect of P-elimination on the frequency of different algae in the raw water

Table 3.5. Elimination of several substances by the PEP

Parameter	PEP – Inflow min.-max $\bar{x}$	PEP – Outflow min.-max. $\bar{x}$	Elimin. %
o-PO_4-P µg/l	1-220 29.3	1-5 1	96.6
Total P µg/l	27-480 116.5	1-13 4.3	96.3
DOC mg/l	0.9-7.3 2.4	0.4-2.2 1.0	58.3
Spect. absorpt. coeff. 254 nm m^{-1}	3.4-20.8 8.1	0.3-4.7 2.4	70.4
COD mg/l	3.7-22.3 11.1	0.1-6.3 2.6	76.6
Turbidity FTU	0.6-48.7 10.4	0.01-0.8 0.06	99.4
Chlorophyll µg/l	1.0-204.3 25.2	0.1-17.3 1.3	94.9
Colony count (22°C) Colonies/ml	285-290 000 11 504	0-17 100 263	97.7
Coliforms bacteria / 100 ml	0-68 000 5979	0-171 8	99.87

from a eutrophic to a mesotrophic state within a period of 1-2 years (Bernhardt & Coworkers, 1985).

These investigations show clearly that the phosphorus loading of the reservoir from all tributaries and precipitation, in spite of phosphorus elimination, amounts to 15-20 µg/l P_{tot} (annual mean), a level just sufficient to allow the desired reduction in biomass production and a shift in the composition of the algal population to occur. When this value is exceeded in wet years due to storm water, i.e. up to 25 µg/l P_{tot} (as in 1980/1982), algae develop in larger amounts again in winter and spring (Fig. 3.7). Obviously, only very efficient processes are capable of maintaining a large lake in a mesotrophic state when the environment around the lake is such that eutrophic conditions would normally occur.

A similar process is employed in Berlin (Germany). In order to control eutrophication of Lake Tegel, precipitation and flocculation followed by sedimentation and secondary flocculation with filtration are used to treat part of the inflow (max 6 m^3/s) of this lake (32 Mio m^3 content, max. depth 16 m, calculated water retention time ~ 120 d). Due to the high P content of the tributaries (~ 1.5 mg/l P) and the high loading with organic matter, a two-stage flocculation with sedimentation and filtration was selected (Hässelbarth, 1979; Grohmann, 1982). These measures allow the P content to be reduced to < 30 µg/l.

3.5.3 *Measures in the lake*

Measures to be used in the lake can be divided into:
- limitation or prevention of the internal loading,
- reduction of the algal development independent of the nutrient influx.

Internal transport processes convey the nutrients to the free water zone where they can lead to extensive algal blooms. This phenomenon can take place particularly in lakes shallower than 10 m and in parts of a lake where large amounts of nutrient-rich sediment accumulate on the lake bottom. Although phosphorus is released from the sediment at considerably higher rates under anaerobic conditions than under aerobic conditions, considerable amounts of phosphates and substances stimulating algal growth are still released from the sediment under aerobic conditions. Such occurrences have been described many times in the literature (Boström et al., 1982; Golterman, 1977; Hoyer et al., 1982; Clasen et al., 1982).

In order to reduce or even to eliminate the effect of the internal loading, the technically realizable solutions are described in the following.

1. *Sediment covering*. A survey of the results obtained by covering the sediment has been given by Cooke (1980) (also Scharf et al., 1984). Sheeting is often employed to cover the sediment but it has to be weighted to prevent it from floating. Sand and gravel cannot be used to keep it in place because they provide a good substrate for macrophyte seedlings after 1-2 years.

The sheeting sould be permeable to gases to prevent them from ballooning due to the release of gases from the sediment. In some lakes, black polythene sheeting, usually of 0.1 mm thickness, has been used to cover the sediment, mainly with the aim of suppressing macrophyte growth. However, polythene has the disadvantages that it is impermeable to gases and therefore has to be perforated which, in turn, favours the growth of the macrophytes through the sheeting. In addition, this material is sensitive to light. Further information on this method is available (Scharf et al., 1984).

2. *Removal of sediment by dredging*. Dredging can only be used to remove the sediment from small and shallow lakes. The greatest problems are the treatment and deposition of the material removed, and the need to pay particular attention to the amount of toxic metals present. Kalbe (1976) reported that the dredged material is suitable for improving the soil in orchards. However, it cannot be used if the material contains more than 10% calciumcarbonate or 15% organic matter.

Recently, experience has been gained on the use of dredging in Lake Baldeney (Germany), the fourth lake in a chain of 5 on the lower course of the River Ruhr. The lake (6.9 million m^3 capacity, 2.64 km^2 surface area) has a maximum depth of 4 m (mean 2.7 m), but it is made up of large shallow areas with depth less than 1 m due to sediment deposition. The rate of inflow varies considerably as it ranges from 20 m^3/s to 2000 m^3/s. The lake, together with the other impoundments, serves to improve the water quality of the nutrient loaded river Ruhr (Koppe et al., 1982).

The sediment was removed to a depth of 2.3 m using a hydraulic dredger with a cutter. One million m^3 of sediment were pumped into sedimentation basins subdivided into compartments covering areas of 22 ha each. In order to prevent sedimentation in the transport pipes, which had a length of 7.2 km, a pumping rate of 3.5 to 4.0 m^3/s had to be maintained. Thus sediment was transported at the rate of 0.6 to 0.8 m^3/s with flush water (at a ratio of 1:6 to 1:10), into one compartment after an other. The sediment was allowed to settle for 1 week and the overlying water was then pumped into the lake. The P content of this water was 0.35 mg/l P_{tot}, which was half the concentration in the River Ruhr. Soon after the sedimentation basin had been filled with sediment, a dense and diverse flora developed on the compacted material.

An example of such a field operation has also been given by Björk (1973) and Cronberg (1982) for the dredging of Lake Trummen. By means of suction pipes, layer of sludge with depth of 60 cm was removed from the lake bed within a period of 2 years. The nutrient-rich sludge was allowed to settle in a sedimentation basin and used directly as an agricultural fertilizer to improve soil. The water in the sedimentation basin was treated with aluminium sulphate as flocculant and again allowed to settle. The remaining water was pumped back into the lake. In this operation, 50 t P (90% reduction of the P-content) and 450 t N were removed from the lake with the sediment. The average amount of biomass decreased from 75 mg/l to 10 mg/l. In particular, algae such as Microcystis that usually produces algal blooms disappeared. Following removal of the lake sediment, the diversity of the plankton population increased. The transparency of the water increased from 0.2 m to 0.8 m (Cronberg, 1982). The lake is now used for recreation.

Without doubt the practice of removing sediment from lakes and from shallow areas is more effective than covering it with synthetic sheeting because the nutrients are actually removed from the lake. It does, however, involve more technical problems and higher costs, and it is only feasible if space is available for the deposition of the sediment removed.

3. *Injection of nitrate into the sediment as an oxygen carrier.* An effective method of inhibiting the internal loading of a shallow lake has been developed by Ripl and consists of mixing a nitrate solution into the sediment (Riplox-method) (Ripl, 1976). In Lake Lillesjon (S), which has an area of 4.2 hectares and a depth of 4.2 m, the pollution load due to wastewater dishcarge was eliminated and the sediment (black FeS, high P content, low Fe concentration in the upper sediment) was treated by employing the following steps consecutively:

– 13 t $FeCl_3$ to increase the available iron to precipitate phosphate and remove H_2S,

– 5 t $Ca(OH)_2$ to neutralize the sediment, and

– 12 t $Ca(NO_3)_2$ to oxidize the organic and inorganic matter (FeS).

These chemicals were ploughed into the sediment. Nitrate is transformed to nitrogen gas, which escapes. After only a few weeks, Lemna disappeared from the lake, the transparency rose from 2.3 m to 4.2 m and the P_{tot}-concentration sank from < 3 mg/l in the hypolimnion to ~ 0.04 mg/l P in the entire water column in the hypolimnion. The lake can now be used for recreation (Björk, 1985).

4. *Removal of water from the hypolimnion.* The removal of water that is low in oxygen content and rich in nutrients from the deeper parts of the lake and its replacement with oxygen-rich water increase oxygen levels at the lake bottom. During the process, the natural outflow is closed. The method is based on an idea of Olszewski (1961) and is practiced in numerous lakes in Switzerland, Austria (Lake Piburg) (Pechlaner, 1971) and in several lakes in the GDR (Klapper & Rudolf, 1979). Recently, Scharf was able to eliminate the meromictic conditions of the volcanic lakes of the Eifel (Germany) using this method (Scharf et al., 1984). The working principle is shown in Figure 3.8 (Klapper, 1980).

The results depend on the management of the water quantity. In lakes with higher flow-through rates and a relatively short times of retention, the desired results occur quickly. For example, 10% of the volume of Lake Piburg (Pechlaner, 1971) was removed and replaced within 6 months. The outflowing water, however, sometimes smells of H_2S.

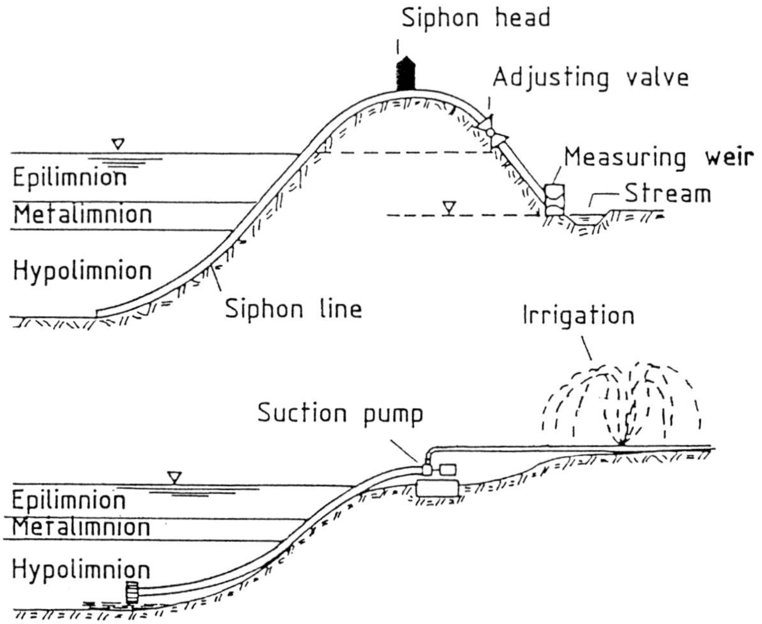

Figure 3.8. Examples of techniques for discharge of hypolimnetic water (from Klapper (1980) & Pechlaner (1971), modified)

Together with effective measures of reducing the phosphorus influx, this method can accelerate the process of oligo-trophication; it may even be the only solution. However, without simultaneous reduction in the P influx, it has little use because its effect lasts only a short time.

5. *Biomanipulation.* During the last years, the principle of biomanipulation has been successfully employed. Such measures are particularly suitable for shallow, natural lakes, for man-made lakes and for reservoirs that are fed with large amounts of water from rivers flowing from catchment areas in which no sanitation measures can be taken due to their multifarious use.

A comprehensive report on biomanipulation has been published by Shapiro et al. (1982). A simplified version of the principle of this method is given in Figure 3.9. Large zooplankton species, e.g. Daphnia, Diaptomus, decimate the population of phytoplankton species due to their grazing. This desired effect can be

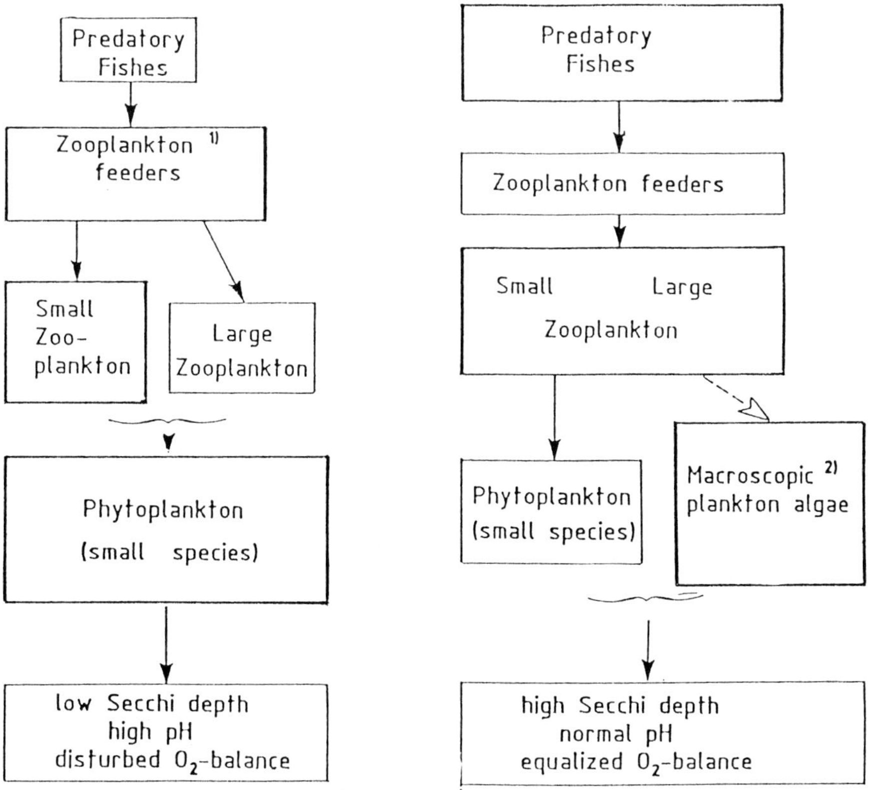

1) Pelagic carnivorous fishes and predatory Zooplankton. (Importance of the latter uncertain)

2) colonies of visible size, e.g. Aphanizomenon, Microcystis a.s.o.

Figure 3.9. Scheme for the manipulation of pelagic food web (aquatic food chain)

adversely affected or even eliminated by white fish, perch and roach that feed primarily on large zooplankton; predatory zooplankton, such as Leptodora kindti or insect larvae such as Chaoborus flavicans can have the same effect. The small zooplankton feeders such as the rotifers, Keratella or Kellicotia, remain in large numbers under these conditions. The effect of their grazing on phytoplankton is, however, slight. Most rotifers feed mainly on microalgae and cannot effectively reduce the phytoplankton population.The number of zooplankton feeders must therefore be kept low, and this result can be effected by maintaining predatory fishes such as pike and pike-perch (Benndorf et al., 1984).

Sport fishing in such lakes should be limited to fish longer than 45 cm in order to protect the predatory fish. This method was employed by Benndorf and co-workers in a shallow man-made lake (average depth 7.4 m) (Benndorf et al., 1984).

Predatory fishes (pike) were introduced to feed on the small fish whose source of food was zooplankton. After a few years the population of small fish was decimated and the zooplankton population increased. The fraction of large zooplankton such as Daphnia and predatory zooplankton (mainly larvae of Chaoborus flavicans and the water flea Leptodora kindti) in the total biomass increased from 1/3 – 1/2 to 2/3 – 3/4. The intensive filtration of water where Daphina prevail leads to the results cited below:

- lower phytoplankton biomass,
- greater transparency (Secchi disc),
- reduced numbers of bacteria,
- a better and more favorable oxygen balance along the depth profile.

Within the first year, the phytoplankton biomass was reduced from an average of 10.2 g/m^3 (maximum 18.5 g/m^3) to an average of 3 g/m^3 (maximum 6.5 g/m^3). However, the composition of the phytoplankton population changed so as to favour an undesirable blue-green alga because filamentous algae, and in particular filamentous blue-green alga, cannot be ingested by zooplankton. Therefore, higher biomass values were obtained in the second year (mean value 17.4 g/m^3, maximum value 40.9 g/m^3) due to the appearance of large quantities of a blue-green alga. The authors suggest keeping a larger stock of silver carp (Hypopthalmichthys molitrix) in the lake to control the growth of filamentous algae.

The problems associated with this technique lie in regulating the composition of the zooplankton with fish. Control of the process is only possible to keep if accurate and continuous information is obtained about the composition of the fish population in the lake. In larger and deeper reservoirs and lakes that are used for drinking water, this method will not yield satisfactory results.

6. *Artificial destratification*. The artificial mixing of a stratified lake can
   - change the composition of the algal population,
   - reduce the number of algae and the algal growth rate,

– aerate the lake and thus compensate for the oxygen deficit resulting from metabolic activity.

Destratification finds application chiefly in artificial reservoirs with a more or less constant depth (Knoppert, 1978; Steel, 1975), and it has only limited use in water bodies formed by damming a river. These reservoirs are deepest at the dam, and with increasing distance upstream they become shallower and the full mixing depth decreases. The compensation depth (the limit of the photic layer which is to be defined soon lately) exceeds the full mixing depth, depending on the extinction coefficient of the water. In these reservoirs, mixing would increase the biomass and not produce the conditions desired.

With artificial circulation of the lake, the oxygen content of the water is increased at all depths, and at the sediment-water interface, aerobic conditions are created, which suppress the release of nutrients from the sediment.

a) *Light limitation effect by destratification.* The light limitation effects have been extensively investigated and elucidated in recent years by a large number of workers (Oskam & Bernhardt, 1978). The algal development is limited, if the artificial mixing holds the algae in the darker and deeper parts of the lake for a sufficient time to reduce their photoassimilation rates (light limitation). This limitation occurs if the mixing depth [z_m], which is usually the actual depth [z], is greater than the compensation depth [z_k]. The compensation depth is the lower limit of the photic layer at which photosynthesis and respiration of algae equal each other. The light intensity at this depth is usually taken to be 1% of the light intensity at the surface, I_0. Since the light intensity decreases as

$$I(z) = I_0 e^{-\varepsilon z}$$

the compensation depth is given by

$$\ln 100 = \varepsilon \cdot z_k \qquad (3.1)$$

or

$$z_k = 4.6/\varepsilon \qquad (3.2)$$

where ε = sum of the extinction coefficients from the water (ε_w) and the algal (ε_c)

$$\varepsilon = \varepsilon_w + C \cdot \varepsilon_c \qquad (3.3)$$

where ε_C = the self-shading coefficient, C = algal concentration.

However, in lakes with clear water (of low extinction coefficient and thus great compensation depth), an artificial full mixing can reduce photosynthesis only if the lake is deep enough such that the depth of mixing layer can be greater than the compensation depth and the mixing proceeds so slowly that the algae are kept in water with low light intensities long enough to inhibit to growth. In turbid lakes or lakes with coloured water, the mixing depth does not need to be so great, and the lakes can be shallower.

Experience has shown that the depth of fully mixed water layer should be at least 5-10 times the Secchi depth of a lake devoid of algae, but not less than 15 m.

Furthermore, other factors which have to be considered in the calculation are the rates of photosynthesis and respiration, the sedimentation of the algae, the loss of algae due to zooplankton feeding, and of course the nutrient content of the water.

The water temperature has a decisive influence not only on the rates of photosynthesis and respiration, but also on the production of zooplankton and, therefore, on the extent of grazing (further information is available in Oskam, 1973; Steel, 1978; Duncan, 1975). By destratification, the cold water in hypolimnion is mixed with the warm water in epilimnion, and the temperature of lake rises as a whole to 15-20°C.

Algal development starts early in the year when the water only has a temperature of 3°C. The first algae to develop are usually diatoms. However, if light can be limited by mixing, the algal development can be delayed; its peak then occurs at the end of April or the beginning of May. At this time, partly as a result of mixing, the water temperature is high enough to allow for the rapid development of zooplankton (Steel, 1978; Duncan, 1975; Steel, 1973). Good results produced by this practice are expected by the more favourable conditions created for zooplankton, e.g. Daphnia. However, small carnivorous (zooplankton feeding) fish should not present in appreciabale quantities.

Oskam (1982) has reported that, following artrificial mixing of a reservoir in Biesbosch (Netherlands), mainly cladocerans (Daphnia longispina, D. magna and Bosmina longirostris) and to a lesser extent copepods (Cyclops sp.) developed, and these drastically reduced the level of phytoplankton (Fig. 3.10).

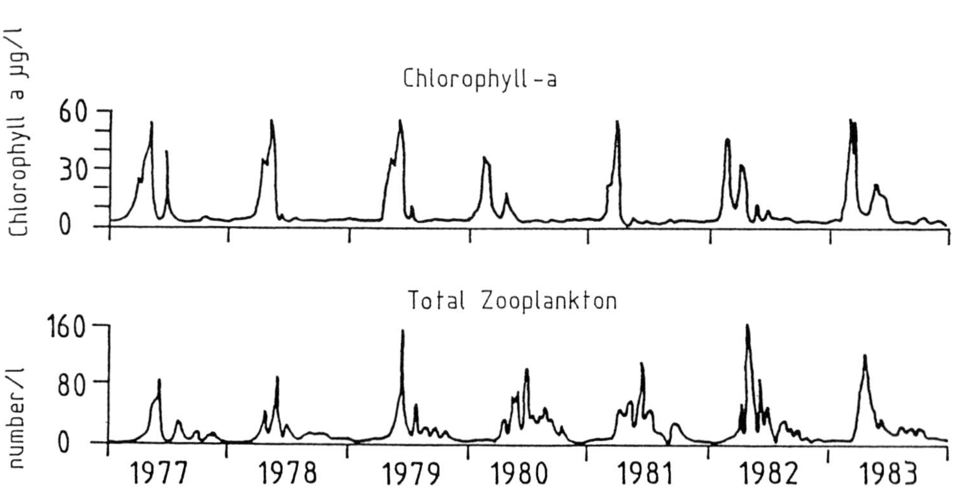

Figure 3.10. Chlorophyll concentration and numbers of Zoo-plankton in the Biesbosch-Reservoir (from Oskam, 1982, supplemented)

b) *Negatiave effect of destratification – Unsuccessful examples of destratification.* However, if this process is not used properly, mass developments of green algae and diatoms can occur. Steinberg (1983) has reported on the effect of destratification using the air-lift principle in a small meromictic lake having a maximum depth of 11 m and a volume of 180 000 m^3 (Lake Fischkalter, Germany). Figure 3.11 shows that following the start of destratification, the already dominant blue-green alga (Oscillatoria sp.) increased. The cause was the

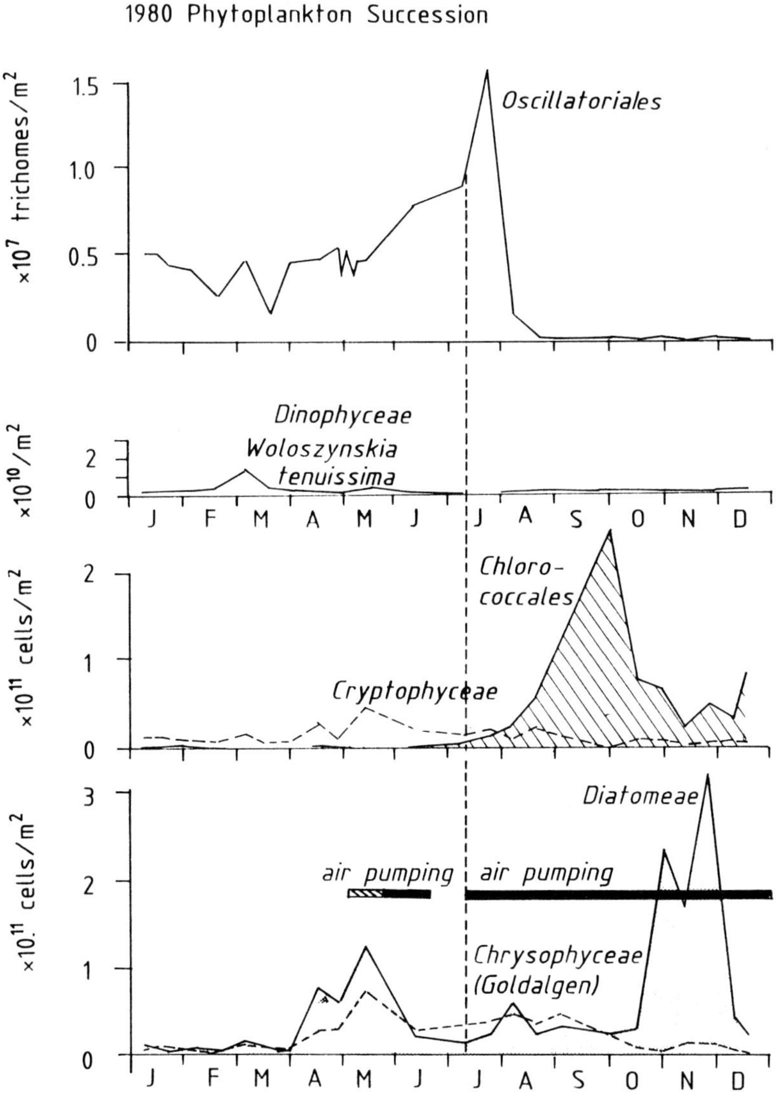

Figure 3.11. Phytoplankton succession prior and during artificial destratification with free ascending air bubbles in Lake Fischkalter. (Black: air pumping) (from Steinberg (1983))

transport of nutrients from the sediment-water interface and the monomolimnion into the trophogenic zone. However, the mass development of blue-green algae collapsed within a few days, after which the lake flora was dominated by chlorococcal algae and later by diatoms. The reason for the development of green algae and diatoms in place of filamentous blue-green algae is the low ability of blue-green algae to tolerate changes in the environment. The artificial destratification generates a water circulation which destroys the milieu in which the filamentous blue-green algae have an advantage over green algae and diatoms.

The destratification of the lake water under the given morphometric conditions in this example resulted in significant increases in the production of biomass. Measurements of chlorphyll in a 1 m^2 water column (integrated value throughout the depth) averaged 0.36 g/m^2 before destratification and 0.68 g/m^2 and 0.58 g/m^2 during the next two years. These investigations show that homogenization of the water quality in relatively shallow lakes, that is, those with a maximum depth of only about 10 m increases the production of biomass.

7. *Hypolimnetic aeration.* The goal for hypolimnetic aeration is to introduce oxygen into the hypolimnion of a standing water body without disturbing the temperature. This measure is particularly suitable for improving the oxygen balance in lakes where the hypolimnion extends at depth greater than 10 m and the ratio of the volume of epilimnion to that of hypolimnion is less than 2.

The hypolimnetic aeration is, however, not suitable for directly reducing the extent of an algal development. The effect is indirect and is due to a suppression of the internal loading. This measure is conceived to overcome the harmful effects of the disturbance of the oxygen balance in a lake, and it should be supported by measures which reduce the nutrient input.

How hypolimnetic aeration should be planned and carried out has been reported extensively in the literature (DVGW, 1985; Lorenzen et al., 1977; Bucksteeg et all., 1985).

The area in the lake, in which hypolimnetic aeration has an effect, depends on the volume of aerated water discharged horizontally. In order to obtain an effect over long distances within a lake, the flow of oxygenated water should theoretically allow for a complete circulation of the hypolimnion within 50-100 days. The hypolimnetic aeration plant must, therefore, be designed to meet this demand.

The oxygen consumption rates in the hypolimnetic water body and in the sediment-water interface are equally important. The daily amount of oxygen which must be brought into the hypolimnion can be calculated from its rate of oxygen consumption ($g/m^3 \cdot d$) in the water and by the sediment (g oxygen removed per m^2 sediment per day). The actual oxygen consumption is usually 50 to 200% higher than the calculated value. As soon as the oxygen concentrations become lower than 5 mg/l at the sediment-water interface, reductions in MnO_2 and nitrate occur and phosphorus is released (Fig. 3.12). In addition, the measur-

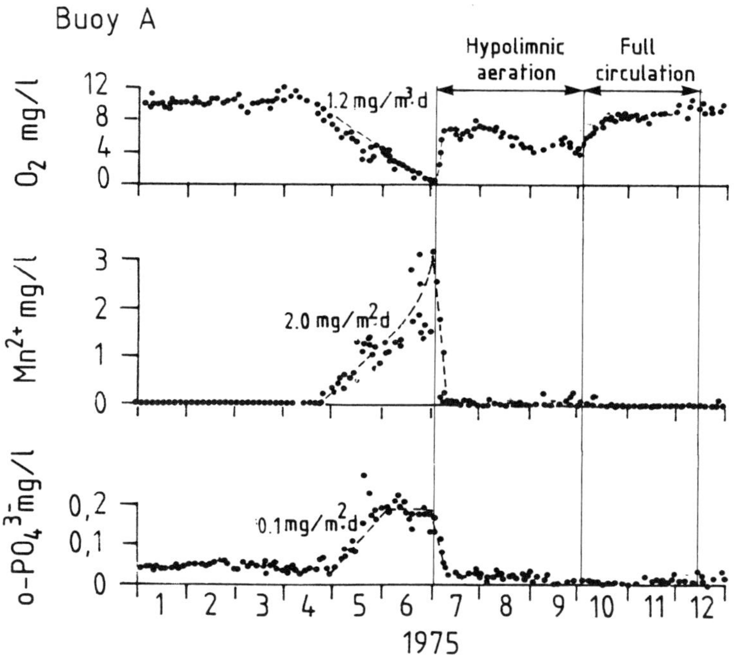

Figure 3.12. Effect of hypolimnetic aeration and artificial full circulation by ascending air bubbles on the concentration of oxygen, manganese and o-PO_4-P in the sediment-water interface of the Wahnbach Reservoir. The O_2-depletion rate bases on the hypolimnion (31 m depth), the Mn- and PO_4-mobilisation rate basis on the microlayer (5 m)

able oxygen consumption rate decreases if the oxygen concentration in the hypolimnetic water is lower than 5 mg/l. Therefore, in practice the capacity of a hypolimnetic aeration plant should be at least 1.5 times greater than measured oxygen consumption rate.

The results from 15 years of use of the hypolimnetic aerator at the Wahnbach Reservoir, developed by the Wahnbach Reservoir Association (Bernhardt, 1978); $Vol_{epi}= 20\cdot10^6$ m³, $Vol_{hyp}= 16\cdot10^6$ m³ and max. depth ~ 45 m) show that (Fig. 3.12):

– In spite of occasional mass developments of algae, the oxygen consumption at the sediment-water interface is compensated for during stratification. No anaerobic conditions occur at the lake bed (> 2 mg/l O_2).

– Unwanted concentrations of iron and manganese do not occur in the hypolimnion. Treatment to remove manganese is generally not necessary.

– The release of o-phosphate from the sediment remains low (limitation of internal loading).

– During stratification, the hypolimnion water remains at temperatures of < 10°C that is below the guide values for drinking water quality (15°C) (EEC, 1980).

8. *Suppression of algal growth by chemical means.* Methods of destroying phytoplankton by applying chemicals to lakes have often been discussed. They should, however, be used with caution.

A technique used by Pütz et al. (1983) was successful in the drinking water reservoirs of Klingenberg and Lehnmühle in Germany. In these reservoirs, growth of a motile alga (Synura uvella) caused markedly unpleasant smell and taste in the drinking water. The distribution of 7 t of lime/10^6 m^3 over the surface of the water destroyed the algal population within a few days. Due to the treatment, the pH of the water at the surface immediately rose to 9.2, and within a few days the pH of the whole water body was 8.3-9.2.

Extensive laboratory studies showed that the motility of these colonies is inhibited if the pH is above 7.5 and that the cells burst within a pH range of 8.5-9.0. This measure can also be classed under the heading of biomanipulation because CO_2 consumption, normally causes the pH in lake to rise to 9-10.

In contrast, measures involving the use of algicides to kill the algae should be rejected because these chemicals are not compatible with the biotope. Also, such measures only lead to temporary improvements because the sudden death of the mass population, which is usual during stratification, causes a breakdown of the oxygen balance. Such a practice is unacceptable particularly for a drinkingwater impoundment and does not allow the requirements for drinking water to be met.

3.6 SUMMARY

Growth of algae in natural and man-made lakes exerts an unfavourable effect on drinking water treatment processes and on drinking water supplies. Therefore, counter measures are necessary. Because phosphorus is the limiting growth factor, technical measures are required that reduce the P input or the P content of the lake.

However, depending on the purpose of the lake, a mesotrophic or weakly eutrophic condition is quite acceptable. Naturally occuring eutrophication in lakes can be desirable for landscaping, and measures undertaken to control it must, therefore, be suited to the lake's purpose and the trophic grade required for the purpose.

The technically realizable measures and their efficiencies in P eliminaiton have been grouped according to their usability in the catchment area, in the lake tributaries, at the inflow to the lake and in the lake itself. Some of these methods are illustrated by means of case studies that include details of the locality, running experience and effects on the lake and water quality. In this respect, shallow lakes are particularly important if the internal loading from the nutrient-rich sediment provides for the growth of the algae. The measures used for this purpose can be a successful counter-measures, if the P-input is simultaneously reduced.

This survey gives primary attention to methods that have been employed on a practical scale and which have given good results. Methods have also been included that reduce the biomass without reducing the nutrient content of the inflowing streams.

REFERENCES

Benndorf, J. & K. Pütz 1984. Process governing the elimination of nutrients in pre-dams. *EARD*, 2: 17-54.

Benndorf, J., R. Hallebach & K. Pütz 1976. Die Leistung von Vorsperren bei der Rückhaltung der Pflanzennährstoffe aus Abwässern und landwirtschaftlich genutzten Flächen. *Limnologica (Berlin)*, 10: 617-622.

Benndorf, J., H. Kneschke, K. Kossatz & E. Penz 1984. Manipulation of the pelagic food web by stocking with predacious fishes. *Int. Revue ges. Hydrobiol.*, 69: 407-428.

Benndorf, J., F. Recknagel & H. Schultz 1984. Food-web manipulation for a man-made lake (Bautzen reservoir). Internat. Post Graduate Course in ecological approaches to resources development. In *Land management and impact assessment in developing countries, Dresden*, VI(3): 20-41.

Bernhardt, H. 1978. Die hypolimnische Belüftung der Wahnbachtalsperre. *Gas-Wasserfach, Wasser-Abwasser*, 119: 177-182.

Bernhardt, H. 1983a. Input control of nutrients by chemical and biological methods. IWSA Specialized Conference Eutrophication and Water Supply. *Water Supply*, 1: 187-206.

Bernhardt, H. 1983b. Aufbereitung von Wasser aus Seen und Talsperren. *DVGW-Schriftenr. Wasser*, 206: 23-1 till 23-40.

Bernhardt, H. 1984. Treatment disturbances with water out of eutrophic reservoirs as consequence of extensive algal development. *IWSA-Kongreß Monastir/Tunis, Special Subj.*, 4: SS4-7 to SS4-15.

Bernhardt, H. & Coworkers 1985. Oligotrophication in lakes by means of chemical nutrient removal from the tributaries. Its demonstration with Wahnbach-Reservoir (Final report). Arch. Hydrobiol./Suppl., 70(4): 481-533. (Monographische Beiträge).

Bernhardt, H. & H. Schell 1979. Die verfahrenstechnische Konzeption der Phosphoreliminierung am Wahnbach mit Hilfe der Flockenfiltration (System-Wahnbach). *Z. Wasser Abwasser Forsch.*, 12: 123-133.

Bernhardt, H. & H. Schell 1982. Energy-input-controlled direct filtration to control progressive eutrophication. *J. Am. Water Works Assoc.*, 74: 261-268.

Bernhardt, H., S.H. Eberle, D. Donnert, H. Strüwe & A. Wilhelms 1981. Anwendung der Aktivtonerdefiltration zur Eliminierung von Phosphaten aus kleinen Talsperrenzuläufen. *Z. Wasser Abwasser Forsch.*, 14: 180-187.

Björk, S. 1973. Restoring lakes in Sweden. *Technol. Rev.*, 76: 53-60. Mass. Inst. of Techn. Cambridge.

Björk, S. 1985. Scandinavian lake restoration activities. *Lakes Pollution and Recovery, European Water Pollution Control Association, Internat. Congress Rome, 15-18 April*, 293-301.

Böhnke, B. 1980. Literaturstudie zur weitergehenden Abwasserreinigung. *Gewässerschutz-Wasser-Abwasser*, 43.

Boller, M. 1985. Practical experience with phosphorus removal and nitrogen conversion. *Lakes Pollution and Recovery, European Water Pollution Control Association, Internat. Congress Rome, 15-18 April*, 128-138.

Boller, M., J. Eugster & A. Weber 1981. Erfahrungen mit der ersten schweizerischen Flockungsfiltrationsanlage auf der Kläranlage Hochdorf. *Gas Wasser Abwasser*, 61: 349-365.

Boström, B., M. Jansson & L. Forsberg 1982. Phosphorus release from lake sediments. *Arch. Hydrobiol. / Ergebn. Limnol.*, 18: 5-59.

Briese, D. 1984. Nutzungskonflikte und ihre Regelung in Einzugsgebieten von Trinkwassertalsperren unter besonderer Berücksichtigung des Wegenetzes. Dissertation Univ. Gieß.

Bruno, G. 1985. Bracciano sewerage system and sewage treatment plant automatic-control. In *Lakes Pollution and Recovery, European Water Pollution Control Association, Internat. Congress Rome 15-18 April*, 194-198.

Bucksteeg, K. & F. Hollfelder 1985.Künstlich erzeugte Strömung – Ein Hilfsmittel zur Seentherapie. 27. *DVWK-Seminar 'Seenströmungen', München, 13-14.X.1985*.

Clasen, J. & Bernhardt, H. (1980). OECD cooperative programme for inland-waters (Eutrophication control – Shallow lakes and reservoirs, final report). The Water Research Centre, Medmenham Laboratory, P.O. Box.16, Medmenham, Marlow, Bucks., UK.

Clasen, J., H. Bernhardt, O. Hoyer & A. Wilhelms 1982. Phosphate remobilization from the sediment and its influence on algal growth in a lake model. *Arch. Hydrobiol./Ergebn. Limnol.*, 18: 101-113.

Cooke, G.D. 1980. Covering bottom sediment as a lake restoration techniques. *Water Resources Bull.*, 16: 921-926.

Cronberg, G. 1982. Phytoplankton changes in Lake Trummen induced by restoration. Long term whole-lake studies and foodweb experiments. *Folia limnologica scandinavia*, 16: 1-119.

Donnert, D. & S.H. Eberle 1982. Die praktische Erprobung und Optimierung des Aktivtonerde-Verfahrens zur Nachreinigung von Abwässern. In D. Gleisberg (ed.) *Weitergehende Reinigung kommunaler Abwässer insbesondere zur Phosphatelimination*, 67-81. Hoechst-Symposium, Hoechst AG Knapsack.

Duncan, A. 1975. Production and biomass of three species of Daphnia coexisting in London Resvoirs. *Verh. Internat. Verein. Limnol.*, 19: 2858-2867.

DVGW (1985). Maßnahmen zur Sauerstoffanreicherung von Oberflächengewässern. Techn. Mitt. – Merkbl. W250 ZfGW-Verlag GmbH, Postfach 901080, Frankfurt/M.

Eberle, S.H., D. Donnert & H.P. Sapandowski 1980. Abtrennung und Rückgewinnung von Phosphat aus biologisch gereinigtem Abwasser mittels Aktivtonerde. *Gewässerschutz, Wasser, Abwasser*, 42: 835-860.

Edmondson, W.T. 1985. Recovery of Lake Washington from eutrophication. In *Lakes Pollution and Recovery, European Water Pollution Control Association, Internat. Congress Rome, 15-18 April*, 228-234.

EEC, Council Directive of 15th July, 1980, Relation to the quality of water intended for human consumption (No. 80/778/EEC), Official Journal No. L 229 of 30th August, 1980.

Flögl, H. & W. Flögl 1985. O-ring-collectors. *Lakes Pollution and Recovery, European Water Pollution Control Association, Internat. Congress Rome, 15-18 April*, 140-166.

Gleisberg, D. (ed.) 1982. Weitergehende Reinigung komnunaler Abwässer, insbesondere zur Phosphatelimination. *Hoechst Symposium*, Hoechst AG Knapsack.

Golterman, H.L. 1977. Sediments as a source of phosphorus for algal growth. In H.L.Golterman (ed.), *Interactions between sediments and fresh water*, 286-293.

Grau, A. 1977. Untersuchungen zur Ermittlung der Leistungsfähigkeit einer Hangversickerung als spezielles Verfahren einer Bodenpassage. *Gewässerschutz-Wasser-Abwasser*, 24.

Grau, A. 1978. Der Einsatz einer Hangversickerung zur Vorreinigung verschmutzter Wässer im Einzugsgebiet einer Trinkwassertalsperre. *Vom Wasser*, 50: 101-123.

Grohmann, A. 1982. Verfahrenskomination zur Phosphatelimination an den Berliner Seen und die Effizienz von Zusatzstoffen. *Z. Wasser Abwasser Forsch.*, 15: 121-131.

Grohmann, A., H.H. Hahn & R. Klute 1985. Chemical water and wastewater treatment. In R.Leschber und G. v.Nieding (ed.) *Schriftenr. des Vereins für Wasser-, Boden- und Lufthygiene*, 62. Stuttgart/New York: Gustav Fischer Verlag.

Hamm, A. 1976. Untersuchungen zur Nährstoffbilanz am Tegernsee und Schliersee nach der Abwasserfernhaltung – zugleich ein Beitrag über die diffusen Nährstoffquellen im Einzugsgebiet bayerischer Alpen- und Voralpenseen. *Z. Wasser Abwasser Forsch.*, 9: 110-121 and 135-149.

Hamm, A. 1983. Input control of nutrients by technical measures. IWSA Specialized Conference Eutrophication and Water Supply. *Water Supply*, 207-215. London, UK: Pergamon Press Ltd.

Hamm, A. & Coworker 1989. Kampendium – Auswirkungen der Phosphate Höchstmengen verordening für Waschmittel auf Kläraulagen und in Gewässen. Academia Verlag Richam GmbH, D5205 St. Augustin, BRD.

Harremoes, P., E. Arvin & M. Henze 1985. Wastewater nutrient removal – A state of the art review. In. *Lakes Pollution and Recovery, European Water Pollution Control Association, Internat. Congress Rome, 15-18 April*, 120-127.

Hässelbarth, U. 1979. Phosphor-Eliminierung aus den Zuflüssen Berliner Seen. *Z. Wasser Abwasser Forsch.*, 12: 133-147.

Hoyer, O., H. Bernhardt, J. Clasen & A. Wilhelms 1982. In situ studies on the exchange between sediment and water using caissons in the Wahnbach reservoir. *Arch. Hydrobiol./ Ergebn. Limnol.*, 18: 79-100.

Kalbe, L. 1976. Restoration of highly eutrophic shallow lakes and possible uses of lake sediments. In *EUTROSYM 76, Proc. Internat. Symp. Eutrophication and Rehabilitation of Surface Waters, Karl-Marx-Stadt and Berlin, GDR, September 20-25*, 5: 59-71.

Klapper, H. 1980. Management of standing water ecosystems. *Ingenieurschule f. Wasserwirtschaft Magdeburg. 03-1955-05-0.*, 17-19.

Klapper, H. & G. Rudolf 1979. Deepwater depletion as a method for oligotrophication of lakes. *Symp. Biol. Hung.*, 19: 197-199.

Klute, R., V. Mertsch & H.H. Hahn 1984. Studie zum Einsatz der weitergehenden Abwasserreinigung durch F:allung/Flockung in der Bundesrepublik Deutschland. *Korr. Abw.*, 10: 808-815.

Knoppert, P.L. 1978. Das Speicherbeckenprojekt Brabantse Biesbosch. DVGW-Schriftenr. *Wasser*, 16: 68-79.

Koppe, P., H. Kornatzki, F.A. Nusch, F. Dietz & G. Reinnarth 1982. 50 Jahre Baldeney-See – Einige Aspekte seiner wasserwirtschaftlichen Besonderheiten. *Gas-Wasserfach, Wasser-Abwasser*, 123: 17-29.

Lorenzen, M. & A. Fast 1977. A guide to aeration circulation techniques for lake management. EPA-600/3-77-004: 115-116.

Martini, P. 1985. The project CO.B.I.S. plants. *Lakes Pollution and Recovery, European Water Pollution Control Association, Internat. Congress Rome, 15-18 April*, 188-193.

Müller, S. 1984. Oberflächenabfluß, Abschwemmung und Abtrag von landwirtschaftlichen

Nutzflächen als mögliche Ursachen der Befrachtung von Trinkwassertalsperren. Dissertation Univ. Gieß.

N.N. 1983. Phosphatelimination in Vorsperren. TGL 27885/02 (DDR).

Olszewski, P. 1961. Versuch einer Ableitung des hypolimnischen Wassers aus einem See. Ergebnisse des ersten Versuchsjahres. *Verh. Internat. Verein. Limnol.*, 14: 855-861.

Oskam, G. 1973. A limnetic model of phytoplankton growth and its use of algal control by reservoir mixing. *Int. Symp. on Man-Made Lakes, their Problems and Environmental Effects, Geophysical Monograph Series*, 17: 629-631.

Oskam, G. 1978. Modellvorstellungen zur Entwicklungskinetik von Algen. *DVGW-Schriftenr. Wasser*, 16: 14-28.

Oskam, G. 1978. Die Vorausberechnung der Algenbiomasse in den Biesbosch Speicherbecken. *DVGW-Schriftenr. Wasser*, 16: 90-107.

Oskam, G. 1982. Quality aspects of the Biesbosch reservoir. *Aqua*, 0497-0504.

Oskam, G. & H. Bernhardt (ed.) 1978. Verminderung der Algenentwicklung in Speicherbecken und Talsperren. *DVGW-Schriftenr. Wasser*, 16: 13-112.

Pechlaner, R. 1971. Die Restaurierung des Piburger Sees (Tirol). *Carintia II., Sonderheft*, 31: 97-115.

Pöppinghaus, K. 1975. Chemisch-physikalische und biologische Maßnahmen zur Stickstoffeliminierung aus dem Abwasser. *Gewässerschutz-Wasser-Abwasser*, 17: 77-105.

Pütz, K. & Coworker 1983. Die Massenentwicklung der Geißelalge Synura uvella in den Trinkwassertalsperren Klingenberg und Lehnmühle – Ihre Auswirkungen auf die Trinkwasserversorgung und ihre Bekämpfung. *Wasserwirtschaft – Wassertechnik*: 135-138.

Ripl. W. 1976. Biochemical oxidation of polluted lake sediment with nitrate – A new lake restoration method. *Ambio*, 5: 132-135.

Sampl, H. 1980. Zum gegenwärtigen Stand der Sanierung der Kärtner Seen. *Österreichische Abwasser-Rundschau.*

Scharf, B., A. Hamm & C. Steinberg 1984. Seenrestaurierung. In Besch, W.K. & Coworkers (ed.), *Limnologie für die Praxis*, 1-71. Verlag ecomed, Landsberg/Lech.

Schulte-Wülwer-Leidig, A. 1985. Der Einfluß unterschiedlicher landwirtschaftlicher Bodennutzung auf die Nährstofffrachten kleiner Wasserläufe der Wahnbachtalsperrenregion. Dissertation Univ. Gieß.

Shapiro, J., V. Lamarra, G. Lindmark, M. Lynch, E. Smeltzer & G. Zoto 1982. Experiments and experiences in biomanipulation – Studies of biological ways to reduce algal abundance and eliminate blue-greens. Interim Report No. 19 of the Limnological Res. Center, Univ. of Minnesota, Minneapolis.

Steel, J.A. 1973. Reservoir algal productivity. Univ. of Newcastle upon Tyne Symposium. In A. James (ed.) *The use of mathematical Models in Water Pollution Control.*

Steel, J.A. 1978. The use of simple plankton models in the management of Thames valley reservoirs. *DVGW-Schriftenr.*, 16: 42-59.

Steel, J.A. 1975. The management of Thames valley reservoir. *Water Research Centre Symposium: The effects of storage on water quality, Medmenham.*

Steinberg, Chr. 1983. Effects of artificial destratification on the phytoplankton population in a small lake. *J. Plankton Res.*, 5: 855-864.

Stumm, W. & J.W. Leckie 1971. *Phosphate exchange with sediments, its role in the productivity of surface waters.* Eidgen. Techn. Hochschule, Separatum Nr. 406. Dübendorf, Schweiz.

Uhlmann, D. & H. Klapper 1985. Protection and restoration of lakes and reservoirs in the German Democratic Republic – Case Studies. *Lakes Pollution and Recovery European*

Water Pollution Control Association, Internat. Congress Rome, 15-18 April, 262-271.
Uhlmann, D. & J. Benndorf 1980. The use of primary reservoirs to control eutrophication caused by nutrient inflows from nonpoint sources. In N. Duncan & J. Rzoska (ed.), *Land use impacts on lake and reservoir ecosystems*, 152-195. Wien: Fakultas-Verlag.
Vollenweider, R. A. & J. Kerekes 1982. Eutrophication of waters monitoring, assessment and control. *OECD, Paris.*
Wilhelmus, B., H. Bernhardt & D. Neumann 1967. Vergleichende Untersuchungen über die Wirkungen von Vorsperren. *DVGW-Schriftenr.*, 16: 140-176.
Yee, W.C. 1966. Selective removal of mixed phosphates by activated alumina. *J. Am. Water Works Assoc.*, 57: 239-247.

Sediment-laden flows: Control of inflow and outflow 4

AKIRA WADA
School of Marine Science and Technology, Tokai University, Shizuoka, Japan

4.1 INTRODUCTION

Sedimentation in reservoirs is a phenomenon that occurs more or less naturally because sediments carried into the reservoir by the river are trapped by the dam. If sedimentation is so rapid that sediments fill a large part of the reservoir, the ability of the reservoir to control floods is reduced, the river-bed upstream rises, and other functions of the dam, such as the production of electric energy, are impaired. Thus the utility of the resource is greatly reduced with comparable reductions in the economic return on the investment. Furthermore, deleterious effects can result from the long-term persistence of turbid water and the lowering of downstream river-beds. In recent years, various organizations have conducted studies of countermeasures to reduce sedimentation and of sand-flash measures to remove sediment. However, their effectiveness is quite unpredictable because the results are affected by natural phenomena and by both topographic and geological conditions. Drastic countermeasures are circumscribed because they must also satisfy economic and environmental conditions. Decision making is all the more difficult because the nature of the problem is such that very little research has been conducted in this field.

This section presents current sedimentation practice; Section 4.2 presents the problems; Section 4.3 the sand-flash technique along with several case studies; Sections 4.4 and 4.5 contain examples of planning and studies of the sand-flash method and dredging. The last named technique is probably the most effective among the various measures for controlling sedimentation.

4.2 CURRENT STATUS OF SEDIMENTATION IN RESERVOIRS

The construction of a dam changes the equilibrium or balance of the sediment transport in natural rivers. Upstream of the dam, it changes the dynamic relation between the inflow of water and the amount of sediment transported, thereby causing sedimentation in the reservoir. This sedimentation progressively reduces the storage capacity and thus limits the primary function of the reservoir (flood control, agriculture, urban and industrial water supplies, power generations, or other) and reduces or eliminates the functions of structures attached to the dam. In addition, a rise of the river-bed upstream of the reservoir can increase the probability of disasters due to flooding, and the lowering of a river-bed downstream of the reservoir can damage river structures.

In 1970, the Ministry of Construction of Japan conducted a survey by means of a questionnaire on current problems caused by sedimentation in existing reservoirs. These problems and the number of cases in which they occur are summarized in Table 4.1. Clearly, many problems exist, both within the reservoirs and in the river channels themselves.

The problem of sedimentation in reservoirs dates back some 5000 years when sedimentation was found to have occurred in an ancient dam (Sadd, El-Kafara Dam) discovered in Egypt. It has occurred with the construction of dams over the millennia, and no effective solution to it has been found so far.

Table 4.1. Sedimentation problems and cases

Area of problem	Concrete problems	Number of cases
Inflow river channel section in the reservoir	Reduction of effective storage capacity	9
	Rise of river bed, etc.	11
	Difference between estimation and actual conditions	8
Near the dam body	Power generation loss, wearing and blocking of intake, blocking of outlet	4
Others	Generation of turbidity, lowering of downstream bed, etc.	–

Table 4.2. Summary of reservoir sedimentation indexes in the world

Country	Reservoir data collected	Annual mean sedimentation ratio		Mean specific sedimentation	
		Number of samples	Mean value (%)	Number of samples	Mean value ($m^3/km^2/year$)
Japan	414	366	1.71	277	533
USA	197	189	1.16	185	580
India	73	59	0.57	59	537
The Republic of China (Taiwan)	12	12	4.51	12	6 465
Other countries	56	35	3.41	41	319
Total	768	677		524	

Note: Mainly based on data concerning reservoirs for power generation as of 1968

About two-thirds of the land in Japan is mountainous, with steep topographical features and geological structures subjected to violent geological action. The country is also subjected to natural disasters caused by typhoons and the associated heavy rains. Hence, the erosion of both surface soil and the underlying ground is intense, some 4-8 times the world average. Accordingly, Japan's sand deposits are near the top among those countries in which the major sedimentation in the world occurs, as shown in Table 4.2. Annual sediment production is estimated to be at least 10^8 m^3, of which the volume of sediment that flows from catchment areas above dams (about 80 000 km^2) into reservoirs has a yearly average of 4.2×10^7 m^3.

4.3 METHODS FOR REMOVAL OF SAND DEPOSITS

4.3.1 *Countermeasures against deposits*

Countermeasures against sedimentation in reservoirs are affected not only by the environmental conditions of the reservoir but also by such factors as the purpose of the dam, the size, configuration and hydraulic characteristics of the reservoir, the fluctuation of water level, and the characteristics of the sediments that flow into the reservoir. An effective sand removal plan must thus take into account a wide variety of factors. The area between flow upstream of the backwater end of the reservoir and the dam is divided into four parts in accordance with sediment deposits and the hydraulic characteristics of the reservoirs, Figure 4.1.

1. *Area upstream of the reservoir.* This area includes the source of sediment and the transport region upstream of the reservoir. Major countermeasures in this region include systematic management of sediment production and soil conservation to control both water flow and erosion.

2. *Backwater region.* In this region the river-bed is sometimes exposed during changes of water-level due to reservoir operation, as, for example, during use of sand-flash channel. Countermeasures include sand storage weirs, screened weirs, slit weirs, and excavation of sand (Fig. 4.2). If a weir is used, a way must be provided to remove the sediment deposited behind it.

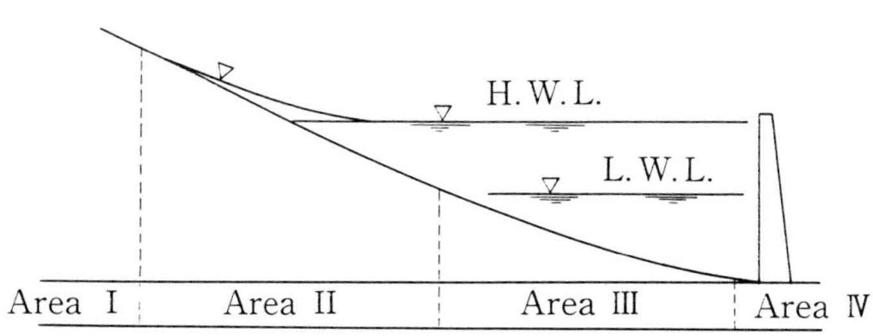

Figure 4.1. Area classification viewed from sand flash measures

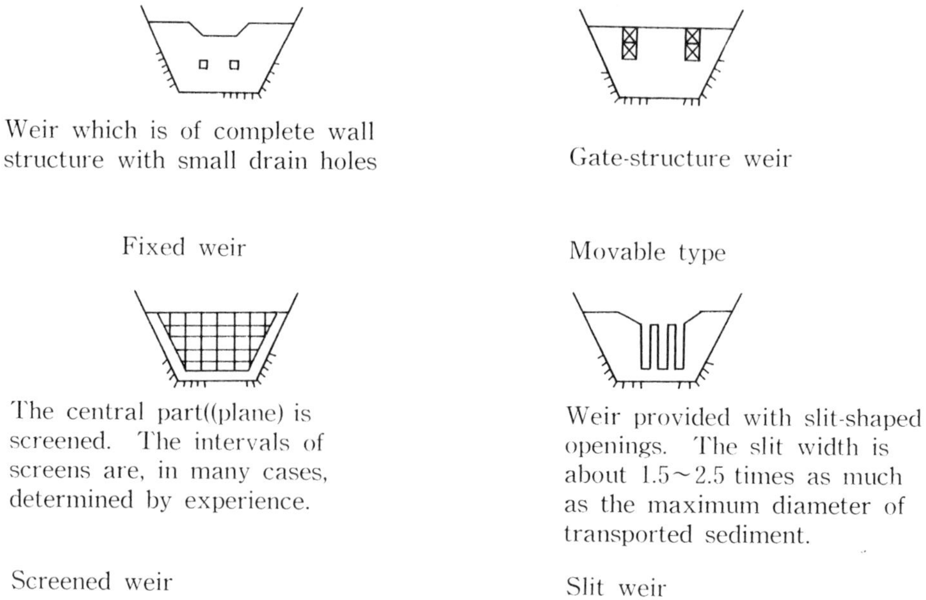

Figure 4.2. Sand deposit measures in the backwater region of the reservoir

3. *Reservoir.* The region between area (II) and a section immediately upstream of the dam is always submerged. Dredging is an effective countermeasure here.

4. *Dam and area immediately upstream.* In the area immediately upstream of the dam, such countermeasures as sand-flash gates and discharge pipes as shown in Figures 4.3 and 4.4 are effective.

Decisions concerning measures to prevent sand deposits in area (IV) depend on whether the sediments accumulated in the reservoir can be discharged directly downstream of the dam. The points to examine are whether the sand-flash level reaches the intake height of the sand-flash facilities and whether the amount of water needed to execute sand-flashing fits into the reservoir operation plan. Unless these conditions are satisfactory, sand-flash facilities are not appropriate. In such cases, countermeasures must be located in areas (I) and (II) upstream of the reservoir.

If sand-flashing is feasible, the method used must reflect the size of the sediment. Sand-flash measures and corresponding sediment grain size are shown in Table 4.3. As the table shows, methods using various types of weirs installed in river channels within area (II) are effective countermeasures against sand and gravel; in area (III) various dredging methods can remove sand, gravel and some pebbles; in area (IV) the sand-flash gate and discharge pipe can remove a wide range of grain sizes, from clay to pebbles.

Typical examples of countermeasures being taken at existing reservoirs are as follows:

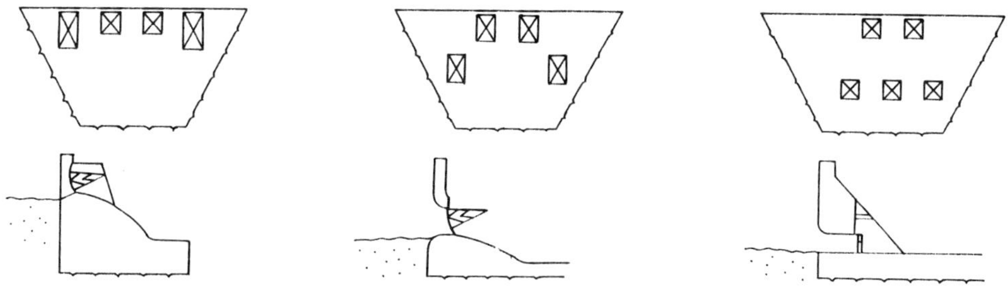

Figure 4.3. Types of sand-flash gates

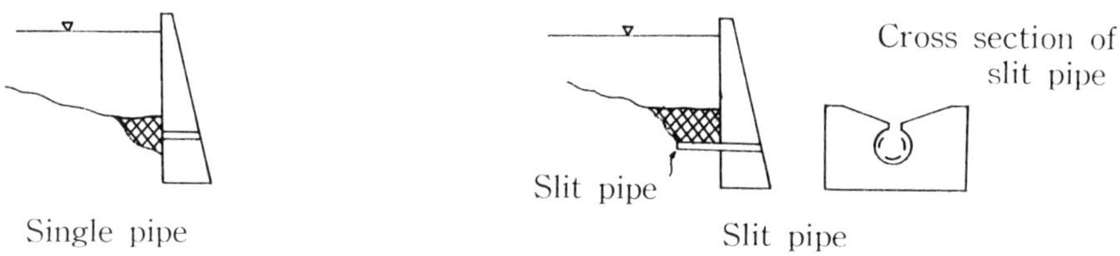

Figure 4.4. Types of sand discharge pipes

Table 4.3. Selection of measures to control sediment of various grain sizes

Sediment classification / Corresponding area & method		Soil engineering classification	Clay	Silt	Sand	Gravel	Pebbles
		Hydraulic classification	Wash-load	Wash-load	Wash-load / Suspended sediment	Bed-load	Bed-load
		Size (mm)	– 0.005	0.005 – 0.01 – 0.05	0.05 – 0.1 – 0.5 – 1	5 – 10	50 – 100 – 500
Area II		Screened weir				—	—
Area II		Slit weir				—	—
Area II		Sand storage weir			—	—	—
Area III	Grab dredging facilities	Grab dredger			—	—	—
Area III	Grab dredging facilities	Dipper, back-hoe dredger			—	—	—
Area III	Grab dredging facilities	Bucket dredger		—	—		
Area III	Grab dredging facilities	Tower excavator		—	—	—	—
Area III	Grab dredging facilities	Pump dredger		—	—		
Area III	Grab dredging facilities	Diving dredger		—	—		
Area IV	Sand-flash Gate	Overflow type	—	—	—	—	—
Area IV	Sand-flash Gate	Curtain-wall type	—	—	—	—	—
Area IV	Sand-flash Gate	Conduit pipe type	—	—	—	—	—
Area IV	Scour pipe	Single-pipe type	—	—	—		

1. *Utilization of spillways*. Crest spillways are currently utilized at the time of flood flows as a method of removing sand deposits. An example of this method is a part of the spillway system for Ohma Dam, shown in Figure 4.5. This concrete

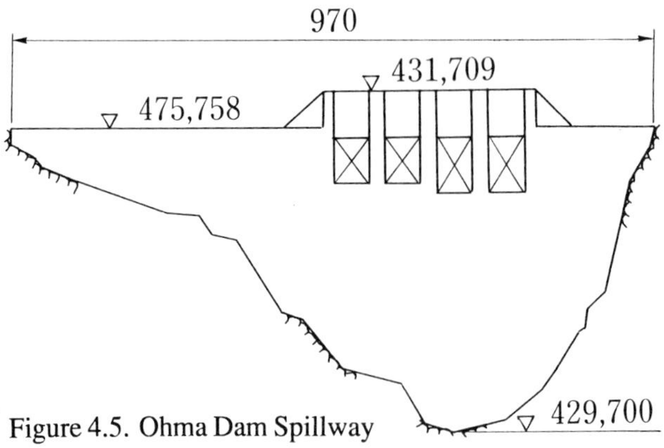

Figure 4.5. Ohma Dam Spillway

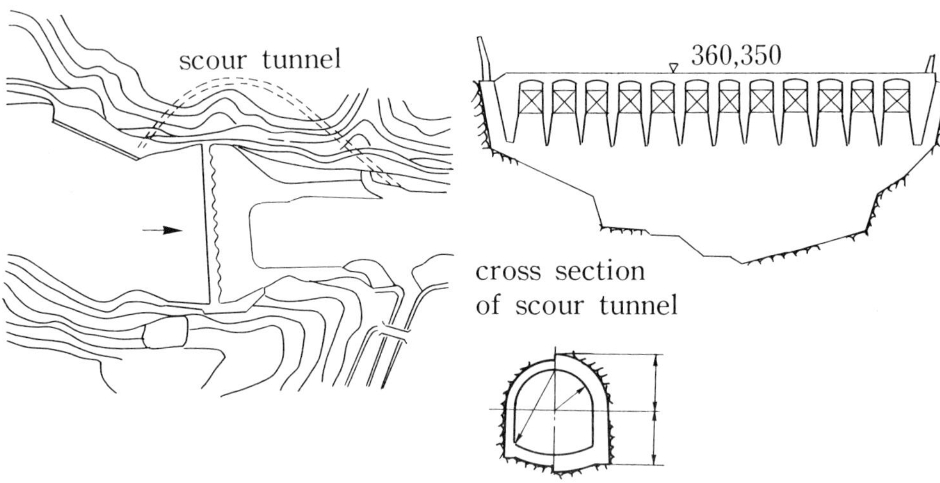

Figure 4.6. Sand-flash tunnel equipment at Yasuoka Dam

gravity dam was constructed in 1935, and the annual sediment inflow of about 40 000 m^3 had almost completely filled its reservoir with sand by 1963. The collapse of mountain sides and other upstream erosion caused the river-bed to rise about 13 m in the neighborhood of the dam. The flood spillways are currently utilized to remove sand through suitable operation of the gates installed at the crest of the dam spillways.

2. *Sand-flash tunnels.* Figure 4.6 shows a typical example of a sand-flash tunnel located at Yasuoka Dam, a concrete gravity dam built in 1936 across the main stream of the Tenryu River. Its reservoir was almost filled with sand by 1944 because of an annual sediment inflow of about 1 030 000 m^3. The sand-flash tunnel was constructed in 1949. Efforts were made to remove about 130 000 m^3 of sediments, all at one time, during the shutdown of power generation at the end

of one year and the beginning of the next year. Although some sediment was removed in the region just above the dam, the sediment inflow was so large for the size of the sand-flash tunnel that the effort to recover effective storage capacity and to lower the river-bed upstream was unsuccessful. Later, beginning in 1957, some 20 000 m^3 a year was dredged from the reservoir by means of a clam shell dredger with a bucket capacity of 1.5 t. The dredging prevented the flow of sediment into the intake channel. This dam also underwent large-scale repairs due to wear-out of concrete in the apron section of the dam body and to scouring near the dam foundation.

3. *Dredging method.* A successful case of dredging occurred at the multi-purpose Miwa Dam. This dam was constructed in 1959 across the Mitsumine River, a tributary of the Tenryu River, and the sediment inflow was estimated at 10^6 m^3 during a flood of about 500 m^3/s, from past records of sand deposits. Because the sand deposition was estimated to be 3×10^5 m^3, it was predicted to reach twice the value of the planned sand deposits by 1985. Dredging was undertaken and 1.85×10^6 m^3 of sand had been removed by ground dredging between the construction of the dam and 1973. The dredging both maintained the storage capacity and provided material for concrete aggregates. However, as the planned volume of the sand-deposits cannot be controlled by this method alone, dredging by pump dredging was started in 1973; these measures have provided satisfactory results so far.

4.4 SAND-FLASH METHOD

Sand-flash channels, which are positioned in the flash gate and discharge pipe, utilize the tractive force of flowing water to remove sand deposits, as shown in 4.3. The sand-flash areas include the dam body, the region immediately upstream and a part of the reservoir. The outline of the sand-flash channel method, the hydraulic design method and their application and design examples are described in the following sections.

4.4.1 *Outline of the sand-flash channel method*

Use of the sand-flash channel method is limited to (a) relatively small storage volumes with large discharges, (b) cases in which dredging and other methods are not suitable, and (c) situations in which the absence of a countermeasure against sand deposits gives a fatal blow to function of the reservoir. The criteria for sand-flash channels that will function properly are as follows:

1. *Cross section.* A diameter of 3-5 m is needed. If the diameter is below 3 m, drift wood and other debris can cause blocking; if the diameter is greater than 5 m, the cost may be too high.

2. *Slope.* The range of suitable slopes is $1/30$-$1/80$. If the slope is greater than $1/30$,

the wear on the surface of the sand-flash channel will be severe; if the slope is below 1/80, local deposits of sand may occur.

3. *Channel length and alignment.* The maximum channel length is some 400-500 m. Greater lengths are too costly. Also, any curvature in the middle channel can cause both local deposits of drift sand, because of secondary currents, and wearing of the channel surface. Therefore straight pipes should normally be used.

4. *Intake heights.* The intake should be as low as is feasible in order to lower the level of sand deposits in the reservoir and to increase the sand-flash volume.

5. *Operational considerations.* The water level at the entrance to a sand-flash channel should be lowered after a flood has passed. Frequent use of the equipment is desirable in order to flush out sediments near the entrance of the sand-flash channel most effectively.

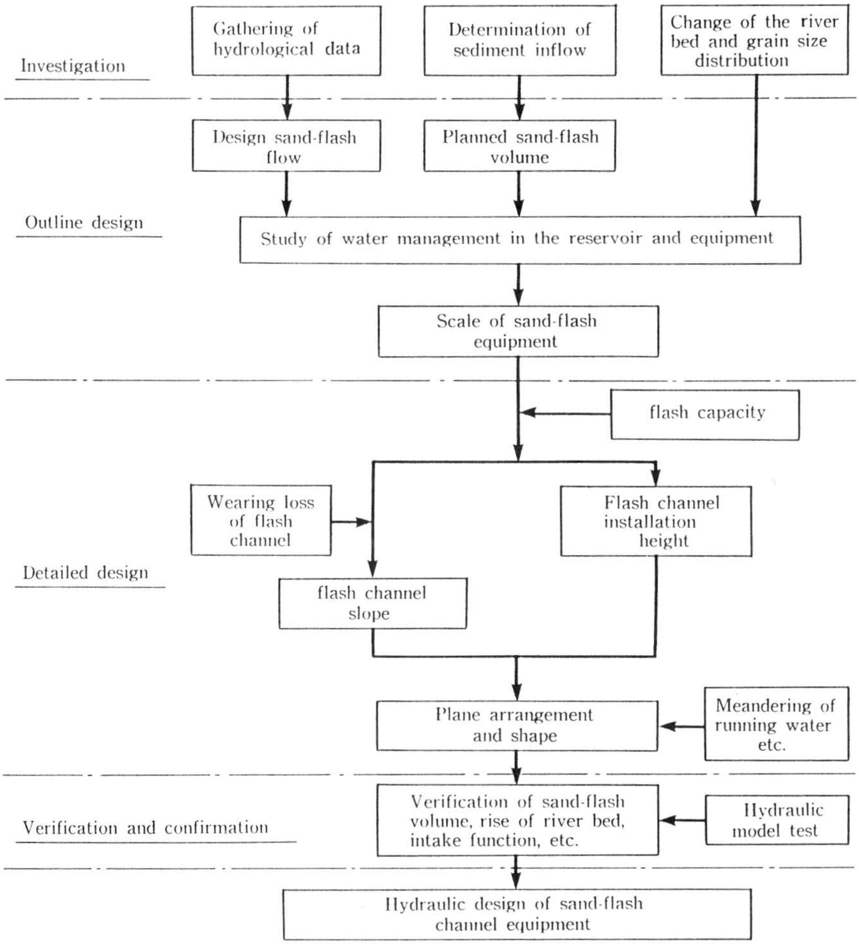

Figure 4.7. Flowchart for hydraulic design of sand-flash channel equipment

4.4.2 *Hydraulic design*

A typical hydraulic design for sand-flash channels is illustrated in Figure 4.7. In order to design a channel, one must first collect data on hydrological conditions and on sediment transport in the area concerned. In addition, one needs to study the deformation of the river-bed and the grain-size distribution of sediments that flow into it. The rates of flows from the reservoir should be adequate to flash the sand through the designed sand-flash channel, and these must be determined beforehand in order to make a preliminary design of the sand-flash channel equipment. Next, the slope of the channel, the channel installation height, and the horizontal position and cross-sectional configuration of the channel should be designed in such a way that water can run through the planned channel safely. Finally, a detailed hydraulic design of channel equipment is made; normal operation of intake equipment, which is the primary function of the reservoir, must be feasible without raising the water-level in the backwater region. The slope of the channel should be such as to minimize wearing damage of the channel wall.

The designers should utilize existing hydraulic information during preliminary design, and they may need to conduct hydraulic model experiments before arriving at a final design. A detailed study of reservoir operating conditions and flash processes should include the inflow region, the reservoir itself, and the sand-flash channel as an integrated system. Also, simulations based on a numerical model are needed to predict the condition of flow in the reservoir, the changes in the level of sedimentation in the reservoir, the amount of bed-load in the flash channel, etc. Although useful, this technique still presents problems, and its effectiveness will depend upon research and future developments.

4.5 AN EXAMPLE OF A SYSTEM FOR SAND REMOVAL

1. *Description of facility.* A dam having height of 76.7 m, a total storage capacity of 9.01×10^6 m^3 and an effective storage capacity of 1 660 000 m^3 was constructed in the Kurobe River. The river carries a large amount of sediment; the annual sediment inflow into this reservoir has been estimated from the volume of sand deposits in the existing reservoir to be about 400 000 m^3. In order to secure the effective storage capacity of the reservoir and to prevent a rise of the river-bed at a second dam located at the backwater end of the reservoir, the downstream dam is provided with two lines of large-scale sand-flash channels (5 m high and 5 m wide) that utilize the energy of flowing water to remove sand deposits, as illustrated in Figure 4.8.

2. *Laws of similarity on the hydraulic model.* Laws of similarity for model tests considering sediments often create experimental difficulties. However, the problem has been reduced in this case by use of a model with a relatively large

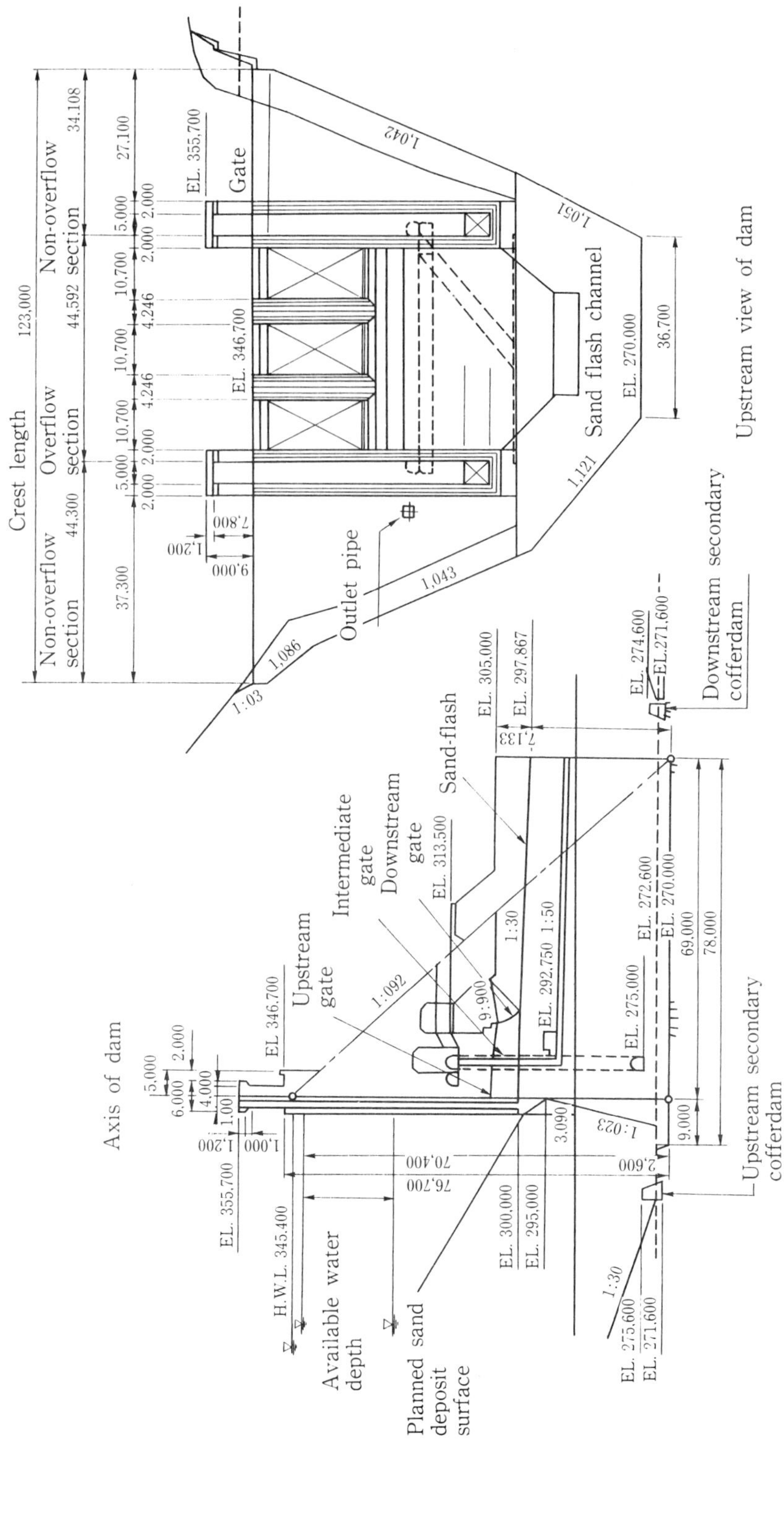

Figure 4.8. Scour channel equipment of A dam

scale (1/50) and by setting the size of the sand used in the test so as to bring it closer to the grain size distribution at the site. In this way some quantitative test results were obtained. The grain size distributions of the test sand and the sand deposited at the site are shown in Figure 4.9. The conversion of various quantities from model to prototype was made in the following manner, in accordance with the Froude law of similarity:

Length $L = nl = 50l$
Time $T = \sqrt{n}\,t = \sqrt{50}\,t$
Quantity of flow $Q = n^{5/2}\,q = 50^{5/2}\,q$

Here, capital letters represent prototype values and small letters model values; n is the scale ratio.

To convert the weight of the sand, W (kg), collected during the test in the volume q_{Bt} (m^3), the specific weight of the sand $\sigma/\rho = 2.65$ and the porosity $\lambda = 0.38$ were used:

$$q_{Bt} = \frac{W}{1000\,(1-\lambda)\,(\sigma/\rho)} = \frac{W}{1640}$$

Also, conversion of the sand removal rate per unit time Q_B (m^3/s) and the total sand removal rate Q_{Bt} (m^3) was made as follows:

$$Q_B = n^{5/2}\,q_B$$
$$Q_{Bt} = n^3\,q_{Bt}$$

Where q_B is the sand removal rate, q_{Bt} is the volume of sand removed for the model, and Q_B and Q_{Bt} are the corresponding prototype values.

The concentration C of sand flowing in the sand-flash channel is as follows:

$$C = 100\,Q_B\,(1-\lambda)\,/\,Q\ (\%)$$

3. *Method of operating the sand-flash channel.* A study was made of how to operate the sand-flash channel efficiently. The most efficient flow rate was 80 m^3/s although the higher the flow rate, the greater the volume of sand removed. Also, the removal of the annual inflow of sediment, using two rows of sand-flash channels, is estimated to take about 4-5 days at a flow rate of 80 m^3/s, about 4 days at a flow rate of 160 m^3/s and about 3 days at a flow rate of 320 m^3/s. These flows are adequate to remove the sand at this dam.

Whenever the flow rate exceeds 320 m^3/s, the water level in the regulating reservoir is high and the velocity of flow in the intake part of the sand-flash channel is low, so that hardly any sand is removed. In order to remove sand efficiently, the dam should be operated in such a way that the sand accumulated at the upstream end of the regulating reservoir is transported to the vicinity of the dam by lowering the level of the water in the reservoir whenever an opportunity to do so arises.

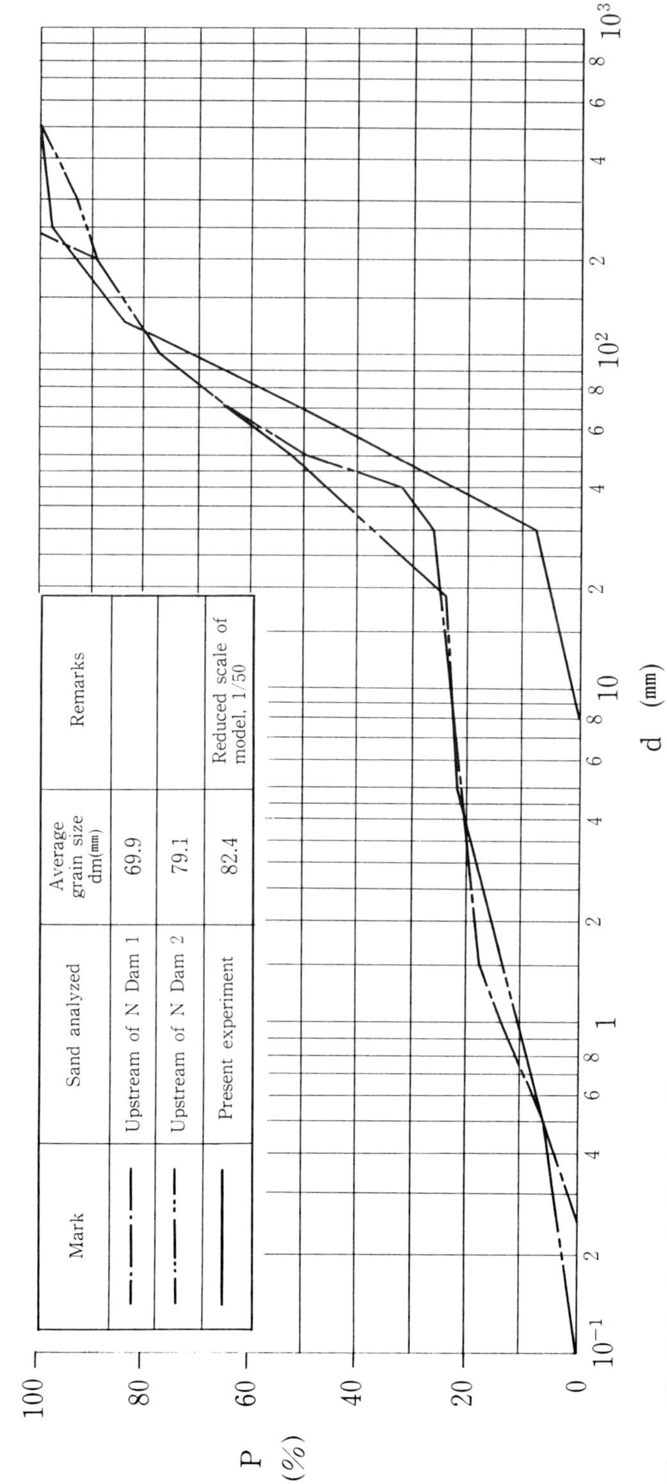

Figure 4.9. Grain-size distribution of river-bed sand

4.6 EXCAVATION AND DREDGING

Countermeasures against sand deposits that may be used are excavation and dredging. Sand deposited near the backwater end of the reservoir can be excavated in the dry during low-water periods, and excavating machines such as power shovels, back-hoe and drag-line, which are normally used to excavate on dry soil, can be utilized.

In contrast, sand deposited within the reservoir is usually excavated underwater by dredging. Such dredging vessels as grab dredgers, pump dredgers and bucket dredgers are available. Use of a pump dredger is usually advantageous. Table 4.4

Table 4.4. Results of studies on the availability of aggregates for four dams

		H Dam	S Dam	HT Dam	I Dam
Volume of sand deposit in the dam	Total volume of sand (1000 m^3)	39 414	75 031	18 898	25 813
	Effective volume of sand deposit in the reservoir (1000 m^3)	2 702	28 221	10 285	15 333
Inflow rate into the dam	Average sand deposit speed after commencement of filling (1000 m^3/year)	2 815	3 900	1 050	1 173
	Average sand deposit speed in the recent 10 years (1000 m^3/year)	67	2 200	611	479
	Sediment inflow and average sand deposit volume (estimated volume in 20-27 years) (1000 m^3/year)	2 930	4 060	830	1 000
Content of grain deposit (in reference to the total volume of sand accretion)	Pebbles (%) (1000 m^3)	(10.0) 3 941	(10.0) 7 503	(50.0) 9 449	(24.0) 6 195
	Sand (%) (1000 m^3)	(80.0) 31 531	(55.0) 41 267	(45.0) 8 504	(66.0) 17 037
	Silt and clay (%) (1000 m^3)	(10.0) 3 941	(35.0) 26 261	(5.0) 945	(10.0) 2 581
Excavatable volume by ground and underwater (Unit: 1000 m^3)	Total volume of sand deposit	38 200	76 300	20 100	27 700
	Ground excavatable volume	4 500	–	4 900	3 700
	Under excavatable volume	21 100	12 000	6 700	6 100
	Unremovable volume of sand	12 600	64 300	8 500	17 900
Result of grain-size distribution test	Appropriateness of grain-size distribution	Appropr.	Appropr.	Appropr.	Appropr.
Result of material test	Appropriateness of aggregate strength	Appropr.	Appropr.	Appropr.	Appropr.
Result of concrete strength test	Appropriateness of product strength	Appropr.	Appropr.	Appropr.	Appropr.

contains the volumes that can be excavated or dredged at four dams, the grain size distributions of the sand deposit and other information.

Disposal of the sand removed also introduces problems. A place to dispose of the sand may be difficult to find; the surrounding area is usually mountainous, and costs are high because the dam site is usually far from regions in which the material could be used effectively. Moreover, utilization of river gravel as aggregate or banking material, which has usually been taken from the downstream part of a river, has caused such problems as lowering of the river-bed; thus one is compelled to rely on the sand accumulated in the reservoir as the future source of supply. The composition of sediments in the four dams shown in the table are as follows: H Dam has mainly sand; S Dam, sand containing silt; HT Dam, pebbles and a large amount of sand; I Dam, sand containing pebbles. Fortunately, these sediments have been found to be appropriate for use as concrete aggregates.

4.7 SUMMARY

Methods are available for controlling sedimentation in at least some reservoirs. The examples presented illustrate the present situation concerning sand deposition, the methods of removing the sand deposited, an example of the design of a sand-flash channel, and an example of sand removal by dredging. Countermeasures against sand deposition involve (1) hydraulic measures, (2) control of land erosion, considering geology of the catchment area and (3) flood control. With so many factors, comprehensive studies are indispensable. The following items require further research if reservoir planning is to become more effective.

1. Prediction of sediment production and inflow from the upstream regions of rivers.
2. Prediction of the process of sedimentation and its removal in reservoirs.
3. Countermeasures against local sand deposition and wearing damage in pipe transport channels.
4. New sand removal methods utilizing the tractive force of flowing water and others.

BIBLIOGRAPHY

About overall countermeasures against sand deposits in dams 1978. Development Section, River Bureau, Ministry of Construction (in Japanese).

Hydraulic formulas 1985 (3rd. ed.). Rivers, 5. Sedimentation, 6. Sediment production and runoff, 7. Section relating to river-bed fluctuation, etc. The Japan Society of Civil Engineers (in Japanese).

Kira, H. 1982. *Sand deposit in dams and its countermeasures.* Morikita Shuppan (in Japanese).

Overall system development investigation reports on removal of sand deposit in dams for power generation 1981-1982. Agency of Natural Resources and Energy, Ministry of International Trade & Industry, New Energy Corporation (in Japanese).

Study manual for countermeasures against sand deposit 1980. Land Development Technical Center (in Japanese).

Study on sand deposit in reservoirs 1976. Dam Maintenance Study Group, Federation of Electric Enterprises, Hatsuden Suiryoku, No. 143 (in Japanese)

Tada, T. 1979. Review and countermeasures against sand deposit problems in dam reservoirs. *11th lecture meeting on dam technology, March 1979*. Japanese Commission on Large Dams (in Japanese).

Water quality in rivers and estuaries

Water purification in rivers 5

MIKIO HINO
Department of Civil Engineering, Tokyo Institute of Technology, O-okayama, Meguro-ku, Tokyo, Japan
Presently: Faculty of Political Science, Chuo University, Hachiohji, Tokyo, Japan
TOMONORI MATSUO
Department of Urban Engineering, University of Tokyo, Hongo, Bunkyo-ku, Tokyo, Japan

5.1 SELF-PURIFICATION IN RIVER

5.1.1 *Elements of self-purification*

Originally mankind depended on the fundamental process of self-purification in rivers. Self-purification is a well-known process in nature to restore and preserve the water quality of a river. This natural process is composed of the following steps (Fig. 5.1).

1. *Reaeration.* Mass transfer through the interface between the surface of a stream and the atmosphere increases the dissolved oxygen concentration (DO) in the stream. The transfer rate is proportional to the deficiency in DO.

2. *Oxidation of organic waste.* With discharge of organic matter into a river, the bacteria begin to grow and consume the DO so as to oxidize the organic waste. The biochemical process of the oxidation is described in the following section. The bacterial population and then the protozoa and fish populations also increase in accordance with the ecological food-chain pyramid if the amount of dissolved oxygen is sufficient.

3. *Production of CO_2.* Increased fauna population (bacteria, protozoa and fish) contributes to the decrease in DO and to the increase in the amount of carbon dioxide (CO_2) dissolved in the water that is necessary for algal photosynthesis. The deficiency in DO causes the process of reaeration in 1.

4. *Supply of oxygen by algal photosynthesis.* Algal growing on the surface of gravel and pebbles on a river bed supply oxygen as a result of photosynthesis. For algal photosynthesis to occur, the stream transparency must be high; i.e. the water should be clean enough for the sun's rays to reach the bottom.

The last step of algal photosynthesis is usually neglected but it can be significant for rivers that are shallow and relatively clean (Aida et al., 1975; Hino, 1977).

This normal biological cycle requires the proper amount of organic waste. Too much waste can destroy the cycle. An abrupt increase in the organic waste causes

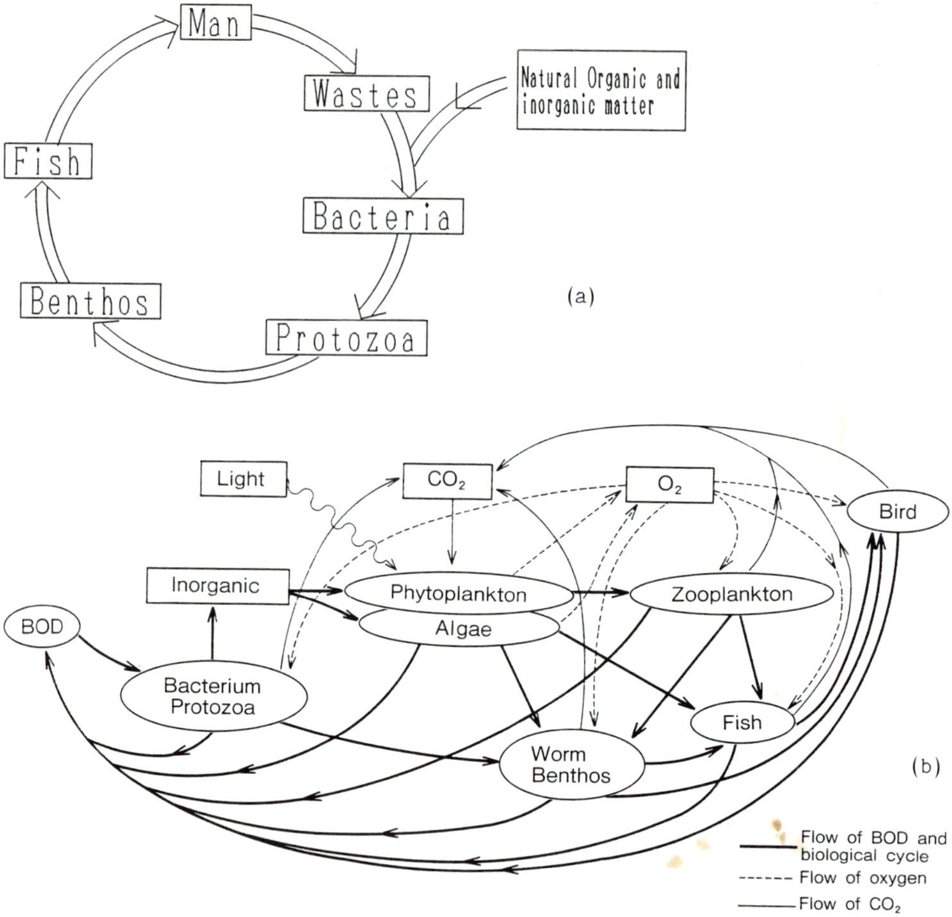

Figure 5.1. a) Natural biological cycle in streams, b) food chain in the ecological system

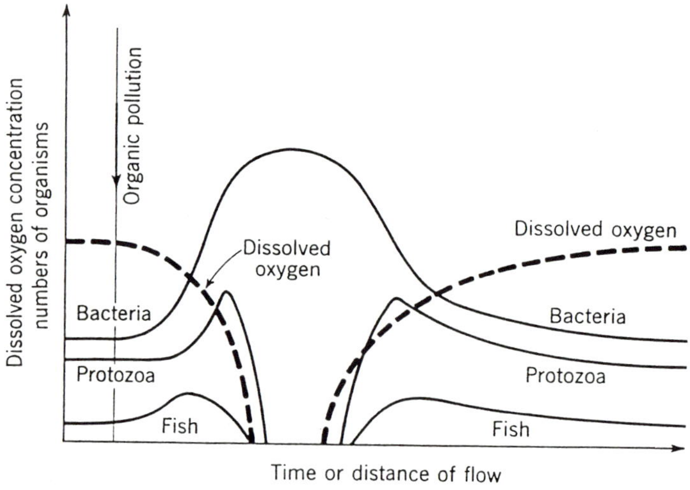

Figure 5.2. Effect of pollution on biological life in a stream (from McKinney, 1962)

an excessive increase in the bacterial population with a greater and greater oxygen demand. As the oxygen level decreases, the higher ranking fauna in the water begin to die off; the final result is an anaerobic environment with odors and dark colours (Fig. 5.2). The complicated behaviours of the ecosystem composed of six components (BOD, inorganic nutrient, phytoplankton, aerobic bacteria, anaerobic bacteria and dissolved oxygen) have been discussed by Kanda and Hino (1988) and Hino (1991), which disclosed the three categories of the aerobic equilibrium, oscillatory state (between aerobic and anaerobic states) and anaerobic equilibrium, depending on the external conditions of the amount of BOD input and the intensity of insolation.

5.1.2 *Biological oxidation*

Bacterial activities in the decomposition of organic waste are for the purpose of producing energy for synthesization of the complex molecules such as the proteins and polysacchorides that are needed to build new cells (Fig. 5.3). The first process is called 'catabolism' or 'catabolic reaction', and the latter 'anabolism'. Also, the energy needed for bacterial activity is supplied by 'autolysis' that involves the oxidation of organic matter converted and stored in the bacteria themselves (Aiba et al., 1973; Mara, 1976).

1. Catabolism

$$\underset{\text{(organic waste)}}{C_xH_yO_zN} + O_2 \rightarrow CO_2 + H_2O + NH_3 + \text{Energy} \tag{5.1}$$

2. Anabolism

$$\underset{\text{(organic waste)}}{C_xH_yO_zN} + \text{Energy} \rightarrow \underset{\text{(bacterial cells)}}{C_5H_7NO_2} \tag{5.2}$$

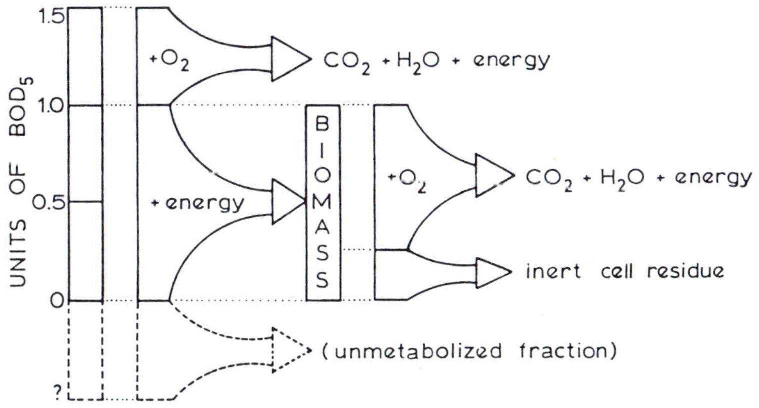

Figure 5.3. The catabolic, anabolic and autolytic reactions of aerobic biological oxidation (from Mara, 1976)

3. Autolysis

$$\underset{\text{(bacterial cells)}}{C_5H_7NO_2} + 5O_2 \rightarrow 5CO_2 + NH_3 + 2H_2O + \text{Energy} \tag{5.3}$$

5.1.3 *BOD – Index of water quality*

The best known and most widely used index of water quality is BOD (Biochemical Oxygen Demand), and it is defined as the oxygen (measured in ppm or mg/l) demanded for the bacteria in water to decompose the organic matter within 5 days at a temperature of 20°C, or for other standard conditions.

A high concentration of organic matter in water means a high demand of oxygen for bacterial growth. A measure of water quality in terms of BOD_5 (mg/l) is as follows:

Classification	BOD_5
Very clean	< 1
Clean	2
Fairly clean	3
Doubtful	5
Bad	> 10

5.2 THEORY ON THE VARIATIONS OF BOD AND DO ALONG A STREAM

5.2.1 *Streeter-Phelps theory*

The theory of self-purification, Equations (5.4) and (5.5), first proposed by Streeter and Phelps in 1925 and advanced by many others (e.g., Dobbins, 1964; Camp, 1965) since then treats the bacterial activity only indirectly, that is, through the deoxygenation coefficient that is taken as a constant

$$\frac{\partial L}{\partial t} = \varepsilon \frac{\partial^2 L}{\partial x^2} - V\frac{\partial L}{\partial x} - (K_1 + K_3)L + w \tag{5.4}$$

$$\frac{\partial D}{\partial t} = \varepsilon \frac{\partial^2 D}{\partial x^2} - V\frac{\partial D}{\partial x} + K_1 L - K_2 D - u \tag{5.5}$$

where L = BOD concentration, D = deficiency in dissolved oxygen, ε = longitudinal diffusion coefficient, V = mean velocity of a stream, K_1 = deoxygenation coefficient, K_2 = reaeration coefficient, K_3 = removal rate by settling, deposition and adhesion, w = BOD input and u = oxygen supply rate by photosynthesis.

If three assumptions are accepted: the flow is steady; the diffusion terms and K_3, w and u are negligible; and K_1, K_2 are constants, the downstream variation of L

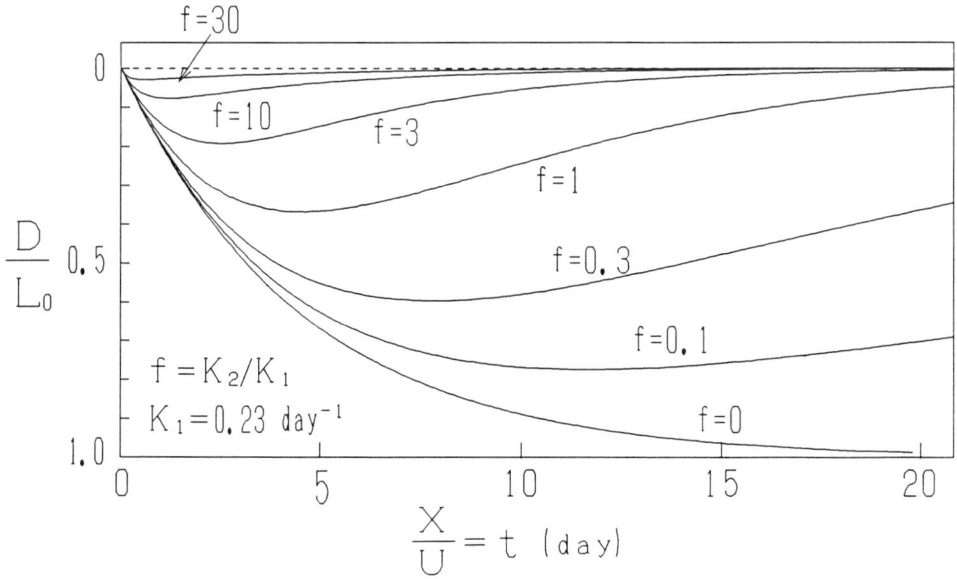

Figure 5.4. Curves of 'oxygen sag'

and D can be derived. The depletion and subsequent recovery of D expressed by Equation (5.7a) of (5.7b) is called 'Oxygen Sag' (Fig. 5.4)

$$\frac{L}{L_0}=\exp\left\{-K_1\frac{x}{V}\right\} \tag{5.6}$$

For $K_1 \neq K_2$:

$$\frac{D}{L_0}=\frac{K_1}{K_2-K_1}\left[\exp\left\{-K_1\frac{x}{V}\right\}-\exp\left\{-K_2\frac{x}{V}\right\}\right]+\frac{D_0}{L_0}\exp\left\{-K_2\frac{x}{V}\right\} \tag{5.7a}$$

For $K_1 = K_2$:

$$\frac{D}{L_0}=K_1\frac{x}{V}\exp\left\{-K_1\frac{x}{V}\right\}+\frac{D_0}{L_0}\exp\left\{-K_2\frac{x}{V}\right\} \tag{5.7b}$$

In the above expressions the subscript 0 indicates the initial state.

The equation of dissolved oxygen, Equations (5.7a) or (5.7b), is obtained by substituting the expression for L from Equation (5.6). Thus, no interactions among BOD, D and bacterial population, explained in the above section (5.1.1), are considered by this approach.

5.2.2 *Thory considering bacterial activity*

From the interactions between the three members of the ecosystem (Fig. 5.5), a set of organic pollutions (BOD), the concentration of dissolved oxygen and the

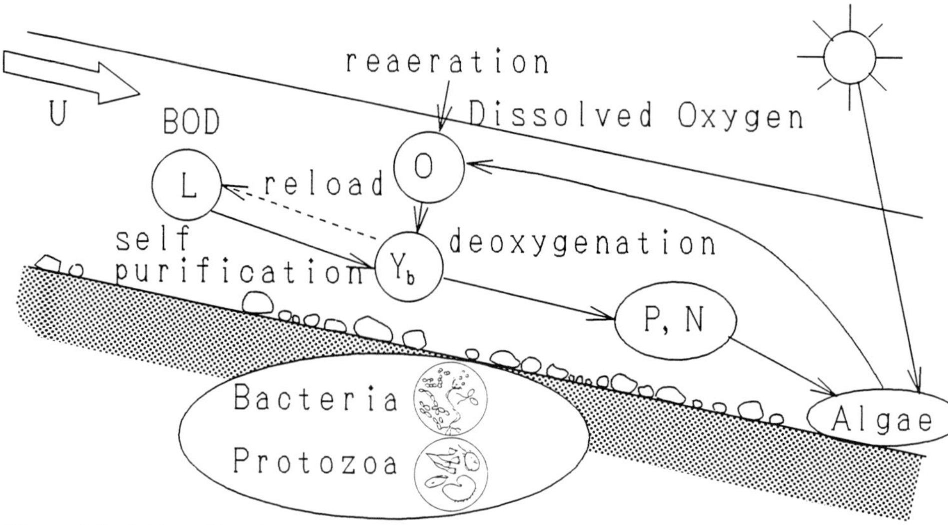

Figure 5.5. Self-purification processes in rivers.

population density of bacteria in the stream or on the river bed can be treated as follows:

$$\frac{\partial L}{\partial T} = \varepsilon \frac{\partial^2 L}{\partial x^2} - V \frac{\partial L}{\partial x} - \mu' Y - \beta' Y + w \tag{5.8}$$

$$\frac{\partial D}{\partial t} = \varepsilon \frac{\partial^2 D}{\partial x^2} - V \frac{\partial D}{\partial x} + (\mu + m) Y - K_2 D - u \tag{5.9}$$

$$\frac{\partial Y}{\partial t} = - V_b \frac{\partial Y}{\partial x} + \mu Y - \beta Y \tag{5.10}$$

where Y = population density of bacteria; mass of bacteria in unit volume of stream flow or on unit area of river bed, V_b = convection velocity of bacteria ($V_b = V$, if they are convected together with stream; $V_b = 0$, if bacteria on river beds are dominant), μ = growth rate of bacteria, m = oxygen consumption rate by bacteria, μ' = coefficient of disintegration of organic substances for bacterial respiration and nourishment, β = decay or extinction rate of bacteria, and $\beta' Y$ = conversion of βY into BOD by reloading of bacteria into flowing water.

These equations have been solved analytically for various cases as discussed by Hino (1979a, b, 1981a, 1991). Figure 5.6 shows the result of the solution of the above equations when the bacterial activity is assumed to be dominant on the river bed. The result shows that the deoxygenation coefficient

$$k_1 = -\frac{1}{L} \frac{dL}{dx} \tag{5.11}$$

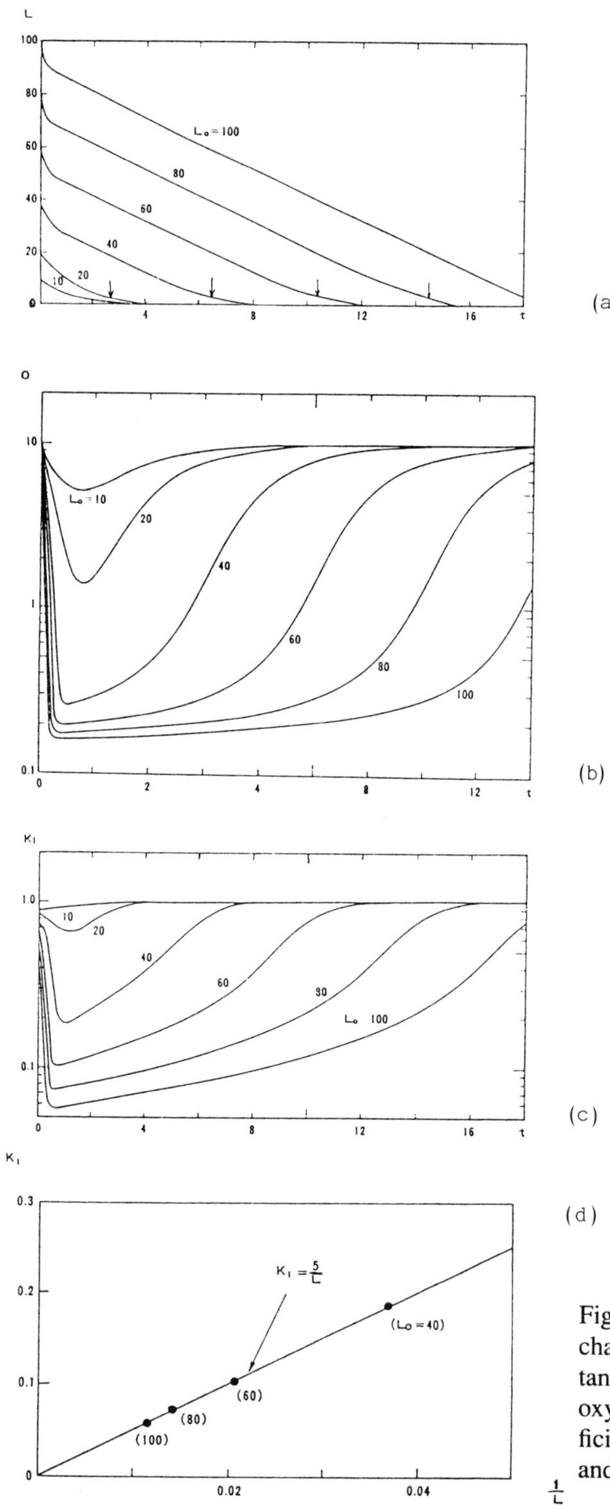

Figure 5.6. Theoretical result on the changes along the downstream distance of a) BOD, b) DO (dissolved oxygen), c) K_1 (deoxygenation coefficient), d) relationship between K_1 and $1/L$

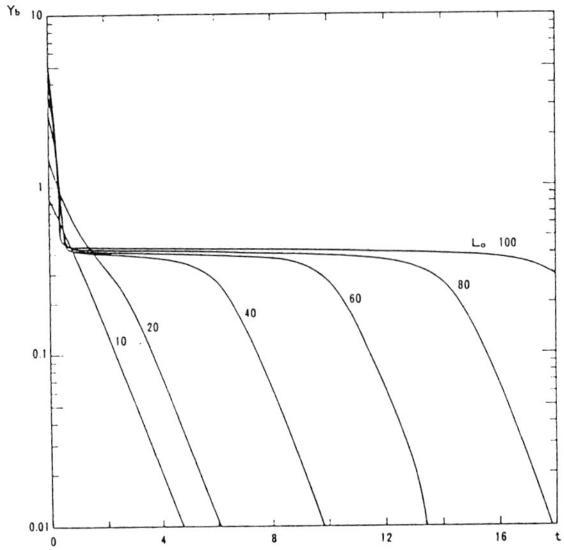

(e)

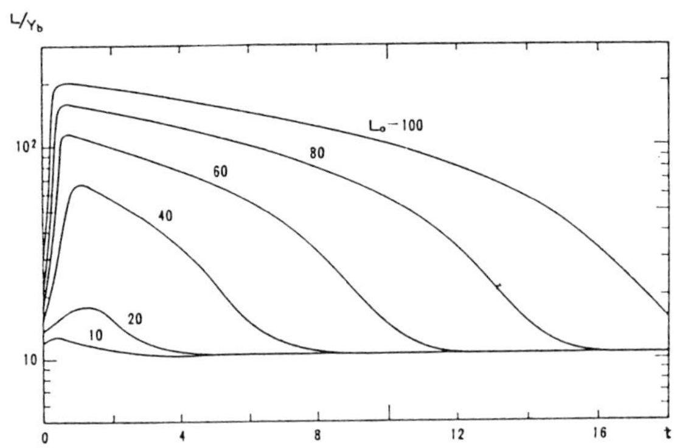

(f)

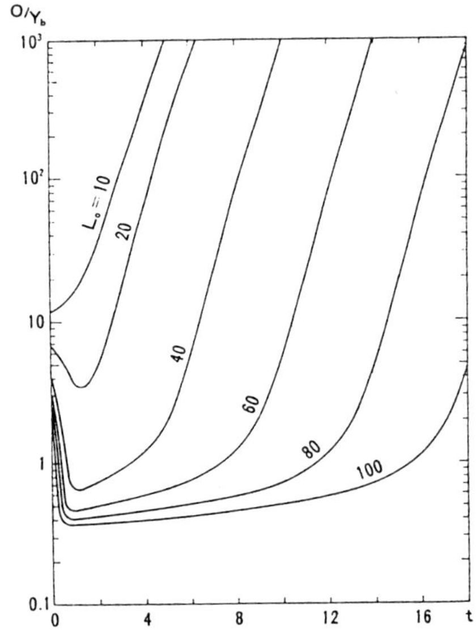

(g)

Figure 5.6 (continued). e) bacterial growth, f) the ratio of L/Y_b (L = BOD and Y_b = bacterial mass on the river bed), and g) the ratio of O/Y_b (from Hino, 1977b) (which is usually treated as constant) is not constant but it depends on the bacterial activity and thus on the oxygen concentration dissolved in stream. The bacterial mass is regulated by BOD. An excessive amount of BOD input in streams causes the explosive growth of bacteria and the increase in the oxygen consumption. As a result, the bacterial activity is suppressed on the contrary due to the deficiency in DO (dissolved oxygen).

In the nature, the ecosystem of the river stream is much more complicated. Figure 5.7, making use of an idea introduced by Bartsch (1948, cited from Hynes, 1974), demonstrates schematically the changes in the physical and chemical characteristics, and the growth of micro-organisms and large animals. The bacteria are at first abundant and then decline as the organic matter (BOD) is consumed; with a lag from the bacterial growth sewage fungus appears, increases

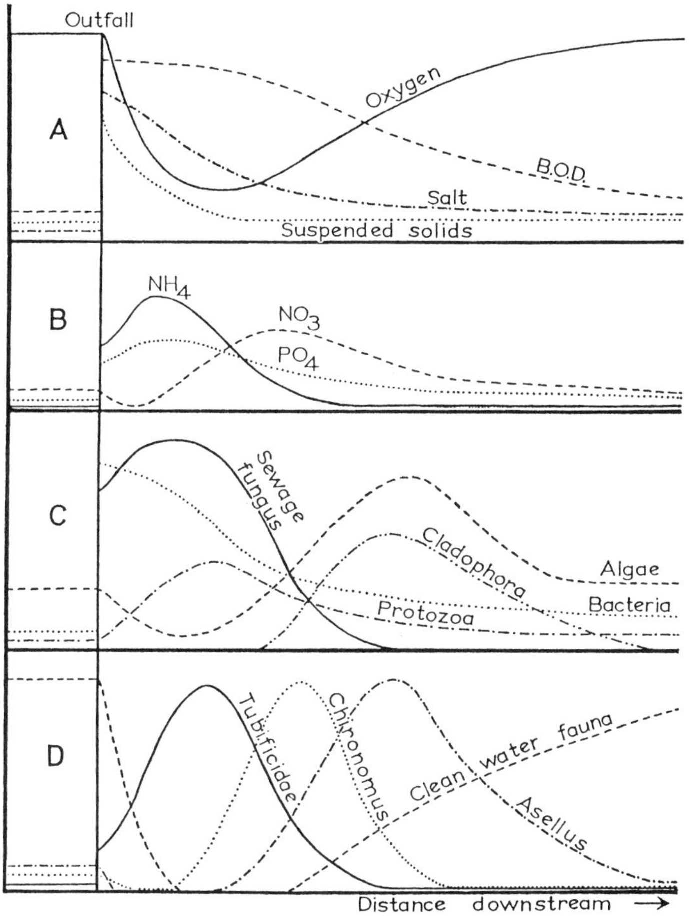

Figure 5.7. The effects of an organic effluent into a stream on A) the physical and B) chemical changes, C) the growth of microorganisms and D) large animals (from Hynes, 1974)

and then declines. The algae at first decrease, but they increase greatly as nutrient salts are released by the activity of bacteria and then decline. Finally, clean water fauna appear and increase as the conditions of BOD and dissolved oxygen are improved.

5.2.3 *Criterion on the critical rate of nutrient loading into lake*

To keep the water quality of a lake or reservoir, the nutrient level of influent rivers must be lower than a critical value. Vollenweider (1975) derived a theory, Equation (5.12), based on the elementary mass balance, for the critical rate of nutrient loading into a lake per surface area ($\mathcal{L}$) and compared it with a large variety of field data with good results

$$\mathcal{L} \leqq \frac{10}{4\xi}\left(1+\frac{1}{10}\frac{h_e}{\tau_r}\right)^2 \qquad (5.12)$$

where $1/\xi$ = a concentration representing a capacity factor of the lakes, h_e = effective depth of a lake or reservoir, τ_r (= V/Q) = mean residence time of water. In Equation (5.12), $\mathcal{L}$ is related to the inflow condition as follows:

$$\mathcal{L} = \text{loading rate } (QX_0)/\text{Lake surface area } (V/h_e) \qquad (5.13)$$

where X_0 = allowable nutrient concentration of influent river, Q = rate of discharge of nutrient into the lake, and V = effective volume of lake or reservoir.

Hino (1981b, 1983), for the three-rank ecosystem equations (nutrient, phytoplankton and zooplankton; Fig. 5.8), derived a theory, Equation (5.14), on the critical nutrient loading rate per surface area $\mathcal{L}$;

$$\mathcal{L} \leqq \left(\frac{h_e}{\tau_r} + \alpha' I\right) \cdot X_c \qquad (5.14)$$

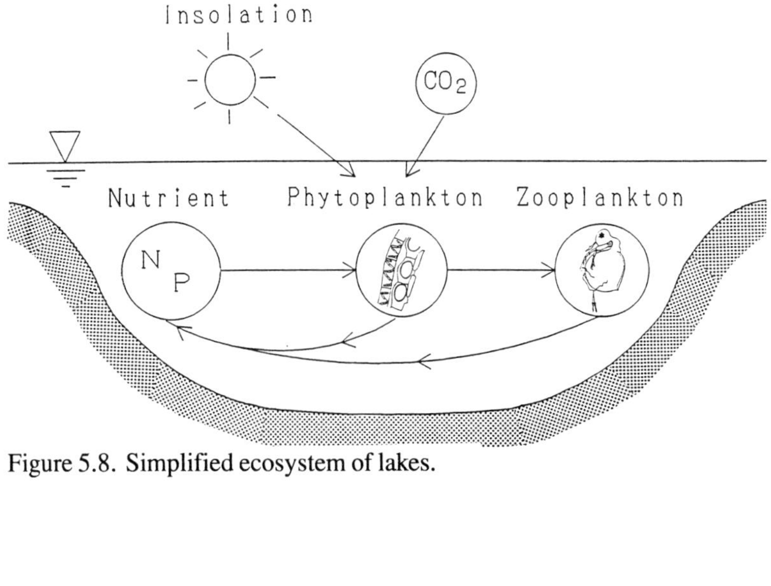

Figure 5.8. Simplified ecosystem of lakes.

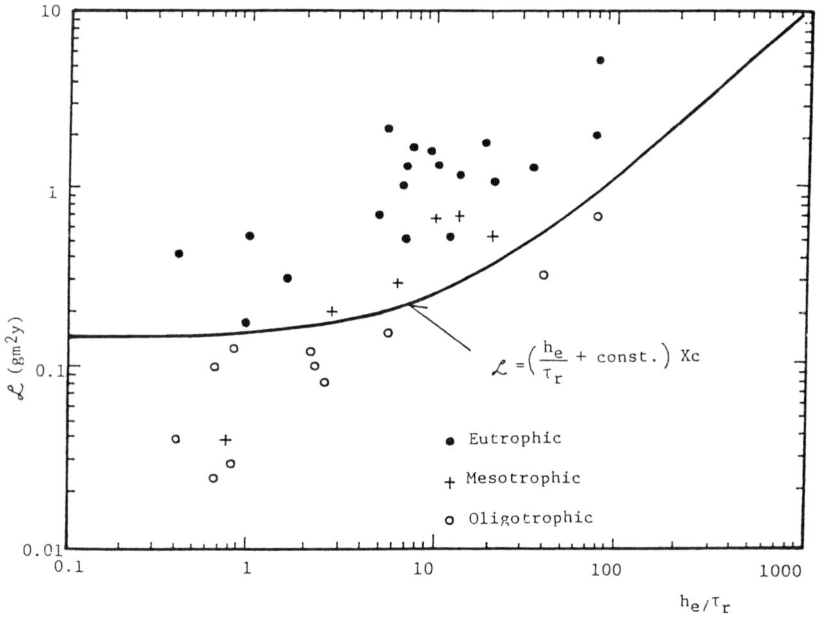

Figure 5.9. Nutrient loading limit into lakes and reservoirs (Hino, 1981b, 1983; Field data from Vollenweider, 1975)

where I = insolation intensity, α' = a constant and X_c = critical concentration of the controlling (most effective) nutrient in the lake (this may be either P or N). If the environmental nutrient concentration is greater than X_c the lake becomes eutrophic (Carlson, 1977). The above formula, Equation (5.14), compares well with the data collected by Vollenweider as shown in Figure 5.9.

5.3 PURIFICATION OF WATER QUALITY BY STABILIZATION POND

5.3.1 *Stabilization pond*

In order to keep rivers clean from pollution by sewage discharge or highly polluted tributaries, counter-measures are neccessally.

Waste stabilization ponds, which are large shallow basins enclosed by earthen embankments have been proposed as raw sewage treatment facilities. Since the treatment operates entirely under natural processes due to both algae and bacteria, such long retention-times as 30-50 days are not uncommon. Stabilization ponds are the most important method of sewage treatment in hot climates where the high temperature stimulates their operation, but they are also used in cold climates such as that of Alaska.

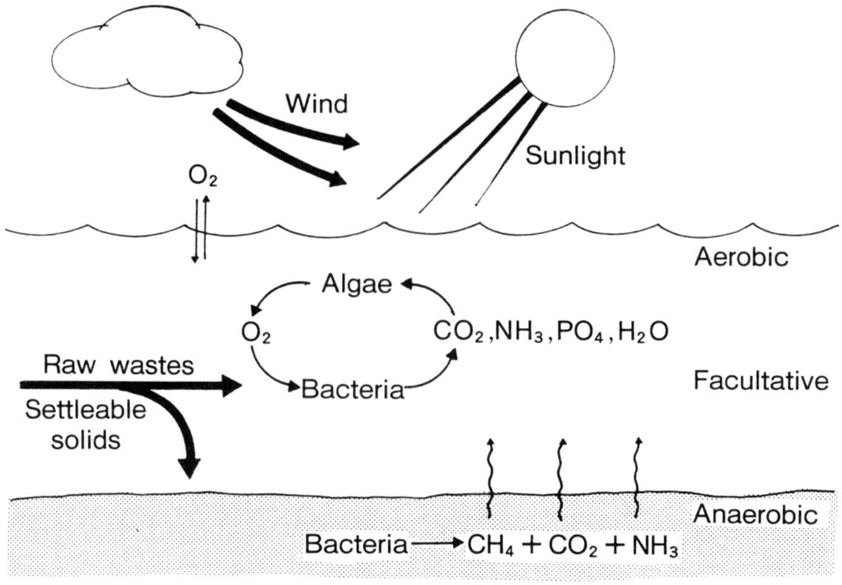

Figure 5.10. Schematic diagram of oxidation pond operations (from McKinney, 1962)

5.3.2 *Estuary lake as stabilization pond*

In case a polluted tributary is confluent with a less polluted main flow or if it flows into a lake or a sea, a small fore-lake or pond that is located upstream of the main body of water effectively reduces the BOD.

In these natural stabilization ponds, the food-chain pyramid has a more complex system including also water plants, benthos and waterfowl. Consequently, the purification process, not only physical (sedimentation, absorption) but also biological and ecological, works more effectively than artificial stabilization ponds (Fig. 5.10). Some reports indicate natural small lakes situated at estuaries work as stabilization ponds. For instance, for a river that discharges into a small lake before entering a large lake (Lake Biwa, Japan), above one third of the pollutants in the river are removed.

A lagoon situated at the estuary of River Nanakita, Sendai City, Japan, is reported to work effectively as a natural treatment system. Kurihara (1980) based on his experience on this lagoon, proposed the artificial lagoon project shown in Figure 5.11.

Even a small reservoir behind a small intake weir contributes to the reduction of BOD by prolonging the retenting time.

5.3.3 *Fish and duck pond*

In an open-air tank with water having a low level of BOD and of dissolved

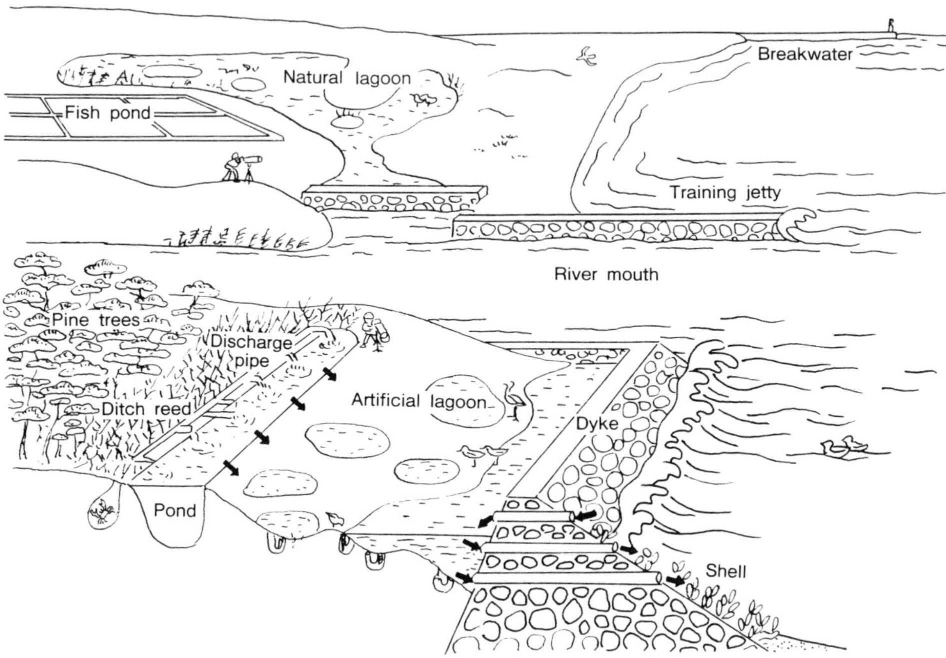

Figure 5.11. Sketch of an artificial lagoon project (from Kurihara, 1980)

inorganic substances such as NH_3 and PO_4, algae and phytoplankton grow and produce oxygen by photosynthesis. This process further stimulates the aerobic bacteria growth, oxidation of BOD and supply of inorganic nutrient (P, N). Zooplankton grow as predators of phytoplankton. Grazing on these algae and zooplankton as a good protein source, fish and ducks can thrive. A dry period for the pond is necessary to decompose by the action of aerobic bacteria NH_3 and H_2S absorbed in the soil of the pond. This type of fish pond was proposed as a method of the secondary sewage treatment by a German biologist, Hofer, in 1907. The final effluent with some residual N and P is used as a fertilizer for farms.

The fish farms in the Isar canal at Munich, Germany, are examples of aquaculture. Thirty ponds – each pond having the dimensions 200×350 m, depth $0.4 \sim 2$ m, and retention time $16 \sim 40$ hr were operated during the warm season from May to October (Fig. 5.12). After nourishment in clean water for two months, grown carp were sold in the Christmas season, thus reducing the running costs. The pond was planned in 1922, began to operate in 1930 and worked until the end of the World War II (Sabata, 1984).

Such a system of water farming is sometimes called 'aquaculture' just as 'field farming' is called agriculture. An example is the integrated water disposal system developed in Papua New Guinea which is a very simple and effective combination of aquaculture and agriculture (Fig. 5.13), (Chan, 1973).

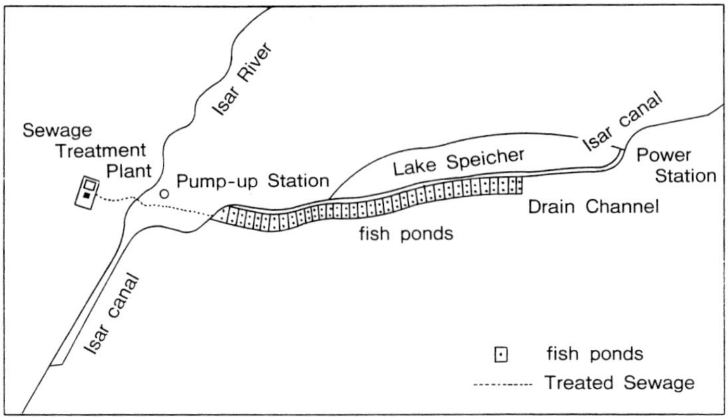

Figure 5.12. Plan of fish ponds of Isar canal (from Sabata, 1983)

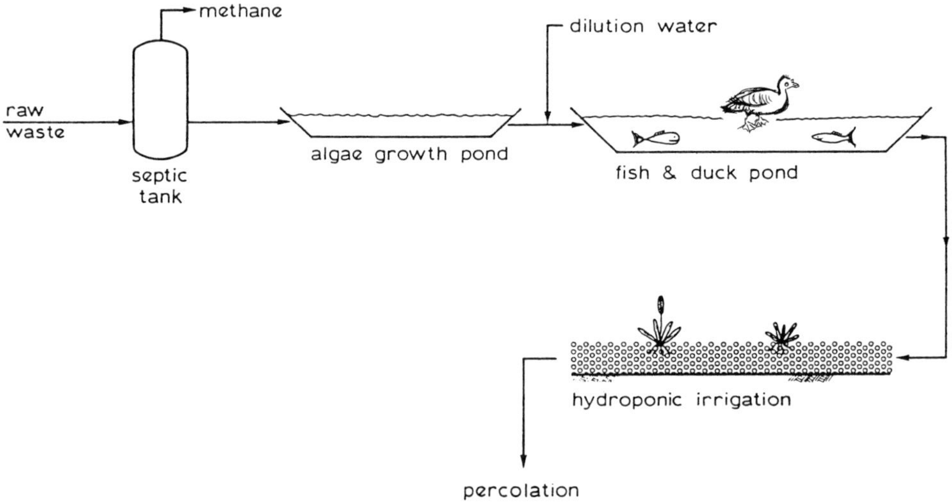

Figure 5.13. Integrated waste disposal system as developed in Papua New Guinea (Chan, 1973)

5.4 PURIFICATION OF WATER QUALITY BY HIGH-RATE BIOFILTRATION

One of the artificial ways to augment and accelerate the self-purification process in nature is to fabricate a structure, usually made of plastic, with a regular geometery like the honeycomb used in the settling chamber of a wind-tunnel. The structures are fabricated so as to provide both a larger surface-area where microbiofilms grow and a high void ratio that not only prevents blockage by excess microbial growth and other solids but also reduces the resistance to flow and thus the operational cost.

5.4.1 *Operational characteristics*

High-rate biofilters are able to remove large amounts of BOD at high BOD loading. Mara (1976) states, 'It is an advantage to conserve the strength of the waste water so as to operate the lowest hydraulic loading possible rather than to dilute the waste water and to operate at higher hydraulic loads and weaker feed strength, even though the BOD_5 load on the filter might be the same. With increasing strength the driving force for BOD_5 removal by the film increase'.

5.4.2 *Design*

The removal process of the biofilter is described by the first order kinetic theories. The retention time t^* is given by

$$t^* = \alpha H/(Q/A)/(xSd) \tag{5.15}$$

where Q = flow rate, A = cross-sectional area of the biofilter, S = surface area of biofilm in the 1 m cube length, d = flow thickness, H = height or length of the biofilter, α = ratio of the effective zigzag path of the biofilter to H ($\alpha > 1$), and x = a measure of non-availability of the theoretical flow area (Sd), ($x < 1$).

Thus, the BOD of the effluent is

$$L_e = L_i \exp\{-KVS/Q\} \tag{5.16}$$

where $K = k_1 \alpha xd$, $V = AH$, Q/V = volumetric hydraulic loading, k_1 = first-order rate constant of BOD removal, which is dependent on temperature.

5.5 OXIDATION BY BIOFILM ON GRAVELS

5.5.1 *Trickling filter*

One of the conventional facilities for the treatment of settled sewage is a biofilter, a kind of biological reactor. The facility ia a circular or rectangular bed of coarse aggregate (Fig. 5.14). Settled sewage is spread over the aggregate bed, and it then

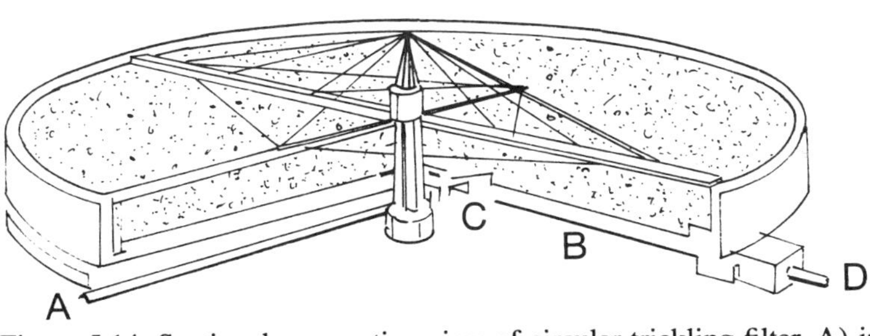

Figure 5.14. Sectional perspective view of circular trickling filter, A) inlet pipe, B) underdrain blocks, C) effluent channel, D) outlet pipe (from Mara, 1976)

trickles down over its surface. Microfilms of bacteria attached on the surface oxidize the organic matter; the oxygen needed for the process is supplied from the atmosphere through gaps in the aggregate.

Some of the microbial films are detached, and they must be removed in secondary sedimentation tanks. Two sets of the biofilter and sedimentation system may be operated in line; a process that is known as alternating double filtration. The order of the filter is reversed before blockage can occur.

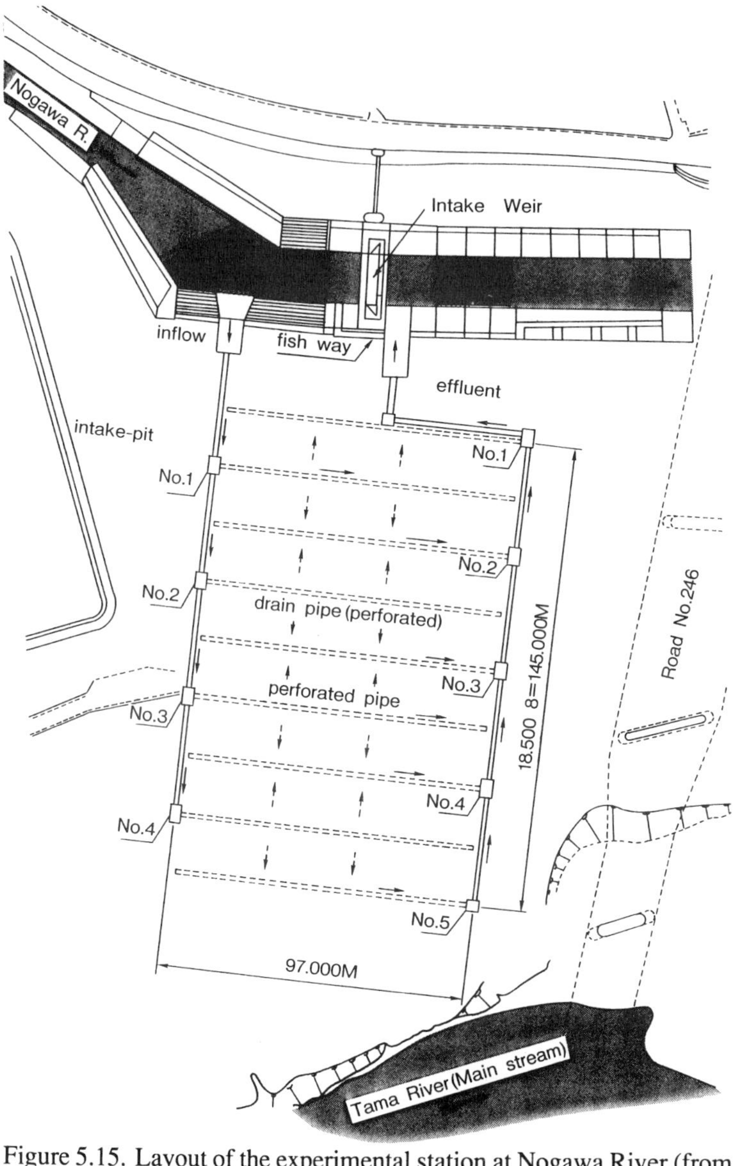

Figure 5.15. Layout of the experimental station at Nogawa River (from Report of Resources Investigation Committee, Agency of Science and Technology, Japan, 1984)

5.5.2 *Inter-gravel contact oxidation*

A new method of river purification has been proposed and is being tested by the Ministry of Construction of Japan (Fig. 5.15). The operational principle of the facility is somewhat similar to that of a trickling filter, since it utilizes the biofilm grown on the surface of gravels.

1. Ditches are dug in the flood plain of a river. They are not paved with concrete so as to reduce the construction cost. Gravel and pebbles from the river bed are packed into the ditch. No additional materials are used.

2. Water behind an intake weir built across the river is drawn into the ditch. No additional energy is required to circulate the polluted water of the river through the ditch if the hydraulic head at the intake weir and the channel slope of ditch are adequate.

3. No sludge treatment is required since within the facilities the sludge is treated simultaneously with the purification process.

Operation of the experimental station of the Nogawa river, a tributary of Tama river, showed the process to be highly effective. The removal rates of BOD and SS were, respectively, 75% and 85%; i.e. BOD 13 (mg/l) decreased to BOD 3.25 (mg/l) and SS 16 (mg/l) to 2.4 (mg/l).

REFERENCES

Aiba, S., A.E. Humphrey & N.F. Millis 1973. *Biochemical Engineering*, Academic Press.

Aiba, S., M. Okada, H. Otake, R. Sudo & T. Mori 1975. Simulation of BOD-DO balance in a shallow polluted river, I and II. *J. of Sewerage Assoc. Jpn.* 12(131): 33-38; (132): 26-37 (in Japanese).

Biswas, A.K. (ed.) 1981. *Models for Water Quality Management*. McGraw-Hill.

Camp, T.R. 1965. Field estimation of oxygen balance parameters. *Proc. ASCE*, 91(SA5).

Carlson, R.E. 1977. A trophic state index for lakes. *Limnology & Oceanography*, 22(2): 361-369.

Chan, G. 1973. In *Workshop on waste recycling systems*. Department of Agriculture, Stock and Fisheries, Papua New Guinea.

Dobbins, W.E. 1964. BOD and oxygen relationships in streams. *Proc. ASCE*, 90(SA3).

Hino, M. 1977. Algal and bacterial growth and self-purification of rivers. Memorandum at IIASA (unpublished).

Hino, M. 1979a. A formula on self-purification of rivers. *Proc. JSCE* (286): 65-75 (in Japanese).

Hino, M. 1979b. Theory on the self-purification by bacterial activity on river beds. *Proc. JSCE*, 290: 67-78 (in Japanese).

Hino, M. 1981a. Ecohydrodynamics. In V.T. Chow (ed.), *Advances in Hydroscience*, 12: 143-193.

Hino, M. 1981b. Note on trophic states and permissible phosphorus loading rates of lakes. Tech. Rept. No. 28, Dept. of Civil Eng., Tokyo Inst. of Tech., 161-166 (in Japanese).

Hino, M. 1983. Prediction of nutrient loading limit to prevent eutrophication of reservoirs by

the three ranks biological system model. *Proc., 20th Congress of IAHR*, V: 510-517, Moscow.
Hino, M. 1991. Ecological models in hydraulic environment, In Environmental Hydraulics, vol. 2, J.H.W. Lee & Y.K. Cheung (ed.), Balkema, The Netherlands.
Hynes, H.B.N. 1974. *The Biology of Polluted Waters*. Liverpool Univ. Press.
James, A. (ed.) 1978. *Mathematical Models in Water Pollution Control*. John Wiley & Sons.
James, A. (ed.) 1984. *An Introduction to Water Quality Modelling*. John Wiley and Sons.
Kanda, M. & M. Hino 1988. Analysis of ecosystem in lakes by mathematical model. *Proc. 33rd Japan Conf. on Hydraulic Res.*, JSCE, 257-262 (in Japanese).
Kurihara, Y. 1980. *Living Lagoon*. Iwanami New Library, Iwanami, Tokyo (in Japanese).
Mara, D. 1976. *Sewage Treatment in Hot Climates*. John Wiley and Sons.
McKinney, R.E. 1962. *Microbiology for Sanitary Engineers*. McGraw-Hill.
Mitchell, R. (ed.) 1972. *Water Pollution Microbiology*. Wiley-Interscience.
Nemerow, N.L. 1974. *Scientific Stream Pollution Analysis*. McGaw-Hill.
O'Kane, J.P. 1980. *Estuarine Water Quality Management*. Pitman Pub. Ltd.
Resources Investigation Committee 1984. Report on water quality improvement by means of purification effects in nature. Agency of Science and Technology, Japanese Government (in Japanese).
Sabata, T. 1983. *Culture of Waterworks – Europe and Japan*. Shincho Selected Library, Shincho-Sha, Tokyo (in Japanese).
Streeter, H.W. & E.B. Phelps 1925. A study of pollution and natural purification of the Ohio river. Public Health Bulletin, No. 146, USPHS.
Uhlmann, D. 1975. *Hydrobiology*. John Wiley & Sons.
Vollenweider, R.A. 1975. Input-outlet models with special reference to the phosphorus loading concept in limnology. *Schweiz, Z. Hydrol.*, 37(1): 53-84.

Improvement of water quality in rivers by aeration at hydraulic structures 6

P. NOVAK
University of Newcastle upon Tyne, UK (retired)

6.1 INTRODUCTION

The potential contribution of hydraulic structures through air entrainment to the improvement of water quality in rivers has long been recognised, but methods for predicting the oxygen uptake have evolved only rather recently. Even though some results of measurements from the field or laboratory were available, predictive equations that would account for at least the most important parameters and some scale effects were difficult to develop.

This text attempts to summarise the present position in methods for estimating oxygen uptake in the case of simple and free overfalls at weirs, multiple jets weirs and cascades, for outflows under gates with hydraulic jumps and for bottom outlets and turbines. Although a brief introduction to the theory of oxygen transfer is given, the main emphasis is on development of methods suitable for hydraulic engineering design; no full analysis of the physics of the process is attempted. However, even if this process is sometimes – because of the complex nature of the phenomena involved – treated almost like a 'black box', an analysis of the hydraulic parameters permits one to formulate a predictive model that should give the designer and operator of the structure some idea of the oxygen transfer likely to occur.

Although air entrainment is nearly always beneficial to water quality, not all environmental effects are positive. Oxygen supersaturation at high-head hydraulic structures can cause fish to die, and design measures may have to be taken to alleviate this danger. A brief note on supersaturation and some of the preventive measures is, therefore, included in this text.

6.2 GENERAL

Air-water flows have three distinct regions: air entrainment, detrainment and air transport by flowing water. This text is concerned only with the first two regions, and with only those cases of local aeration with free surface flows. In these cases local air entrainment is always connected with some form of surface discontinuity and resultant shear planes in the flow with air entrainment in distinct vortices.

The parameters that affect local air entrainment are the geometry, turbulence of the upstream flow and the Froude, Reynolds and Weber numbers of the flow. Local air entrainment may be governed by certain limiting conditions, i.e., the inception limit (requirement of minimum velocity), entraining limit (minimum approach Froude number), air supply limit (e.g., in aeration devices) and transport limit governed by downstream conditions; in this treatment again only the first two conditions are important, and in the practical cases dealt with here they are always satisfied. The order of magnitude for the inception limit can be characterised by the velocity v_c

$$\frac{v_c^3 \rho}{g \mu} \cong (0.5 \text{ to } 1) \times 10^5 \tag{6.1}$$

giving for water jets $v_c \cong 0.8$ to 1 m/s (Ervine et al., 1980).

For hydraulic jumps air entrainment commences at upstream Froude numbers exceeding about 1.7. The relative air entrainment $\beta = Q\text{air}/Q\text{water}$ for a given boundary condition, fully turbulent flow and given fluid properties, is independent of Reynolds number and thus can be shown to be directly proportional to Fr^2 (Kobus, 1985): generally

$$\beta = f(\mathrm{Fr}) \tag{6.2}$$

The physics of air bubble formation and transport have been dealt with by Kobus (1985).

Because the flow geometry and model scales affect the air entrainment process, no generally applicable design criteria can be formulated; thus each major hydraulic configuration has to be dealt with separately. As many of the data available were obtained in the laboratory, the range of scales from which they were obtained is important, and it can be exceeded in application only with due consideration of the scale effects likely to be involved.

In determining the transfer of oxygen in flows with local aeration, one must consider not only the mechanics of air entrainment but also the mass transfer between air and water, i.e., the transfer of oxygen from the air bubbles formed downstream of local entrainment into the solution. The parameters affecting this transfer are mainly the oxygen concentration difference between air and water, contact area, contact time, salinity, temperature, water quality and pressure.

According to Fick's first law of diffusion, the rate of diffusion $\partial M/\partial t$ of a gas across a gas liquid interface of area A into a quiescent body of liquid is a function of the coefficient of molecular diffusion D_m of the gas and the gas concentration gradient $\partial C/\partial x$:

$$\frac{\partial M}{\partial t} = D_m \cdot A \cdot \frac{\partial C}{\partial x} \tag{6.3}$$

The process of distributing gas throughout the body of liquid by diffusion alone is extremely slow, but it can be speeded up by mixing, which physically distributes the solute and fully exploits the diffusion capabilities by maintaining the maximum concentration gradient across the interface.

Various theories have been proposed to describe the process of gas transfer (Danckwerts, 1951; Higbie, 1935; Lewis & Whitman, 1924).

Equation 6.3, which forms the basis of equations developed to describe the oxygen transfer from air to water, is more generally expressed in the form:

$$\frac{\mathrm{d}M}{\mathrm{d}t} = AK_L\,(C_s - C) \tag{6.4}$$

where K_L is the liquid film coefficient and C_s the dissolved oxygen concentration at saturation. K_L is clearly a function of the gas diffusivity (a function of temperature T, the size of diffusing particles and the liquid viscosity ν) and the turbulence or degree of mixing in the liquid medium (time of exposure t_c of the fluid elements). In the case of air entrainment, K_L is further dependent on bubble size. The presence of surface active agents in the liquid may alter the physical properties of the liquid and thus the size d_b of entrained bubbles. Furthermore, films of surface active agents formed at the air/liquid interface can inhibit the diffusion process; however, some evidence indicates that in highly turbulent situations this film and any associated effects on aeration performance are effectively destroyed (Mancy & Okun, 1968).

Thus for a particular diffusion gas such as oxygen in a situation of turbulent air entrainment (mean bubble diameter d_b):

$$K_L = f(D_m, t_c, d_b) = f(T, \nu, t_c, d_b) \tag{6.5}$$

Equation 6.4 may be expressed as an instantaneous rate of change of concentration:

$$\frac{\mathrm{d}C}{\mathrm{d}t} = \frac{1}{V} \cdot \frac{\mathrm{d}M}{\mathrm{d}t} = K_L \cdot \frac{A}{V} \cdot (C_s - C) \tag{6.6}$$

(V is the volume of liquid into which mass $\mathrm{d}M$ of gas diffuses in time $\mathrm{d}t$).

Integrating between the limits of 0 and t for time and C_o and C_t for gas concentration, one arrives at the most widely used equation of gas transfer:

$$\frac{C_s - C_t}{C_s - C_o} = \frac{1}{r} = e^{-K_L \cdot \frac{A}{V} \cdot t} \tag{6.7}$$

$$r \left(= \frac{\text{upstream deficit}}{\text{downstream deficit}} \right)$$

is commonly referred to as the deficit ratio.

Equation 6.7 expanded into a series becomes:

$$r-1=\sum_{1}^{n}\frac{\left(K_L\cdot\frac{A}{V}\cdot t\right)^n}{n!} \tag{6.8}$$

in which

$$(r-1)=\frac{C_{\text{downstream}}-C_{\text{upstream}}}{\text{downstream deficit}}$$

For oxygen transfer from bubbles only (neglecting the effect of the free surface) with n the number of bubbles:

$$\frac{A}{V}=\frac{6\pi\, d_b^2\, n}{Qt}=\frac{6\pi\, d_b^2\, Q_a\, 6t}{\pi\, d_b^3\, tQ}=\frac{36Q_a}{d_b\, Q}=\frac{36\beta}{d_b}$$

Thus (from Eq. 6.2)

$$\frac{A}{V}=f(\mathrm{Fr}, d_b) \tag{6.9}$$

For oxygen diffusion into water at a particular temperature (constant ν and T), Equations 6.5, 6.7 and 6.9 result in

$$r=f(\mathrm{Fr}, t_c, d_b, t) \tag{6.10}$$

The temperature dependence of Equation 6.7 is generally described by a temperature function of the form:

$$\left(K_L\cdot\frac{A}{V}\right)_{T_2}=\left(K_L\cdot\frac{A}{V}\right)_{T_1}\cdot\theta^{T_2-T_1} \tag{6.11}$$

Numerous values of θ have been published, the differences among them being attributable to variations in mixing conditions or turbulence. The greater the turbulence, the smaller is the value of θ; thus extrapolation to higher degrees of turbulence will probably result in θ tending to unity. The effect of increasing temperature is to reduce the gas solubility and thus the concentration difference (the driving force), and also to increase the diffusion rate.

Two phases of gas transfer can be visualised:

1. On exposure to an air interface, the water molecular boundary layer is instantaneously saturated with oxygen irrespective of the initial oxygen deficit.

2. The distribution of oxygen into the body of liquid is effected either by diffusion or by physical mixing or by a combination of these two processes.

In Phase 1 temperature is important in that it determines the saturation concentration. In Phase 2 the diffusion processes are accelerated by temperature increases whereas the mixing process is not affected by temperature (Imhoff & Albrecht, 1972). Thus the relative temperature influence in Phase 2 will be determined by the degree of mixing. The greater the mixing, the lower will be the dependence on diffusion for the distribution of oxygen throughout the fluid and thus the lower the effect of temperature.

Clearly the application of Equations 6.6 to 6.11 to practical situations (overfall, outflow with jump etc.) is difficult, and recourse has to be usually made to experimentation to establish aeration possibilities in a given situation. Thus, the analysis of scale effects or the observance of the limits of experimental work – as already emphasised – is important. From the above dicussion it is clear that one of the main difficulties in the modelling of flow with entrained air and of oxygen transfer stems from the fact that the size of entrained air bubbles remains – for a given water quality – constant irrespective of model scale, and furthermore a limit exists below which the scale of the model should not be decreased in order to preserve the same mode of oxygen transfer as in the prototype for a particular aeration system.

6.3 OXYGEN TRANSFER AT OUTFLOWS UNDER GATES WITH A HYDRAULIC JUMP

Most of the experiments used as a basis for predictive equations in this section and Section 4 were carried out in the hydraulics laboratory of the Department of Civil Engineering at the University of Newcastle upon Tyne (Avery & Novak, 1978; Avery 1976).

As a result of constant recirculation the water was aerated to saturation; it was necessary, therefore, to deoxygenate the water used in the aeration experiments to the extent that even after aeration, the oxygen level was still significantly below saturation. This condition was effected by controlled dosing of a sodium sulphite solution in the presence of a cobalt catalyst into the water upstream of the aeration apparatus. Precautions were taken to ensure sufficient reaction time so that residues of the sulphite were not carried forward to the experimental apparatus.

Oxygen levels were measured by means of calibrated dissolved oxygen meters each incorporating a temperature compensated electrode and a resistance thermometer.

Oxygen levels were expressed as percentage saturation, and oxygen transfer was expressed as the deficit ratio corrected to 15°C using a previously published temperature correction (Gameson et al., 1958) ($\theta = 1.018$ in Equation 6.11). Use of this correction was justified since the energy expenditures in the currently reported tests were comparable to those for which the above value of θ was derived and the temperature of the laboratory water did not vary substantially from 15°C.

As the laboratory water was treated with sodium nitrite ($N_a\,NO_2$) to prevent corrosion, a variation of the concentration of $N_a\,NO_2$ provided a welcome means of varying the bubble size d_b. Three $N_a\,NO_2$ concentrations were used – 0, 0.3%

and 0.6% with a mean bubble size of 2.53, 1.72 and 1.57 mm (Avery, 1976).

The hydraulic jump formed immediately downstream of an underflow sluice gate in a horizontal rectangular flume 100 mm wide and 3 m in length. The experimental range for the supercritical Froude number $Fr_1 = v_1/\sqrt{(gy_1)}$ was $2.8 < Fr_1 < Fr_1$ 8.5 and that for the specific discharge was $0.0145 < q < 0.0808$ m^2/s; the results could best be correlated by the equation (Avery & Novak, 1978)

$$r_{15} - 1 = k_1 \left(\frac{\Delta E}{y_1}\right)^{0.8} \left(\frac{q}{q_o}\right)^{0.75} \tag{6.12}$$

where k_1 is a function of salinity and $q_o = 0.0345$ m^2/s. All the data are compared with a plot of Equation 6.12 in Figure 6.1.

The results can also be described by the equations:

$$r_{15} - 1 = k_2 \, Fr_1^{2.1} \left(\frac{q}{q_o}\right)^{0.75} \tag{6.13}$$

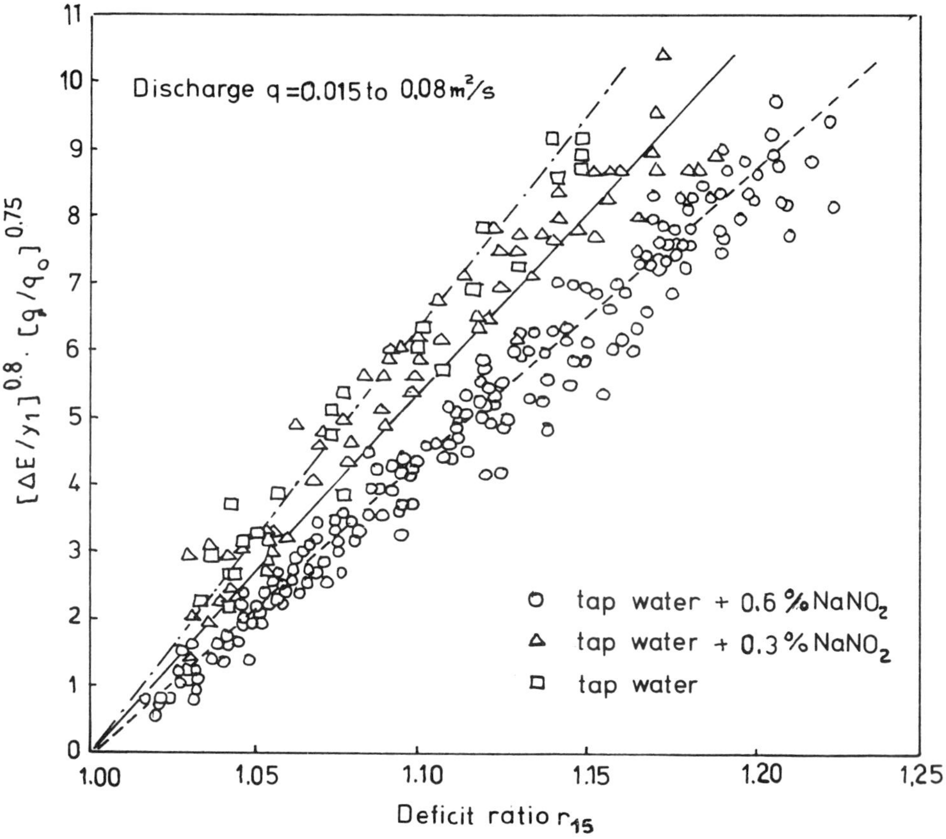

Figure 6.1. Oxygenation in hydraulic jumps: Correlation of measurements with Equation 6.12 (Novak, 1980)

and

$$r_{15} - 1 = k_3 \, \mathrm{Fr}_1^{2.1} \, \mathrm{Re}^{0.75} \tag{6.14}$$

where Re = q/ν is the Reynolds number, and k_2 and k_3 are again functions of $N_a\,NO_2$ concentration. The values of k_1, k_2 and k_3 for the range of salinity tested and for $\nu = 1.143 \times 10^{-6}\ m^2/s$ are as follows:

Sodium nitrite concentration %	k_1 (Eq. 12)	k_2 (Eq. 13)	k_3 (Eq. 14)
0	0.0158	0.00230	1.0043×10^{-6}
0.3	0.0186	0.00285	1.2445×10^{-6}
0.6	0.0230	0.00355	1.5502×10^{-6}

Use of the Reynolds number in Equation 6.14 is not necessarily indicative of viscous effects; Re is used here merely to get a dimensionless expression using discharge q. Other expressions have been published (Apted & Novak, 1973; Holler, 1971) and the more reliable of these results have, after checking, been incorporated in the more generally valid Equations 6.12 to 6.14.

In accordance with Equations 6.7 to 6.9, a reduction in bubble size results in an increase in the oxygen uptake, and $(r - 1)$ is approximately inversely proportional to bubble size.

An attempt was made to compute K_L from Equation 6.7 from the experimental data; Equation 6.9 was used for the value A/V, Equation 6.14 for r, the Kalinske & Robertson (1942) equation for Q_a/Q and tracer experiments gave the time of contact t in the zone of the jump The range of K_L established in this way was (Avery, 1976) $1.6 \times 10^{-4} < K_L < 4.5 \times 10^{-4}$ and the corresponding range of Sherwood numbers $\mathrm{Sh} = K_{Ld_b}/D_m$ was $120 < \mathrm{Sh} < 620$. The following relationship fits the experimental results (Avery & Novak, 1977)

$$\mathrm{Sh} = 0.266 \left(\frac{Q}{A} \frac{\Delta E}{\nu} \right)^{1.2} \tag{6.15}$$

Smutek (1974), using prototype measurements, obtained a value of $K_L \cong 11 \times 10^{-4}$ for a hydraulic jump; hence, some scale effect may occur. Nevertheless all these values are substantially larger than the values of K_L estimated for the free surface of normal open channel flow.

Field verification of Equations 6.12 to 6.14 is very limited (Burley, 1978). Nevertheless, from measurements at a treatment plant with discharges up to 3 times bigger than those used in the laboratory but at rather low Froude numbers (Fr < 3.6) results agreeing reasonably well with the predictive equations were obtained.

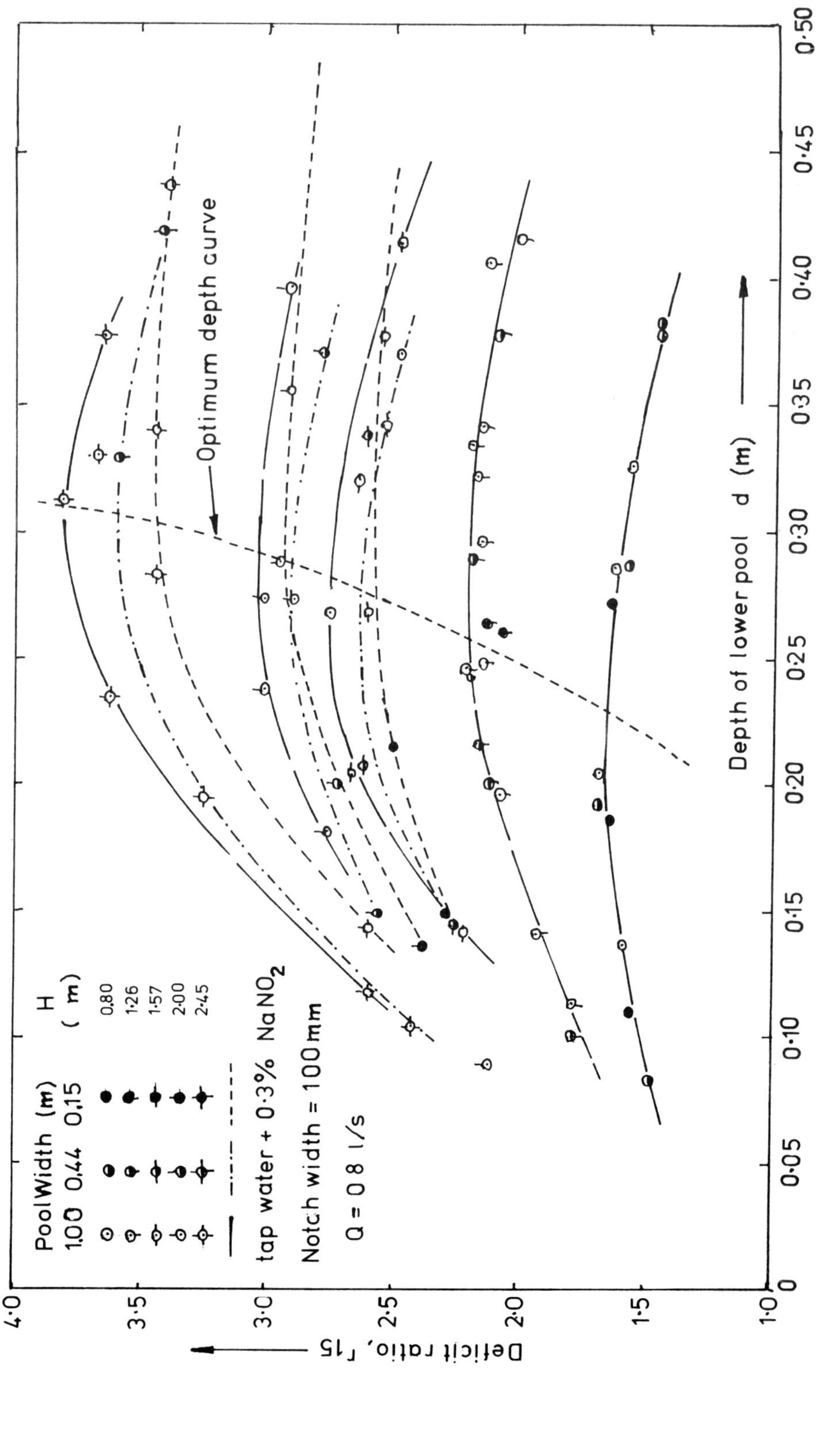

Figure 6.2. The effect on oxygen transfer of the lower pool geometry for a fixed discharge and various elevations of this pool

6.4 OXYGEN TRANSFER AT OVERFALLS

Although rates of oxygen transfer at overfalls have been measured fairly frequently both in the laboratory and in prototype, the diversity of conditions has impeded the process of reaching generally valid conclusions. The differences included structure and jet geometry, downstream conditions and water quality. Also many authors attempted to correlate the results in terms of height of fall only and ignored other parameters, particularly the discharge. A comprehensive summary of published approaches to this problem (Gameson, 1957; Gameson et al., 1958; Holler, 1971; Jarvis, 1970; Nakasone, 1975; Notes on Water pollution; 1973) was carried out by Avery & Novak (1978). They also clearly demonstrated that the plunge pool depth *d* and to a lesser degree its shape and width affect the oxygen uptake. For a given difference in elevation between the weir crest and pool bed *H*, an 'optimum' depth corresponding to the maximum penetration depth of air bubbles into the pool occurs (Fig. 6.2). For a depth *d* smaller than the optimum the contact time of the air drawn into the pool is smaller than for the maximum penetration depth; for a bigger depth this contact time is not increased and the height of fall (and thus the amount of entrained air) is decreased.

The contact time of air bubbles in the downstream pool is thus only one factor affecting oxygen uptake. Equally important is the amount of air drawn into the pool at impact, a quantity that depends on the shape, size and state of the jet and on its velocity at impact.

Because of the complicated and sometimes conflicting effects of bubble contact time and quantities of air entrained, Avery & Novak used a jet Froude number at *impact into the pool* in the form

$$\mathrm{Fr_J} = \frac{v}{\sqrt{gD}} = \left[\frac{\pi\sqrt{(2gh^5)}}{Q}\right]^{1/4} \tag{6.16}$$

where *D* is the jet diameter, *h* the height of fall (distance between the upstream and downstream water level) and *Q* the discharge. For jets with a non-circular shape at impact they used the jet discharge q_J per unit jet perimeter *P* ($q_J = Q/P$):

$$q_J = R\sqrt{(2gh)} \tag{6.17}$$

resulting in

$$\mathrm{Fr}_J = \left(\frac{gh^3}{2q_J^2}\right)^{0.25} \tag{6.16a}$$

These equations apply only if the jets at impact are solid, not disintegrated. Hoření (1956) established experimentally for $0.013 < q < 0.08\ \mathrm{m^2/s}$ the distance L_o from a rectangular notch along the jet centreline to the point of total disintegration as a function of the specific discharge at the crest *q*:

$$L_o = 31.19\,q^{0.319}\ \mathrm{(cm)} \tag{6.18}$$

or

$$L_o = 5.89\, q^{0.319} \cong 6\, q^{1/3} \text{ (m)} \tag{6.18a}$$

(In Equation 6.18 q is in cm^2/s, in Equation 6.18a in m^2/s).

As the minimum velocity for inception of air entrainment is approximately 1 m/s, corresponding to a height of fall 0.05 m, the limits of applicability of Equations 6.16 and 6.17 (and of the subsequent equations for the deficit ratio) are

$$0.05 < h < 6\, q^{1/3} \text{ (m)}$$

Solid jets from a rectangular notch either converge rapidly to a circular cross-section, Equation 6.16, or diverge to flat rectangular shapes, Equation 6.17. The criterion for a jet to diverge was established as (Burley, 1978; Novak, 1980)

$$\frac{H'}{b} > 1.288 \tag{6.19}$$

where b is notch width, H' the head on the notch crest and the height of fall at which convergent jets ($H'/b < 1.288$) become circular as

$$h = 430\, b^{0.36}\, Q^{0.8} \text{ (m)} \tag{6.20}$$

in which b is in metres and Q in m^3/s.

The limits of experiments on jets from rectangular notches carried out in the laboratory at Newcastle upon Tyne that were used to derive the equations for r were:

$$0.1 < b < 0.3 \text{ (m)}$$

$$0.6 < Q < 5.8 \text{ (l/s)}$$

$$0.25 < h < 2.1 \text{ (m)}$$

Under these conditions the following equations were established for the optimum depth (d_{op} in cm) of the downstream pool (h in cm, q_J in cm^2/s in Equation 6.21a:

$$d_{op} = 0.41\, \mathrm{Re}^{0.39}\, \mathrm{Fr}_J^{0.24} \tag{6.21}$$

or

$$d_{op} = 3.3\, h^{0.18}\, q_J^{0.27} \tag{6.21a}$$

and for r_{15} at $d \geq d_{op}$ and for solid jets:

$$r_{15} - 1 = k_4\, h^{1.34}\, q_J^{-0.36} \tag{6.22}$$

(h in cm, q_J in cm^2/s). Or

$$r_{15} - 1 = k_5\, \mathrm{Fr}_J^{0.72}\, h^{0.8} \tag{6.23}$$

(h in cm). Or

$$r_{15} - 1 = k_6 \, \mathrm{Fr}_J^{1.78} \, \mathrm{Re}^{0.53} \tag{6.24}$$

The coefficients k_4, k_5 and k_6 are given as functions of salinity by the following values (Novak, 1980).

$N_a NO_2$ %	k_4 (Eq.22)	k_5 (Eq.23)	k_6 (Eq.24)
0	0.0108	0.0036	0.627×10^{-4}
0.3	0.0151	0.0050	0.869×10^{-4}
0.6	0.0214	0.0071	1.243×10^{-4}

Note: k_4 has dimensions $cm^{-0.62}/s^{0.36}$ and k_5 $cm^{-0.8}$

Figure 6.3 shows some of the experimental results leading to Equation 6.23. In the use of Equations 6.16a and 6.21 to 6.24 for weirs (as opposed to free falling jets from notches) two cases arise: if there is no access of air under the nappe (solid weir) $q = Q/b = q_J$, if the nappe springs clear of the weir with access of air underneath it $q = 2q_J$.

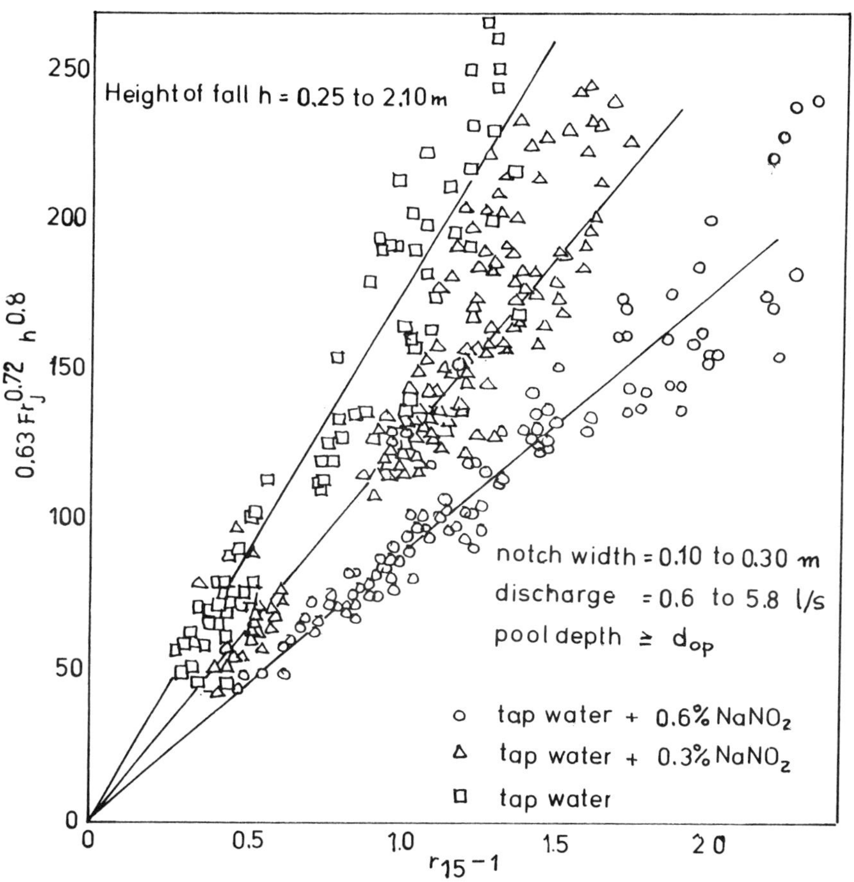

Figure 6.3. Weir oxygenation: Correlation of measurements with Equation 6.23 (Novak, 1980)

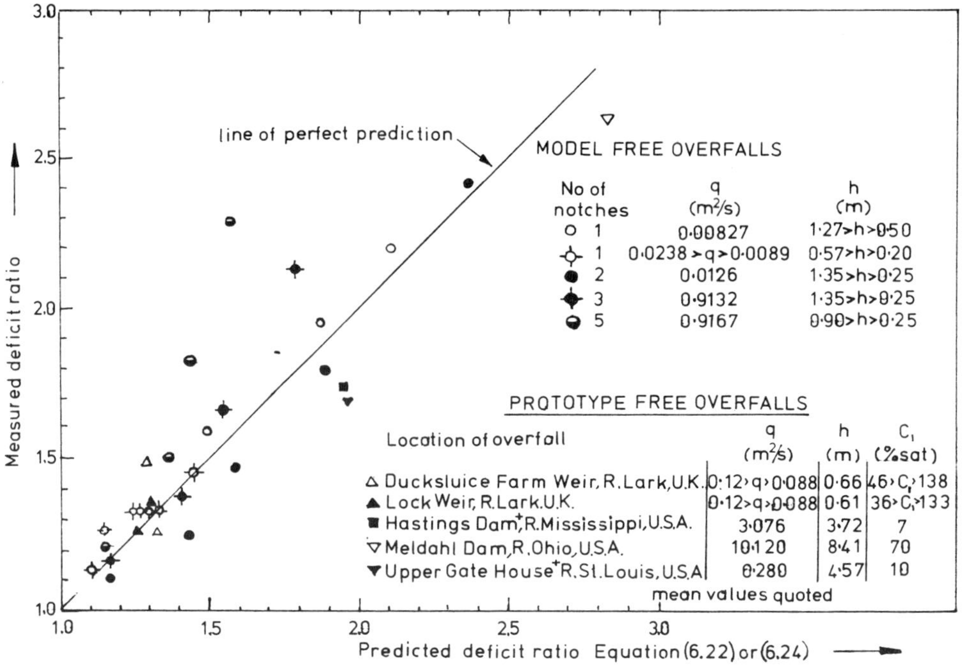

Figure 6.4. Correlation of Equation 6.22 to 6.24 with previously published oxygen transfer measurements for various model and prototype structures

Equations 6.22 to 6.24 were successfully correlated with a number of prototype structures and all published laboratory measurements for which sufficient data were available ((Holler, 1971; Notes on Water Pollution; 1973; Van Der Kroon, 1969) (Fig. 6.4). As can be seen from the figure, a large range of discharges and heights of fall as well as a variety of pollution loads (as evidenced by the initial dissolved oxygen levels) are encompassed in this correlation.

For all the data in Figure 6.4 zero salinity levels were assumed, in the absence of information to the contrary. Pool depths beneath the prototype free overfalls were sufficient to justify the assumption of optimum depth conditions, and jet widths were taken to be equal to the crest widths. Deficit ratios were corrected to 15°C by means of the temperature correction discussed earlier (Gameson, 1958). Once again, the energy expanded in the tailwater was found to lie within the same range as in the laboratory tests from which the above temperature correction was developed (Avery, 1976).

The good correlation obtained with the laboratory tap water experiments tends to confirm that the effect of pollutants on the liquid film coefficient is indeed negligible in highly turbulent flows; nevertheless, some contaminants may increase the air/water interface created (as observed earlier).

Nakasone (1975) incorporated the effect of h, q and d into his equations which are in the form

$$\ln r = K\, h^a\, q^b\, d^c \tag{6.25}$$

in which c is a constant 0.31, and K, a and b vary according to whether $h \lessgtr 1.2$ m and $q \lessgtr 0.065$ m^2/s as follows

	h	q	K	a	b
1	< 1.2	< 0.065	0.0785	1.31	0.428
2	< 1.2	> 0.065	5.39	1.31	– 0.363
3	> 1.2	< 0.065	0.0861	0.816	0.428
4	> 1.2	> 0.065	5.92	0.816	– 0.363

The correlation with (the simpler and physically more justified) Equations 6.22 to 6.24 (if the conditions for $d \geq$ optimum and $h < 6\,q^{1/3}$ are satisfied) is acceptable. The values of a and b in Case 2 correspond closely to the powers in Equation 6.22.

Markofsky & Kobus (1978) developed a diagram for sharp-crested aerated rectangular notches discharging into a deep pool that correlates $(r_{15} - 1)$, Re and Fr (Fig. 6.5). The agreement with Equations 6.22 to 6.24 is fairly good even though the authors state that the graph is not intended for prediction of weir reoxygenation and should be used with caution (Kobus, 1985).

6.5 OXYGEN TRANSFER AT MULTIPLE JET OVERFALLS AND CASCADE WEIRS

Equations 6.22 to 6.24 are also applicable to multiple jet weirs and cascades with pools.

If one circular jet of Froude number Fr_J is split into N jets of equal diameters, the resultant Froude number will be $N^{1/4}\,\mathrm{Fr}_J$. The advantage afforded is obvious, a higher Froude number can be attained for the same discharge and height of fall by splitting the jet and therefore greater aeration is achieved. This result has been experimentally confirmed (Avery, 1976; Tekle, 1983). The application of the equations to multiple jet weirs is valid subject to two conditions:

1. That the jets do not interact (and disintegrate) during their free fall;
2. That the air entrainment and mixing process due to any jet is not unduly affected by that of any neighbouring jets.

Successful use of Equations 6.22 to 6.24 for multiple jets is also demonstrated in Figure 6.4 with the exception of the 5-notch weir for which jet interaction obviously resulted in a greater oxygen transfer than predicted by the equations.

If a cascade weir consists of a series of free overfalls with pools strung together, the oxygen transfer can be computed from Equations 6.22, 6.23 or 6.24, if applied consecutively to each step of the cascade. The process is simplified if the cascade consists of a series of n equal steps; the deficit ratio then would be r^n where r is the deficit ratio for one step.

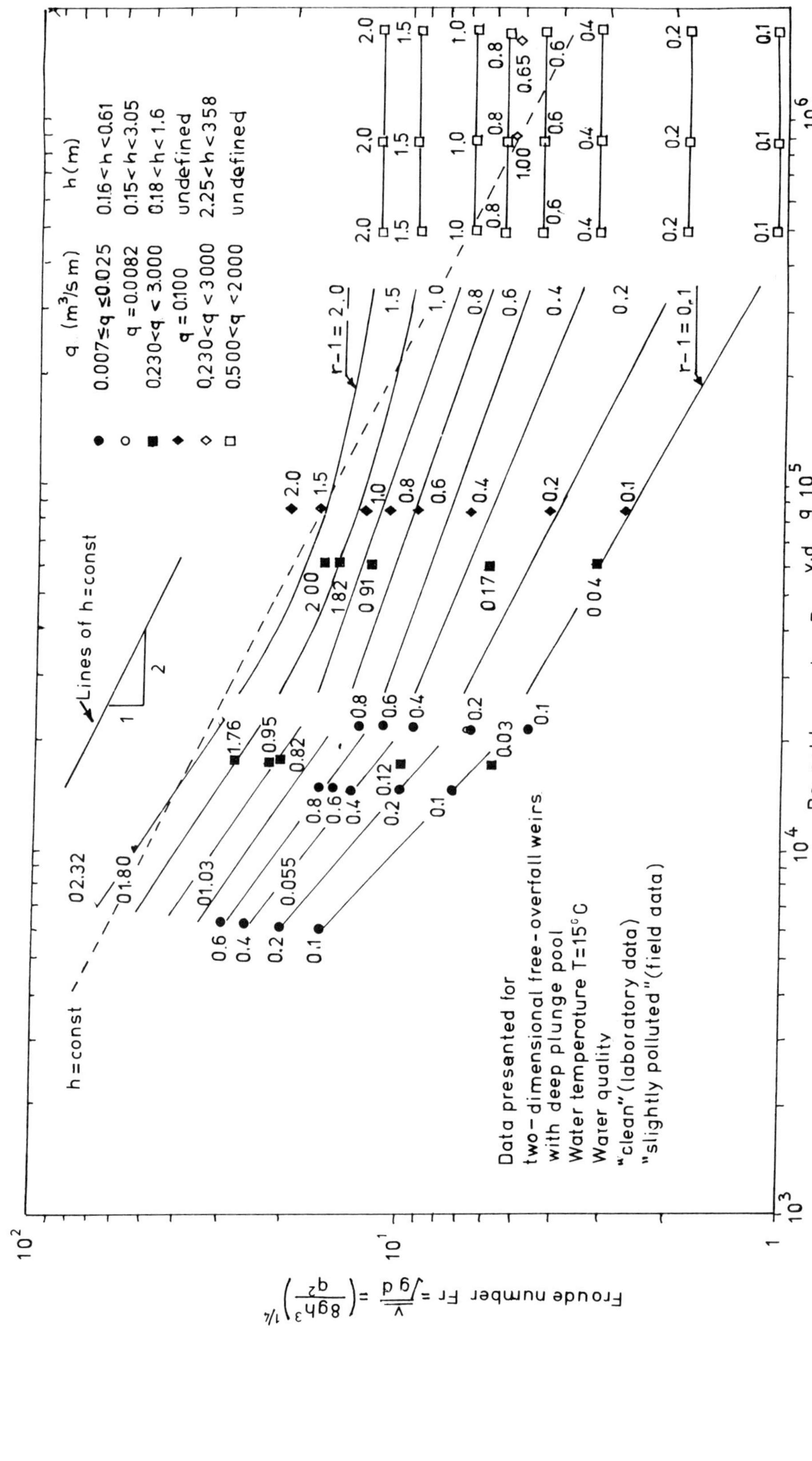

Figure 6.5. Unified presentation of weir aeration data (rectangular weirs with deep plunge pools); Markofsky and Kobus, 1978

The deficit ratio computed in this way can be compared with the ratio for a cascade without pools using the graph shown in Figure 6.6 from Tebutt et al. (1977) summarising the results of work carried out at the University of Birmingham. Experiments were carried out on cascades with three slopes (1:1, 1:2.5, 1:5) individual step heights of $0.05 < H < 0.50$ m and discharges of $0.0116 < q < 0.1447$ m^2/s. The graph relates the cascade slope, number of steps, specific discharge, height of steps and the deficit ratio. Thus, e.g., a cascade 1.00 m high, with $q = 150$ cm^2/s and with 5 steps 20 cm high would achieve a deficit ratio *r*\=\ 2.23 (practically irrespective of slope). The application of Equation 6.24 for the same discharge and a single step 1.00 m high with a pool of optimum depth yields *r*\=\ 2.2 i.e., a value identical to that for a cascade without pools. A cascade of 5 steps 20 cm high with a 20 cm pool at each step (Fig. 6.7) would, on the other hand, give an increased value of $r = 1.23^5 = 2.815$ ($r = 1.23$ is the deficit ratio for a single 20 cm high step).

The efficiency η of aerators is generally gauged as the quantity of oxygen transferred relative to the power expended (kg O_2/kwh):

$$\eta = \frac{3.6\,(C_s - C_1)\left(\frac{r-1}{r}\right)}{gh} \tag{6.26}$$

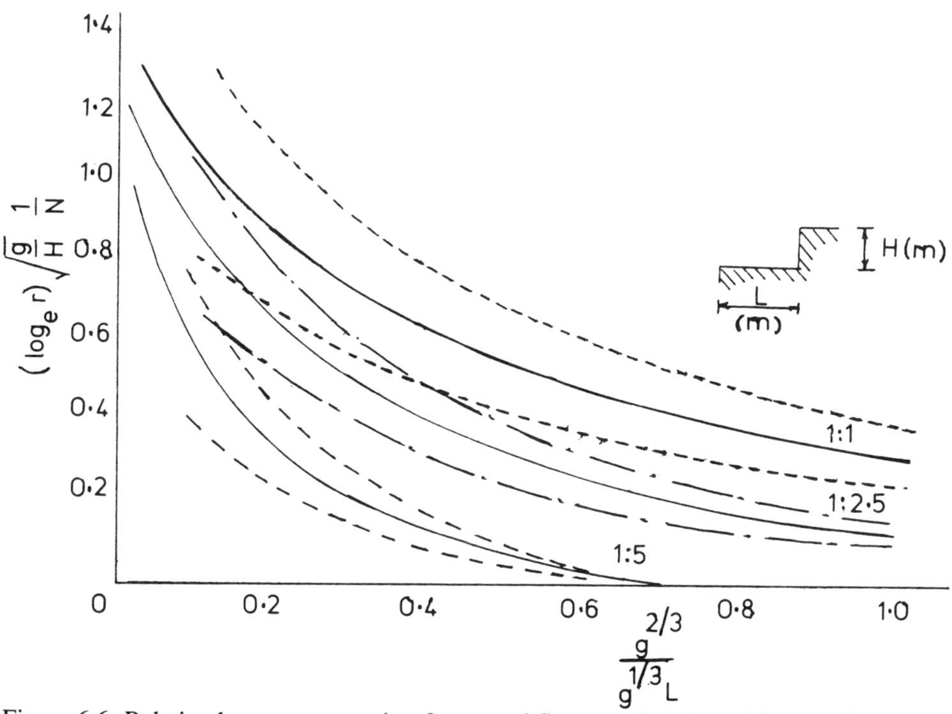

Figure 6.6. Relation between reaeration factor and flow number; dotted lines indicate 95% confidence limits (Tebbutt et al., 1977)

For a weir the efficiency is a function of the Froude and Reynolds numbers as well as of the initial oxygen deficit, which should be specified if efficiencies are compared. Figure 6.7 compares the efficiency of the hydraulic jump, free overfall and the illustrated cascade (with pools with optimum depth conditions) for varying heights of fall and a particular specific discharge.

The high efficiency of the free overfall at small heights of fall illustrates why a cascade of overfalls is more efficient than a single free overfall. The efficiences of

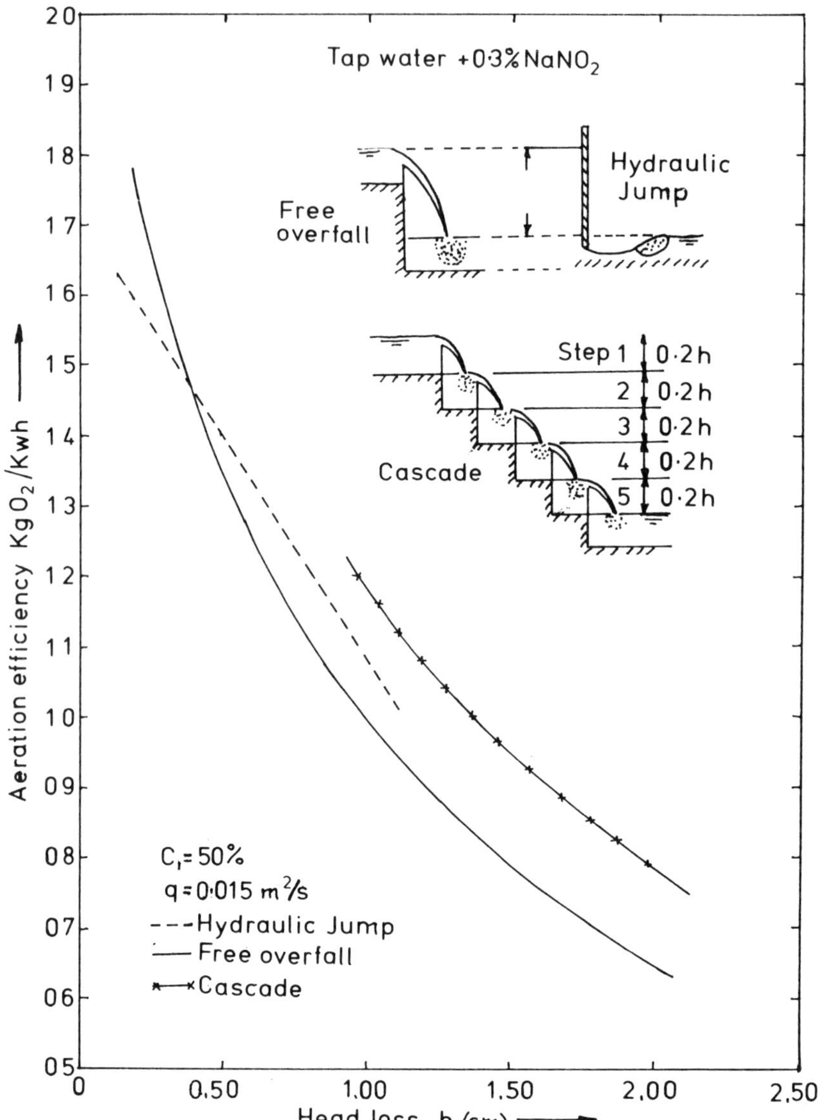

Figure 6.7. A comparison of aeration efficiency for a hydraulic jump, free overfall weir and a cascade weir

the free overfall and hydraulic jump are comparable, whilst the latter has the advantage of requiring the smaller tailwater depth.

The efficiencies for the prototypes in Figure 6.4 varied from 0.2 to 0.9 kg/kwh for a 50% initial deficit (Avery, 1976). The above efficiencies are not uncompetitive when compared with commercial aerators; for the latter the following are examples:

- Turbine aerators 0.45- 1.41 kg/kwh (Imhoff & Albrecht, 1972);
- Surface aerator 0.79-2.00 kg/kwh (Kaplovsky & Sosewitz, 1964);
- U-tube aerator 0.61-2.13 kg/kwh (Speece et al., 1969).

The efficiency of the single free overfall could, of course, be improved substantially by conversion to a cascade, or by serrating the crest to induce the formation of multiple jets.

6.6 OXYGEN TRANSFER AT BOTTOM OUTLETS AND TURBINES

The previous sections have amply demonstrated that the mass transfer and oxygenation of water are greatly affected by the degree of turbulence during mixing and length of contact time of air bubbles with water in highly turbulent flow. Thus even higher values of the deficit ratio and better efficiencies of aeration can be achieved at bottom outlets using valves (with air supply), which create highly turbulent aerated flow in the downstream region with long contact times. Typically this can be done by producing a hydraulic jump inside a conduit or better still be an internal ring hydraulic jump (e.g., downstream of a cone dispersing valve) discharging in the outlet itself (rather than into atmosphere). Haindl (1984) shows some outlets operating in this fashion and quotes oxygenation (C_2) in the region of 10-13 g/m^3, i.e., about double the values associated with normal dam bottom outlet valves (e.g., hollow jet valves discharging into the atmosphere and a downstream pool).

For turbines discharging oxygen deficient water, it is possible to introduce air into the negative pressure zones (in the draft tube) and thus to improve water quality; however this usually decreases the turbine output. It is, therefore, preferable to provide air supply into the outlet from draft tubes of turbines and to use this artificial air supply seasonally as required.

6.7 SUPERSATURATION AND ITS PREVENTION

So far attention has focussed on oxygen transfer at low head structures for which reductions in the oxygen deficiency benefit the environment. If, however, the upstream water is saturated with oxygen, or nearly so, then further oxygen enrichment can lead to oxygen supersaturation that may have detrimental effects on the environment. This situation is more likely to occur at high head structures with their associated high velocities of flow. Most of the research on the subject of supersaturation at hydraulic structures and its environmental effect and on prevention measures has been carried out in the USA (Johnson, 1984; Nebeker & Brett, 1976; Weitkamp & Katz, 1973) and Norway (Tekle, 1983; Tekle et al., 1984).

A liquid is supersaturated if it contains more dissolved gas than the amount that it can normally dissolve at the ambient pressure and temperature. Thus water below dam spillways and power inlets, where a substantial amount of air may be entrained, can be supersaturated; the degassing then takes place in the stilling basin, tailrace and downstream river stretch. This de-gassing, which begins rapidly, becomes slower with time without substantial mixing and turbulence. Dilution and mixing with water at natural saturation can also reduce supersaturation.

Supersaturation leads to de-gassing, or bubble formation, inside the blood vessels in fish, crabs, etc. and results in the gas bubble disease. The lethal supersaturation for fish is some 5-15% above saturation level depending on the time the fish is exposed to the supersaturated water (Nebeker & Brett, 1976; Weitkamp & Katz, 1973).

Supersaturation cannot readily be prevented below long spillways where self-aeration occurs, but it can be alleviated by the intensive turbulence in the stilling basin and by deflecting the flow at entry into the basin for low flows to the free downstream water surface (Johnson, 1984) (for large flows this deflection would be detrimental to the stilling basin operation). Power station intakes with drop shafts are particularly susceptible to supersaturation; therefore, a good hydraulic design should minimise air entrainment. Tunnels at mild slopes, intakes with level controls that keep the shaft filled and the intake submerged (Tekle et al., 1984) and degassing chambers are some of the possible design measures. At least some understanding of air transport under the roof of lined and unlined mild slope rock tunnels is also essential in the design of the power tunnels in case the prevention of air entrainment from a dropshaft intake is uneconomical or impossible (Tekle, 1983).

Normal Pelton turbines provide almost complete degassing of supersaturated water whereas Francis turbines do not provide any effective de-airing; in the latter case the best measure, if feasible, is to use a shallow free-surface tailrace tunnel. At outlets into lakes, reservoirs and fjords, a jet outlet in deep water and an air curtain are measures that can provide sufficient dilution and mixing. The power station can also be operated so as to control supersaturation within acceptable limits. Fish farms can be protected by an air curtain or boom (Tekle et al., 1984).

6.8 SIMILARITY AND FINAL CONSIDERATIONS

Kobus (1985) suggests a functional relationship between air entrainment β, and the Froude and Reynolds numbers that emphasises the importance of the inception limit (plunging jets) and entrainment limit (hydraulic jumps). When modelling the mass transfer process, water quality and surface tension also affect the results. Even, however, if the Reynolds numbers are high enough ($Re > 10^5$) and the models large enough (to overcome the effects of viscosity and surface tension) the problem that the bubble sizes (for given water salinity) are the same in model and prototype remains.

Thus some scale effects are likely to affect the predictions of oxygen uptake that are based on model experiments alone. After all, a simple analysis of the scales (M) in section 6.3 shows (Avery & Novak, 1977) that for a jump from Equations 6.12 to 6.14

$$M_{r-1} = M_l^{9/8} \tag{6.27}$$

where M_l is the length scale; similarly in section 6.4 from Equations 6.22 to 6.24

$$M_{r-1} = M_l^{0.8} \tag{6.28}$$

Equations 6.27 and 6.28 show that to obtain a prototype value of $(r - 1)$ from results of physical model experiments these results have to be scaled up approximately according to the geometrical scale of the model.

Care must evidently be exercised in using the developed predictive equations outside their original experimental limits, which for the case of the hydraulic jump were $2.8 < Fr_1 < 8.5$ and $0.0145 < q < 0.0808$ m^2/s and in the case of free overflow $0.1 < b < 0.3$ m, $0.6 < Q < 5.8$ l/s and $0.25 < h < 2.1$ m. Furthermore it must be emphasised that Equations 6.22 to 6.24 can be used only if $0.05 < h < 6\,q^{1/3}$ (to overcome the limit of inception of air entrainment and not to exceed the limit for total jet disintegration) and for plunge pool depths $d \geq d_{op}$ where d_{op} is given by Equation 6.21, or 6.21a. Nonetheless, the application of Equations 6.12 to 6.14 to available limited field data for the hydraulic jump and Equations 6.22 to 6.24 to a substantial amount of field data in the case of overfalls, resulted in good agreement. Thus, although the experimental limits of the derived predictive equations have been stressed, they probably give acceptable estimates of the likely oxygen deficit ratio even for conditions well outside these limits. In view of the substantial effect of structural details of overfalls in the case of weirs (possible jet interference etc.) and particularly of any pollutant influencing air bubble size, engineering judgement will always be necessary. In many cases of treatment plant outfalls and smaller structures predictions of oxygen uptake based on the discussed equations may be, however, used with reasonable confidence.

REFERENCES

Albrecht, D. 1969. Schätzung der Sauerstoffzufuhr durch Wehre und Kaskaden. *Die Wasserwirtschaft*, 11.

Apted, R.W. & P. Novak 1973. Some studies of oxygen uptake at weirs. *Proceedings, XV Congress, International Association for Hydraulics Research*, paper B23: 177-186.

Avery S. & P. Novak 1975. Oxygen uptake in hydraulic jumps and at overfalls. *Proceedings of the XVI Congress, International Association for Hydraulics Research*, Paper C38: 329-337.

Avery, S.T. & P. Novak 1978. Oxygen transfer at hydraulic structures. *ASCE Journal of the Hydraulics Division*, HY11.

Avery, S.T. 1976. The transfer of oxygen from air entrained by jets entering a free water recipient. Thesis presented at the University of Newcastle upon Tyne, UK, in partial fulfillment of the requirements for the degree of Doctor of Philosophy.

Avery, S. & P. Novak 1977. Modelling of oxygen transfer from air entrained by solid jets entering a free water recipient. *Proceedings, XVII Congress, International Association for Hydraulics Research, Baden-Baden, Germany*, Paper A59.

Burley, G.H. 1978. The effect of jet shape on oxygen transfer at free overfall rectangular weirs and the prototype study of oxygen transfer at hydraulic jumps. MSc Thesis, University of Newcastle upon Tyne.

Danckwerts, P.V. 1951. Significance of liquid film coefficients in gas absorption. *Industrial and Engineering Chemistry*, 43: 1460-1467.

Ervine, D.A., E. McKeogh & E.M. Elsawy 1980. Effect of turbulent intensity on the rate of entrainment by plunging water jets. *Proc. ICE London*, Part 2.

Gameson, A.L.H. Weirs and the aeration of rivers. *Journal of the Institute of Water Engineers*, 11: 477-490.

Gameson, A.L.H., K.G. Vandyke & C.G. Ogden 1958. The effect of temperature on aeration at weirs. *Water Engineering*, 62: 489-492.

Haindl, K. 1984. Aeration at hydraulic structures. In P. Novak (ed.), *Developments in Hydraulic Engineering*, 2. London: Elsevier Applied Science Publishers.

Higbie, R. 1935. The rate of absorption of a pure gas into a still liquid during short periods of exposure. *Transactions of the American Institute of Chemical Engineers*, 31: 365-389.

Holler, A.G. 1971. The mechanism describing oxygen transfer from the atmosphere to discharge through hydraulic structures. *Proceedings, XIV Congress, International Association for Hydraulics Research,* Paper A45: 373-382.

Hoření, P. 1956. Disintegration of a free jet of water in air. *Výzkumný Ustav Vodhospodářský Práce a Studie*, Sešit 93, Praha-Podbaba.

Imhoff, K.R. & D. Albrecht 1972. *Influence of temperature and turbulence on the oxygen transfer in water*. Advances in water pollution research, Jerusalem, Israel.

Jarvis, P.J. 1970. A study in the mechanics of aeration at weirs. Thesis presented to the University of Newcastle upon Tyne, UK, in partial fulfilment of the requirements for the degree of Master of Science.

Johnson, P.L. 1984. Prediction of dissolved gas transfer in spillway and outlet works stilling basin flows. In *Gas transfer at water surfaces*. D. Reidel Publ. Co.

Kalinske, A.A. & J.H. Robertson 1942. Entrainment of air in flowing water – Closed conduit flow. *Proc. ASCE*, 68: 1141-1153.

Kaplovsky, W. & Sosewitz 1964. Artificial aeration of canals in Chicago. *Journal Water Pollution Control Federation*, 36: 463-474.

Kobus, H.E. 1985. *An introduction to air-water flows in hydraulics*. Mitteilungen, Institute fur Wasserbau Universität Stuttgart, Heft 61.

Lewis, W.K. & W.G. Whitman 1924. Principles of gas absorption. *Industrial and Engineering Chemistry*, 16: 1215-1220.

Mancy, K.H. & D.A. Okun 1968. Discussion of paper by Van der Kroon. *Fourth International Conference on Water Pollution Research*, 231-239.

Markofsky, M. & H. Kobus 1978. Unified presentation of weir aeration data. *ASCE Journal of Hydraulics Division.*

Nakasone, H. 1975. Derivation of aeration equation and its verification. *Study on the aeration of falls and spillways, transactions, J.S.I.D.R.E.*, 42-48.

Nebeker, A.V. & J.F. Brett 1976. Effects of gas supersaturated water on survival of pacific salmon and steelhead smolts. *Transactions Am. Fishery Society*, 105(2).

Notes on water pollution 1973. Department of the Environment, No. 61.

Novak, P. 1980. Luftaufnahme und Sauerstoffeintrag an Wehren und Verschlüssen. *Natur und Modellmessungen zum Künstlichen. Sauerstoffeintrag in Flüssen, DVWK*, Heft 49. Paul Parey Verlag Hamburg.

Owens, M. & R.W. Edwards 1963. Some oxygen studies in the River Lark. *Proceedings of the Society of Water Treatment Examinations*, 12: 126-145.

Quirk, Lawler & Matusky 1966. *Study of the assimilation capacity of the St. Louis River from Cloquet, Minnesota to Lake Superior.* Report, Northwest Paper and Wood Conversion Company.

Speece, R.E., J.L. Adams & Wooldridge 1969. U-Tube aeration operating characteristics. *Journal of the Sanitary Engineering Division*, 95(SA3): 563-574, paper 6621.

Smutek, R. 1974. Využití vodního skoku v potrubí pro sycení kapaliny plynem (Use of hydraulic jump in a pipeline for gas enrichment of a fluid). *Vodohospodářský časopis CSAV*, XXII(2).

Tebbutt, T.H.Y., I.T.S. Essery & S.K. Rasaratnam 1977. Reaeration performance of stepped cascades. *Journal of the Institution of Water Engineers and Scientists*, 31(4).

Tekle, T. 1983. Some design problems related to air entrainment and absorption in hydro power systems with secondary dropshaft intakes. *Proceedings, XXth Congress, International Association for Hydraulic Research, Seminar 3, 1983.*

Tekle, T. et al. 1984. Supersaturation of dissolved air in the waterways of hydroelectric power plants – Casual relationships, detrimental effects and preventive measures. Report No. 284075 of Norwegian Hydrodynamic Laboratories, Trondheim.

Van der Kroon, G.T.M. & A.H. Schram 1969. *Weir aeration – Part 1: Single free fall, H_2O(2)*, 22: 528-537.

Weitkamp, D.E. & M. Katz 1973. *Resource and literature review on dissolved gas supersaturation and gas bubble disease.* Seattle Marine Lab., Seattle, Wa.

Wilhelms, S.C. 1975. Reaeration through hydraulic structures. *Symp. on Reaeration Research, Hydraulics Division, ASCE, Gatlinberg, Tenn.*

NOTATION

A	area of air/water interface (L^2)
b	width of rectangular notch, length of weir (L)
C	gas or oxygen concentration in liquid (ML^{-3})
C_1, C_s, C_t	dissolved oxygen concentration prior to aeration, at saturation, at time t (ML^{-3})
d	tailwater depth below weir (L), thickness of jet at pool level (Fig. 5) (L)
d_{op}	optimum tailwater depth for oxygen transfer (L)
d_b	air bubble diameter (L)
D	jet diameter at pool level (L)

D_m	coefficient of molecular diffusion or diffusivity (L^2T^{-1})
ΔE	energy loss across hydraulic jump (L)
$Fr_1 =$	$\frac{v_1}{\sqrt{gy_1}}$ supercritical Froude number (hydraulic jump)
Fr_j	jet Froude number, Eq. (21)
h	difference between water levels above and below a weir (L)
H	difference between notch crest and downstream bed (L)
H'	head on a rectangular notch crest (L)
$k_1 - k_6$	functions of dissolved sodium nitrite concentration, Eq. (12)-(14), (22)-(24)
K_L	liquid film coefficient (LT^{-1})
L_o	distance measured along jet centreline to the point of total disintegration (L)
M	mass of gas transferred (M), scale (prototype/model)
n	number of steps in cascade
N	number of jets formed by weir crest
q	water discharge per unit channel or crest width (L^2T^{-1})
q_o	a reference water discharge per unit channel width (L^2T^{-1})
q_j	jet water discharge per unit jet perimeter at impact (L^2T^{-1})
r, r_{15}	oxygen deficit ratio at T°C, 15°C
R	hydraulic radius (L)
t_c	time of exposure of fluid elements (T)
T	temperature (°C)
v_c	inception limit velocity (LT^{-1})
V	volume of body of liquid (L^3)
y_1	initial (supercritical) hydraulic jump conjugate depth (L)
η	efficiency of oxygen transfer (M/ML^2T^{-2})
θ	temperature coefficient
β	relative air entrainment $\frac{Q_a}{Q}$

Control of water quality in river systems 7

SABURO IKEDA
Institute of Socio-Economic Planning, University of Tsukuba, Tsukuba, Ibaraki, Japan

7.1 INTRODUCTION

Modern industrial, urban (household) and agricultural activities have produced a tremendous amount of wastes and by-products that are eventually discharged into such bodies of water as streams, lakes, and estuaries. Two major types of wastes dissolved in rivers cause the deterioration of water quality:

1. Organic materials in the form of suspended particles and sediments that can cause oxygen deficits and low transparency;
2. Inorganic substances such as nitrogen and phosphorus that can cause outbreaks of algae bloom due to 'water eutrophication'.

The traditional objective in controlling water quality with respect to organic pollutants is to decrease the level of BOD (Biological Oxygen Demand), or to increase the level of DO (Dissolved Oxygen) to some predetermined standard values. The standards for BOD and DO depend on the water quality required for particular water uses. Current practice is to construct sewage treatment facilities so as to reduce organic pollutants before they are discharged into streams. Also aeration by injecting or mixing air into a stream is often used to increase the amount of oxygen dissolved in water (James, 1978; Rinaldi et al., 1979).

The second objective, the control of inorganic pollutants, has emerged, after a certain level of sewage treatment is reached, as a serious issue in most urban areas. The biological treatment of the organic wastes in sewage plants reduces the amount of organic material in the effluent, but it may not reduce the level of inorganic matter. Abundant inorganic materials such as nitrogen oxides (NO_3) and phosphates (PO_4) in effluent water may trigger large-scale algae blooms in lakes, streams and estuaries. Subsequently, the decay of bloomed algae or benthic flora consumes large quantities of dissolved oxygen during the natural process of bacterial degradation.

The lack of economically efficient and suitable techniques for controlling inorganic wastes may add further problems in reaching a balance between the costs and the benefits of improvements in water quality. The management objective of controlling water quality in terms of target or standard levels of pollutants requires the use of 'real-time' monitoring and forecasting of water quality for both organic and inorganic pollutants. The systems control theory can serve as one of the most powerful methodologies in dealing with the measures involved in the planning, designing and operating phases.

This chapter contains some of the new applications of systems control theory to these various water quality problems. The first part is a general view of water quality control models in rivers; it includes inputs from pollution sources and their control by in-stream aeration. The second part deals with the identification of a distributed-source function by means of on-line monitoring data at specific points. A numerical identification algorithm is given with emphasis on the 'non-well posed nature' of the identification scheme. The third part is a theoretical development that utilizes optimal impulse control theory to attack the problem of in-stream aeration. The final part is devoted to the multi-objective optimization problems in river basin management that assist with decisions in water quality management.

7.2 BASIC EQUATIONS OF WATER QUALITY CONTROL

Figure 7.1 is a schematic illustration of pollution sources, pathways, sewage treatment, and transport or reaction processes in a river stream system. Four major pollution sources are encountered in urban areas:

1. Industrial use of water;
2. Municipal use of water;
3. Urban run-off; and
4. Agricultural use of water.

Pollutants from Sources 1 and 2 are collected via a drainage network into sewage plants that may vary in both intensity and magnitude of treatment. Some pollution sources are treated by centralized urban sewage systems, and some that are attached to industrial firms or private houses are individually operated. The outputs from such sewage systems are treated as point sources of pollution in rivers. However, in many Japanese municipalities, a considerable amount of waste water discharged from households is not connected to urban sewage systems. This untreated waste water is one of the major causes of the deterioration of water quality. The run-off of urban precipitation and of irrigation water are other sources of untreated waste water. They contain various pollutants, especially inorganic matter and toxic chemicals or heavy metals. These pollutant sources are typically distributed sources $w(t, x)$ that flow directly into rivers or into the ground water that eventually flows into rivers.

The balance of water quality variables, designated as $S(t, x)$, can be described for a unit reach of a river:

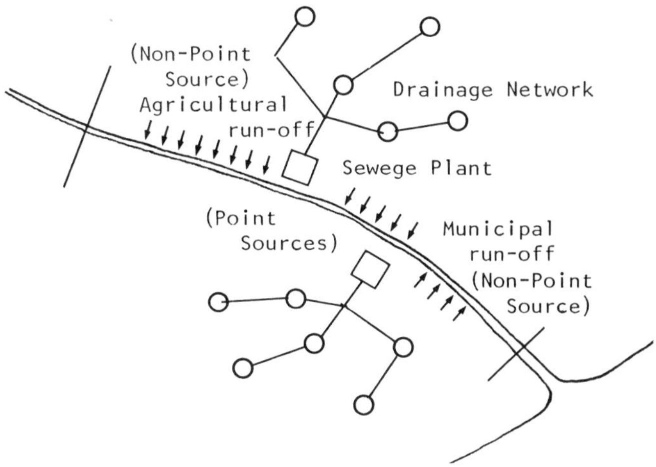

Figure 7.1. Pollution sources for a river

$$\partial S(t,x)/\partial t = D_e \partial^2 S(t,x)/\partial x^2 - v\partial S(t,x)/\partial x + f(t,x,S)$$

(rate of change) = (dispersion) – (advection) + (reaction and decay)

$$+ w(t,x) + u(t,x,S) \tag{7.1}$$

+ (pollutant inputs) + (control)

where $S(t, x)$ = a water quality vector in a river at a location x, t = time, x = downstream distance, D_e = a uniform dispersion coefficient, $f(t, x, S)$ = a reaction function of water quality S at x, $u(t, x, S)$ = a control vector operated in a stream at x, v = an advection coefficient (a mean flow rate), $w(t, x)$ = an input vector of pollutants into a stream at x.

The two transport processes for the material S are dispersion (mixing by turbulent eddies and molecular diffusion) and advection (transport due to the mean river velocity). Deposition of the material and chemical reaction of materials for the decomposition are a third process of reaction function $f(t, x, S)$. The pollutant input $w(t, x)$ represents a series of pollutant inflows along the river stream at various locations x, and the control term $u(t, x, S)$ denotes a remedial measure such as the in-stream aeration $q(t, S)$ at $x = x_d$:

$$u(t,x,S) = q(t,S)\,\delta(x - x_d) \tag{7.2}$$

The well established model for Equation 7.1 is the Streeter-Phelps equation in which BOD and DO are used as state variables of water quality $S(t)$, (Gromiec et al., 1983):

$$\partial S_1/\partial t = D_e \partial^2 S_1/\partial x^2 - v\partial S_1/\partial x - k_1 S_1 + w(t,x) \tag{7.3}$$

$$\partial S_2/\partial t = D_e \partial^2 S_2/\partial x^2 - v\partial S_2/\partial x - k_2 S_1 + k_3(S_s - S_2) + u(t,x) \tag{7.4}$$

where S_1 = Biological Oxygen Demand (BOD), S_2 = Dissolved Oxygen (DO),

k_1 = a decay coefficient of BOD, k_2 = a consumption coefficient of DO, k_3 = a coefficient of surface aeration, and S_s = saturated level of DO,

Several simplified versions of Equation 7.1 are available. One is the assumption of no dispersive transport for cases in which advective flow is dominant over dispersion flow:

$$\partial S(t,x)/\partial t = -v\partial S(t,x)/\partial x + f(t,x,S) + w(t,x) + u(t,x,S) \tag{7.5}$$

Further simplification of this plug-flow model leads to an ordinary differential equation. For example, application of the characteristic transformation method to the convection term of Equation 7.5:

$$\mathrm{d}x/\mathrm{d}\zeta = v, \text{ and } \mathrm{d}t/\mathrm{d}\zeta = 1 \tag{7.6}$$

leads to a set of ordinary differential equations along the characteristic curve $x = v(t-\zeta)$

$$\mathrm{d}S_1(\zeta)/\mathrm{d}\zeta = f_1(v(t-\zeta), S_1) + w(v(t-\zeta), \zeta) \tag{7.7}$$

$$\mathrm{d}S_2(\zeta)/\mathrm{d}\zeta = f_2(v(t-\zeta), S_2) + u(v(t-\zeta), \zeta) \tag{7.8}$$

In the following discussion, this set of BOD-DO equations is used as an example for the mathematical description of control problems.

Another simplification of Equations 7.3 and 7.4 is a stationary version in which the term for time variation $\partial S/\partial t$ is zero.

$$D_e \mathrm{d}S_1^2/\mathrm{d}x^2 - v\mathrm{d}S_1(x)/\mathrm{d}x + f_1(x,S) + w(x) = 0 \tag{7.9}$$

$$D_e \mathrm{d}S_2^2/\mathrm{d}x^2 - v\mathrm{d}S_2(x)/\mathrm{d}x + f_2(x,S) + u(x) = 0 \tag{7.10}$$

These equations are usually used for the allocation of a cascade of sewage plants $w(x)$ for regulating total output or for designing the distribution of aeration devices $u(x)$ along the river.

7.3 OPTIMAL CONTROL SCHEME IN WATER QUALITY

A set of ordinary differential Equations 7.7 and 7.8 is a standard way of representing water quality within a unit reach of a river, and its use makes possible the application of modern control theory to water quality management. Hence, the system equation of water quality vector S is written as

$$\mathrm{d}S/\mathrm{d}t = F(t)^t S + W(t) + U(t) \tag{7.11}$$

where $S = (S_1, S_2)$, $F = (f_1, f_2)$, $W = (w(t), 0)$, $U = (0, u(t))$, and superscript t denotes a transpose of the vector involved.

The possible control variables include two major inputs, W and U, that can affect the water quality of the river system. One is the pollutant input W from a sewage plant, and the other is the direct aeration input U in the river stream. The aeration input U is taken as a control variable, and the pollutant input W is designated as a given function. In addition, for simplicity, t is again taken as an auxiliary variable of the characteristic curve.

The control problem is to determine an optimal control $U^o(t)$ that minimizes the following performance functional J over the time period (0 – T):

$$J = \text{minimize } \{\int_0^T G(S(t), U(t))dt\} \tag{7.12}$$

subject to Equation 7.11.

The linear quadratic function G, commonly used in process control, represents the squared deviation of the state variables $S(t)$ from the target values of $S^*(t)$, and the control cost associated with the squared magnitude of the in-stream aeration $U(t)$:

$$G = (1/2)(S - S^*)^t Q_1(S - S^*) + (1/2)U^t Q_2 U \tag{7.13}$$

where Q_1 and Q_2 are constant non-negative matrices, and superscript t denotes a transpose of the matrices involved. The optimal control $U^o(t)$ can be theoretically derived by the variation principle of the Hamiltonian-Jacobi function (Appendix A1).

The linear control theory (the regulator problem) indicates that, if the system equation (Equation 7.11) is linear in both variables S and U (as is the case for BOD-DO equation), and if the performance function G is quadratic, then one can construct a feedback control U^o with the specific gain matrix R (Riccati Coefficient Matrix):

$$U^o = -Q_2^{-1} R(t)S(t) \tag{7.14}$$

where the gain matrix $R(t)$ is the solution of the following Riccati equation, and superscript –1 denotes the inversed form of the indicated matrices:

$$-dR/dt = F(t)^t R + R^t F(t) - R\,Q_2^{-1} R + Q_1 \tag{7.15}$$

$$R(T) = 0 \tag{7.16}$$

Figure 7.2 illustrates the simple feedback control system in a river for in-stream aeration. However, a number of computational and operational difficulties arise in solving these variational equations of the control function. The case of optimal impulse aeration at a specific control point is discussed in section 7.5.

The following important theoretical problems in applying modern control theory are discussed in the subsequent sections:

1. How to identify the position and temporal change of pollution input, $w(t, x)$, from the observed data of the water quality;

2. How to construct an optimal control scheme of impulse type (i.e. on-off control) feedback control; and

3. How to help the decision makers (policy makers) in river basin management to make trade-offs concerning the 'multiple objectives' problem of water quality control (water quality and control cost).

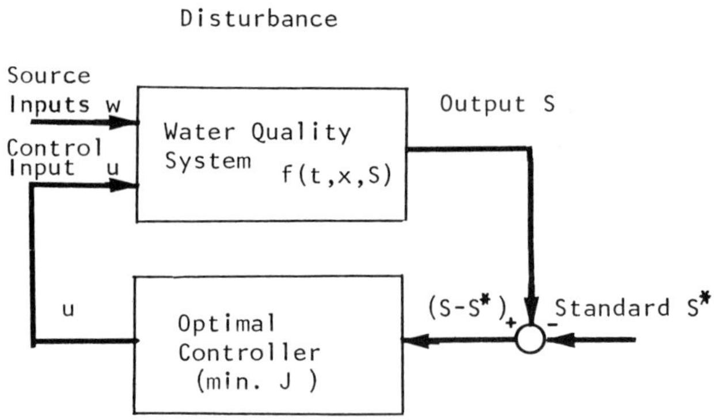

Figure 7.2. Optimal feedback control scheme for water quality system

7.4 IDENTIFICATION OF POLLUTION SOURCE

An important problem is to identify the unknown pollution source $w(t, x)$ from the observed data with respect to the water quality at a given point in a river. In the simplified Streeter-Phelps equations, Equations 7.7 and 7.8, the unknown $w(t, x)$ is expressed in the following form:

$$w(t, x) = f(t)z(x) \tag{7.17}$$

where $f(t)$ is a given function of time t, and $z(x)$ is a unknown pollutant function at location x along the river reach between a and b. The underlying assumption is that the known function $f(t)$ represents a variation with time of the pollutant (BOD) discharge (such as daily variation of discharge) that occurs uniformly at any point $x \in [a, b]$. The function $z(x)$ involves some unknown source information not only on the location of pollutant sources but also on the intensity of discharging pollutants into the river. The points of observation are at the upper and lower ends of the unit reach ($x = a$ and $x = b$ as shown in Figure 7.3).

7.4.1 *Case of no dispersion*

In the case of no dispersion ($D_e = 0$), the new variables $x' = x/v$ and $S_2' = (S_s - S_2)/v$ are introduced for computational simplicity, and rewritten by x and S_2 again, instead of x' and S_2'. Then the Streeter-Phelps equations for the plug-flow case are reduced to:

$$dS_1/d\zeta = -k_1 S_1 + f(\zeta)z(x) \tag{7.18}$$

$$dS_2/d\zeta = k_2 S_1 - k_3 S_2 \tag{7.19}$$

along the characteristic curve:

$$x = t - \zeta \tag{7.20}$$

where ζ means an auxiliary variable.

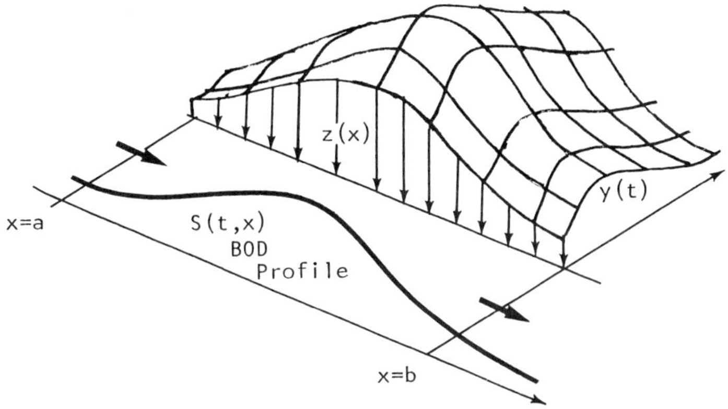

Figure 7.3. A schematic profile of pollution source $w(t, x)$ in river

The measurement of S_1 (BOD) is usually carried out by laboratory monitoring devices rather than by on-site field devices. S_2 (DO) can be measured continuously and automatically at a monitoring station on the shore of a river. Since the costs of measurement of the two water quality (DO and BOD) differ both in operational costs and monitoring efficiency, it would be beneficial to establish both identification schemes from two different data sources of water quality (DO and BOD). Two identification schemes (A or B) of unknown source distribution $z(x)$ are subsequently considered, separately from observations of either S_1 (BOD) or S_2 (DO) (Ikeda et al., 1974).

A. *Identification of z(x) from BOD data*:
Since the BOD Equation 7.18 has no other external variable than the BOD, the solution $S_1(\zeta + b, b)$ at $x = b$ can be directly obtained by integrating Equation 7.18 from $t = \zeta + a$ to $t = \zeta + b$ along the characteristic curve, which is linear in this case:

$$S_1(\zeta + b, b) \exp(k_1 b) = S_1(\zeta + a, a) \exp(k_1 a) + \int_{\zeta+a}^{\zeta+b} z(t - \zeta) f(t) \exp(k_1 t)\, \mathrm{d}t \qquad (7.21)$$

This equation can be rearranged in the form of integral equation (called the Fredholm equation of the first kind), using the observed data S_1,

$$\tilde{S}_1(\zeta) = \int_a^b K(\zeta, s) z(s) \mathrm{d}s \equiv T \cdot z(s) \qquad (7.22)$$

where

$$\tilde{S}_1(\zeta) = S_1(\zeta + b, b) \exp(k_1 b) - S_1(\zeta + a, a) \exp(k_1 a) \qquad (7.23)$$

and T is an integral operator for which the kernel K is defined as

$$K(\zeta, s) = f(\zeta + s) \exp(k_1 s) \tag{7.24}$$

The problem of identifying $z(s)$ is reduced to solve the integral Equation 7.22. Theoretically, this Equation 7.22 can be solved by obtaining the 'inverse operator' T^{-1},

$$z(s) = T^{-1}\tilde{S}_1(\zeta)$$

However, this integral equation has the specific feature of being 'non-well posed' in terms of stability associated with determination of an inverse operator T^{-1} (refer to 7.4.3).

B. *Identification of $z(x)$ from DO data*:
The DO Equation 7.19 contains two state variables, S_1 and S_2. The coupling of the BOD and DO seems to make this identification problem more complex than it was for case A. However, the linearity of the coupling equations enables one to get the same type of integral equation as in case A. For computational simplicity, the upper point is designated as $a = 0$, that is, $0 \le x \le b$.

Integration of the DO Equation 7.19, which is accomplished in the same manner as for the former case, results in:

$$S_2(\zeta + b, b) \exp(k_2 b) - S_2(\zeta, 0) = k_3 \int_o^b \exp(k_2 s)\, S_1(\zeta + s, s)\mathrm{d}s \tag{7.25}$$

The value of S_1 in the integrand follows from Equation 7.21 as

$$S_1(\zeta + s, s) = S_1(\zeta, 0) \exp(-k_1 s) + \int_o^s \exp\{k_1(p - s)\} f(\zeta + p)z(p)\mathrm{d}p \tag{7.26}$$

After some mathematical manipulation, the following integral equation, also a Fredholm equation of the first kind, is obtained from Equation 7.25:

$$\tilde{S}_2(\zeta) = \int_o^b K(\zeta, p)z(p)\mathrm{d}p \equiv T \cdot z(p) \tag{7.27}$$

where $\tilde{S}_2(\zeta)$ is a time function that is known from the observed values of DO and BOD at $x = 0$ and $x = b$:

$$\tilde{S}_2(\zeta) = (1/k_3)\,[S_2(\zeta + b, b) \exp(k_2 b) - S_2(\zeta, 0) - S_1(\zeta) \int_o^b \exp\{(k_2 - k_1)p\}\, \mathrm{d}p] \tag{7.28}$$

and the integral kernel $K(\zeta, p)$ has a more complicated form than that for the BOD case:

$$K(\zeta, p) = \{1/(k_2 - k_1)\}\,[\exp\{(k_1 - k_2)b\} - \exp\{(k_1 - k_2)p\}] \cdot \exp(k_1 p) f(\zeta + p) \tag{7.29}$$

7.4.2 *Case of dispersive flow*

The solution of the dispersion Equation 7.3 for the simple BOD model can also be obtained in the form of an integral equation of the first kind. For a case with no forcing term $w(t)$ in Equation 7.3

$$\partial S_1/\partial t = D_e \partial^2 S_1/\partial x^2 - v\partial S_1/\partial x - k_1 S_1 \tag{7.30}$$

with the following initial and boundary conditions, $S_1(0, x) = 0$, $t = 0$, $x \in [0, b]$; $S_1(t, 0) = 0$, $t \in [0, T]$, $x = 0$.

For Equation (7.30) a fundamental solution U exists, which represents a kind of density distribution function over a domain $s \in [0, b]$ associated with a dispersive phenomenon from a point s to the point x (Ito, 1966):

$$U(t, x, s) = \{1/(2\sqrt{D_e t})\} \exp\left\{-\frac{(x-s)^2}{4D_e t} + \frac{v}{2D_e}x - \left(\frac{v^2}{4D_e} + k_1\right)t\right\} \tag{7.31}$$

From the fundamental solution (7.31), the solution for the equation with pollution source $w(t, x) = f(t)z(x)$ of (7.3), is again the same type of integral equation of the first kind,

$$\tilde{S}_1(t) = \int_o^b K(t, s)z(s)ds, \quad (0 \le t \le T) \tag{7.32}$$

where $\tilde{S}_1(t)$ is the known function defined by the data observed at $x = 0$ and $x = b$ and the kernel K is assigned in the form

$$K(t, s) = \int_o^t U(t - \zeta, b, s)f(\zeta)d\zeta \tag{7.33}$$

In this case, the kernel K has the strong smoothing property of an exponential decay function, which contrasts with the high frequency noise involved in the observed data $\tilde{S}_1$ on the left hand side of (7.32). The numerical examples illustrate this 'non-well-posed' nature of the integral equation for this identification scheme.

7.4.3 *Identification algorithm by regularization method*

The so-called 'inverse problem' of finding $z(s)$ on the basis of a given function of the integral Equations in 7.22, 7.27 and 7.32 is known to be 'non-well posed' in the sense of Hadamard. The concept of 'well-posed' means that:

1. The solution $z(s)$ exists;
2. It is unique; and
3. It depends continuously on the observed data $S(t)$'s.

If the above conditions are not met, the problem is called 'non-well posed' or 'improperly posed' (Appendix 2). A number of attempts have been made to relax the difficulty with an 'improperly posed problem'. These include methods of 'generalized inverses' (Rao and Mittra, 1971), 'penalty function' (Phillips, 1962) and others. The approach used herein is Tihonov's regularization (1963, 1977).

The Fredholm integral equation is transformed into an optimization of the following quadratic cost functional M^{α} to be minimized with respect to z:

$$M^{\alpha}(z) = \int_o^t \{ \int_o^b K(\zeta, s)z(s)ds - \tilde{S}_1(\zeta)\}^2 d\zeta$$
$$+ \alpha \cdot \int_o^b \left\{ z^2(s) + \left(\frac{dz(s)}{ds} \right)^2 \right\} ds \qquad (7.34)$$

The first term on the right hand side of the above equation represents the square of deviations of the true solution from the measured value S_1; and the second term, called the regularization term with a small positive parameter α, prevents the solution $z^{\alpha}(s)$ from diverging beyond some admissible limit, which is subject to the parameter α.

The solution $z^{\alpha}(s)$ is obtained by solving the following Euler equation as a necessary condition for minimizing the functional $M^{\alpha}(z)$ (7.34):

$$\int_o^T \int_o^b K(p, s)K(p, \zeta)z(\zeta)dpd\zeta + \alpha z(s) = \int_o^T K(\zeta, s)\tilde{S}_1(\zeta)d\zeta \qquad (7.35)$$

The finite-difference scheme of the above integral equation becomes:

$$(K^t K + \alpha I)Z = K^t \tilde{S}_1 \qquad (7.36)$$

where discrete matrices (kernel K, source Z, and observed data $\tilde{S}_1$) are defined in the interval $0 \le s \le b$ and $0 \le t \le T$ as follows:

$$K = \{K^{ij}\} = K[(i-1)\Delta t, (j-1)\Delta s]$$

$$Z = \{z_j\} = z[(j-1)/2 \cdot \Delta s]$$

$$\tilde{S} = \{S_i\} = S[(i-1)/2 \cdot \Delta t]$$

$\Delta s = (b-a)/n$, $\Delta t = T/m$, $1 \le i \le m$ and $1 \le j \le n$

I : unit matrix.

The accuracy of the solution $z(s)$ does, of course, depend on the magnitude of the regularization factor α. The larger the parameter, the more stabilized and smoothed is the solution $z^{\alpha}(s)$.

7.4.4 *Example of source identification*

The validity of the scheme (Equation 7.36) in identifying the pollution source $z(s)$ was tested by means of the experiments undertaken in the model river shown in Figure 7.4 (Sawaragi & Ikeda, 1975, Matsuoka et al., 1977). A chlorine solution was injected at the several points as a pollutant source in the following impulsive way:

$$f(t) = \begin{cases} 1.0, & 0 \le t \le \Delta t; \\ 0.0, & \Delta t \le t \le T \end{cases}$$

$$z(s) = \begin{cases} k, & s_i \le s \le s_{(i+1)}; \\ 0, & \text{otherwise} \end{cases}$$

Figure 7.5 displays the observed data $\tilde{S}_1(t)$ at the measurement point D. Chemical reactions along the river system are ignored, i.e. $k_1 = 0$. The estimated dispersion coefficient is $D_e = 54.4 \text{ cm}^2/\text{sec}$. Numerical results for $z(s)$ from the observation data $\tilde{S}_1(t)$ are illustrated in the Figure 7.6 for values of the regularization parameter α from 10^{-19} to 3×10^{-3}. In cases (a), (b) and (c), the model river had

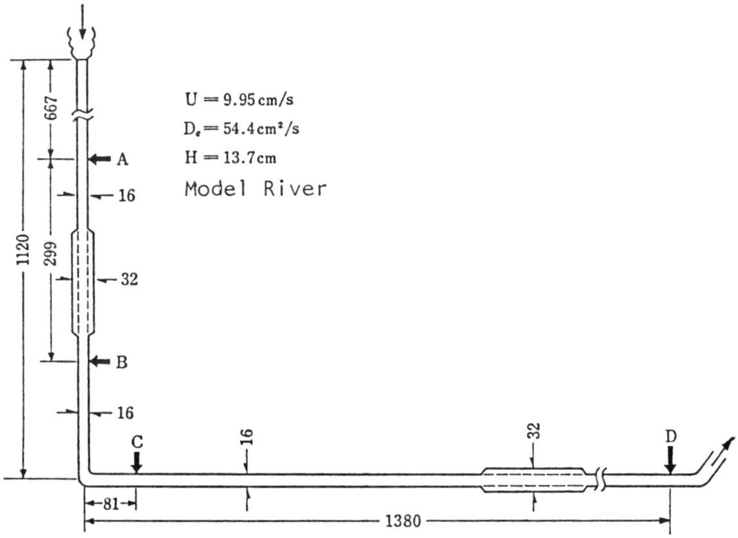

Figure 7.4. Illustration of the model river for source identification and monitoring

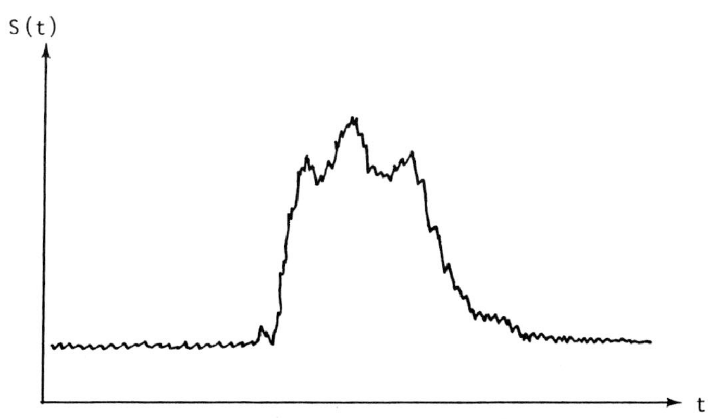

Figure 7.5. Monitoring data observed at point D in the model river

two pollution sources A and B. Case (d) had three sources A, B and C. In all cases the smaller the regularization parameter α was, the more accurate were the estimates of the peak points. However, negative values were obtained for some pollution sources. In contrast, the larger values of the parameter α resulted in positive and smoothed peaks for the source *z*(*s*), but the peak values were lower than those of the former case.

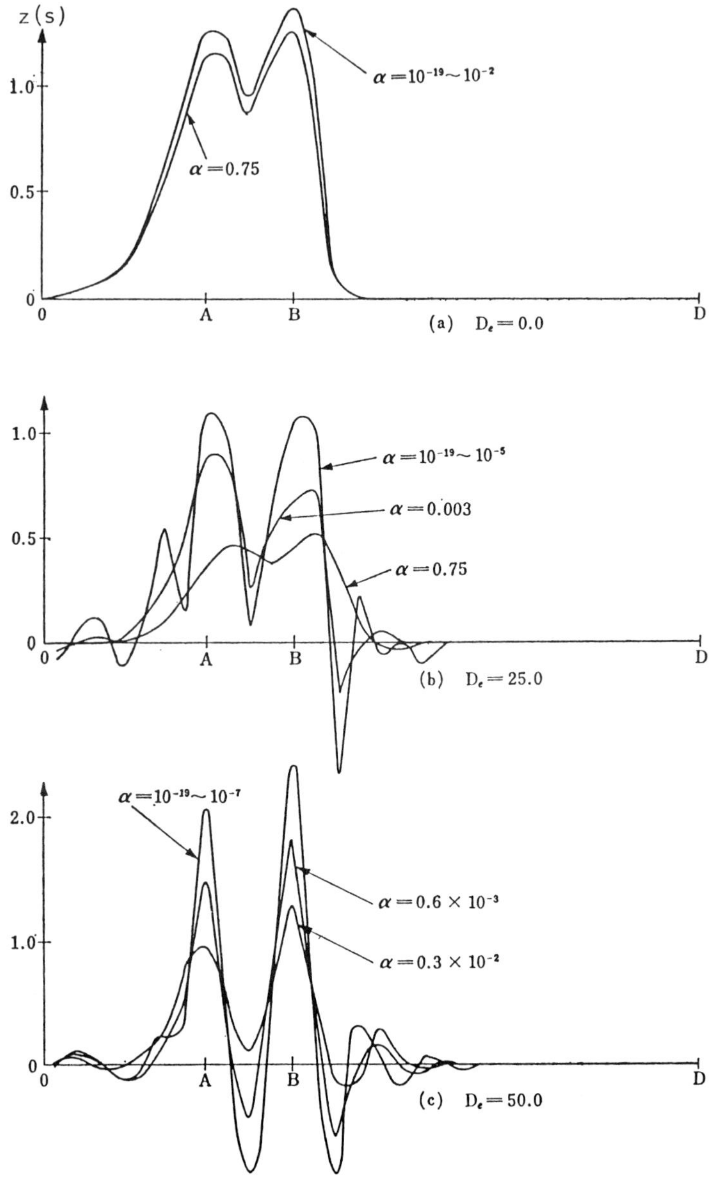

Figure 7.6. Numerical results of identifying pollution inputs

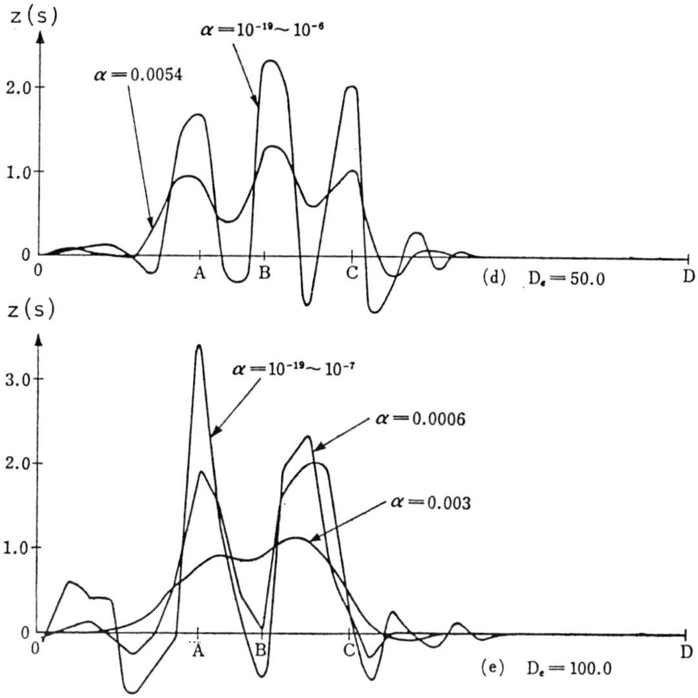

Figure 7.6 (continued).

7.5 OPTIMAL IMPULSE CONTROL OF RIVER STREAM

This section describes an optimal aerator system. The simplified Streeter-Phelps Equation 7.15 is slightly modified assuming the constant flow rate v:

$$\mathrm{d}S/\mathrm{d}t = F(t)S + w(t) + u(t) \tag{7.37}$$

where t stands for $t = x/v$, $w(t)$ is a known pollution source function, and the control function $u(t)$ consists of a series of aerators that has the aeration capacity q^i of on-off control type working on the sequential time $t = t^i$, where t^i is a time when air is impulsively injected into the stream:

$$u(t) = -\sum q^i \delta(t - t^i),\ (1 \leq i \leq N),$$
$$0 \leq t^1 < t^2 \ldots. < t^{N-1} < t^N \leq T \tag{7.38}$$

In other words, a series of aerators with maximum power q^i are located at $x^i = vt^i$ along the river stream. (The constant flow rate v provides for the interchange of the time t $(0 < t < T)$ with river distance x $(0 < x < x_T)$ from $x = vt$).

The performance function J for this optimal aerator problem is a combination of the system's final state $h(T, S(T))$ at the point farthest down stream ($x^T = vt^T$), with the cost of impulse aeration operations c and the integrated value over the interval $[0, T]$ of the transient state function g of the water quality:

$$J = h(T, S(T)) + \sum_{i=1}^{N} c\,(t^i; S((t^i; 0, S_0), q^i) + \int_o^T g(t; S(t; 0, S_0)))\, dt \qquad (7.39)$$

where $S(t; 0, S_0)$ is a solution of Equation 7.37, starting from the initial state $(0, S_0)$, and $g(t, S)$ is a function of water quality $S(t)$. In addition, the functions h, c and g are assumed to be smooth enough so as to guarantee the existence of an optimal solution (i.e., functions that are twice continuously differentiable and that increase monotonously). One of the simplest forms of $g(t, S)$ would be the square of the deviation from the fixed standard values of BOD and DO.

The optimal control problem is to determine timing of aerator to inject air into stream so as to minimize the performance function J (Appendix 3).

Simple example: The above procedures can be illustrated schematically by an example. The DO Equation 7.8, impulse control function $u(t)$, and the performance function J are defined simply as follows (Shima 1978):

$$\mathrm{d}S/\mathrm{d}t = -S + 1 + u, \quad (k_1 = 1, w(t) = 1) \qquad (7.40)$$

$$u = -\Sigma q^{\mathrm{i}} \delta(t - t^{\mathrm{i}}), \quad (q^{\mathrm{i}} > 0, i = 1,..., N) \qquad (7.41)$$

$$J = (1/2)\,\{S(T)\}^2 + cN + \int_o^T (1/2)\,\{S(t)\}^2 \mathrm{d}t \qquad (7.42)$$

where c is unit operation cost of the impulse aeration q^{i} $(1 = 1, 2,..., N)$. From the necessary condition for optimality, an optimal impulse control q must satisfy the following relation:

$$q = S + [1 - \exp\{2(T-t)\}]/[1 + \exp\{2(T-t)\}] > 0 \qquad (7.43)$$

Figure 7.7 illustrates some examples of the paths of various solutions in the domain (t, S) for a simple case $T = 1$ and $c = 0.5$. For example, any solution $S(t; (t,S))$ with the optimal impulse aeration (Equation 7.43) can reach the point $(t, S - q)$ on the curve G^* by using the impulse aeration q:

$$(t, S - q) = (t, S - [1 - \exp\{2(T-t)\}]/[1 + \exp\{2(T-t)\}]) \qquad (7.44)$$

where G^* represents a specific path for the solution that satisfies:

$$S(t) = (1 - e^{2(T-t)})/\,(1 + e^{2(T-t)}) - q$$

$$S(T) = 0$$

That is, the level of the DO reaches zero at the point farthest downstream. The solutions starting from various initial states on the vertical line of $t = 0$ are shown in the figure. The curve AP_1B is the path of the solution $S(t)$ with no control, and it is tangent to the curve G at the point P_1, where the curve G is defined in Appendix. Apparently, all paths that start at the initial states below the point A do not need

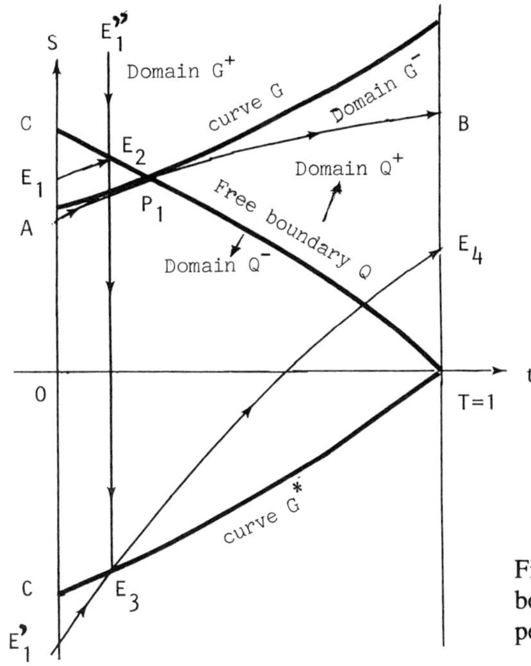

Figure 7.7. Optimal trajectories and free boundary for impulse aeration (refer Appendix 3)

impulse control, as shown by the curve $E_1' E_3 E_4$, because the solutions belong to the domain G^-. This means that the initial water quality $S(0)$ is clean enough to absorb the continuous pollution, $w(t) = 1$, along the whole stream [0,T] without aeration.

The solutions starting from the initial points between A and C, for example from E_1, will intersect the 'free boundary' Q at the point E_2, and then they need impulse control q until arriving at the curve G*. The optimal path of this solution becomes $E_1 E_2 E_3 E_4$. Other paths for the solutions starting over the points C or the states belong to the domain $G^+ \cup Q^+$ (Union (overlapping region) of G^+ and Q^+) need to take the impulse control up to reaching the curve G*, from where no impulse control is needed. The optimal path is shown as $E_1'' E_2 E_3 E_4$.

7.6 WATER QUALITY CONTROL PROBLEMS IN RIVER BASIN MANAGEMENT

The management of the water quality in a river basin involves a variety of theoretical frameworks in order to deal with conflicting societal goals of water utilization and preservation. One of the approaches recently developed is the 'trade-off' analysis for weighing the economic costs of abating pollutants or controlling water quality versus the economic benefits of having cleaner water for both production and aesthetic purposes.

In terms of control theory, this approach is called 'multi-objective optimization' in a decision support framework. A multi-objective optimization can be described in the following way:

$$\max J(s) = \max\,[J_1(s), J_2(s), \ldots, J_p(s)] \tag{7.45}$$

subject to,

$$S = \{s: f_j(s) < 0, j = 1, 2, \ldots, m\} \tag{7.46}$$

where J_i is the i-th competing objective (or attribute), ($i = 1, 2, \ldots, p$), and S is the feasible set of the solution subject to equality or inequality constraints with respect to control variables (s) concerning water quantity and quality (Haimes et al., 1975).

The set of optimal solutions satisfying Equations 7.45 and 7.46 is called 'Parato Optimal' (or non-inferior), and it defines a hypersurface on which none of the objectives improves without lowering at least one of others. The final optimal solution must be singled out of this 'Parato' set on the basis of the decision maker's preference for some combination of various objectives. In order to select the final solution, therefore, one needs to obtain a preference structure from the decision maker (Figure 7.8).

In practical cases, however, information on the decision maker's preference is not necessarily given explicitly. Possibly, the needed function can be extracted from the partial information about the preference attitude of the decision maker by determining his reaction to each alternative solution near a 'Parato' optimal solution. This approach is called the 'interactive method' in multi-objective optimization (Geoffrion et al., 1972; Wierzbicki, 1980; Nakayama et al., 1980).

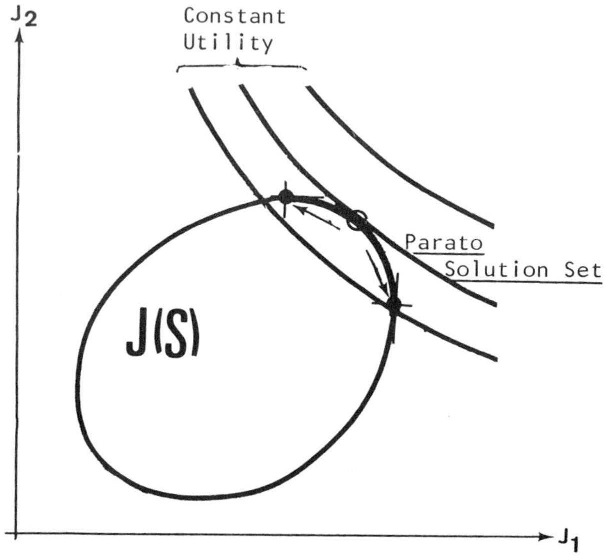

Figure 7.8. 'Parato-Solution Set' and the preference structure as the utility function of the decision maker

Example: The water-quality control problem in river basin management (Sawaragi et al., 1978) serves to illustrate this 'interactive method'. An example is the Yodo River basin located in the western center of the Japanese main island. It has three major tributaries (Katsura River, Uji River and Kizu River) as upper streams, and one main stream (Yodo River, that flows into Osaka Bay, Figure 7.9). The inter-regional conflict arises between the 'upper region' or the two major tributaries (Katsura River and Uji River) and the 'lower region', or the main river. For simplicity, each river is assigned a constant flow rate and a constant amount of intake water, and a constant BOD load is generated in each river basin (Figure 7.9). In addition, this river basin is considered to have no explicit natural decaying process because of the short time required for water to run through the reach in question.

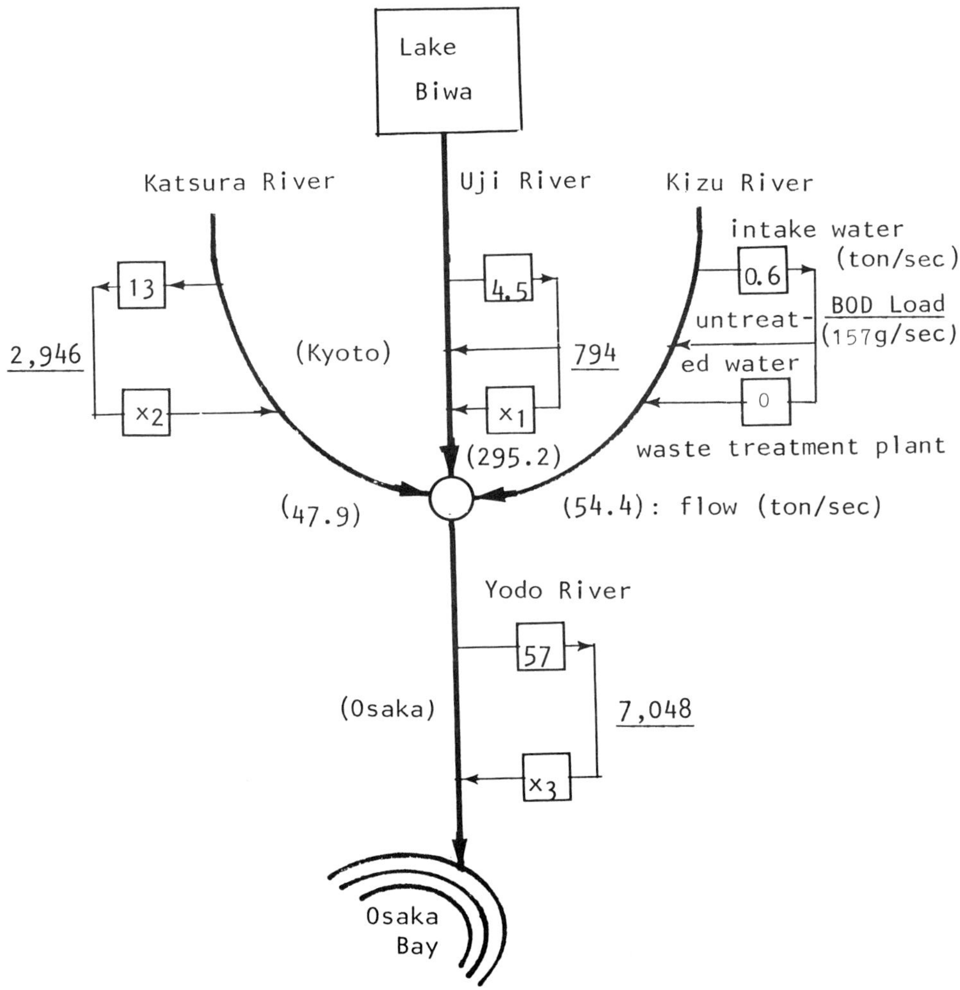

Figure 7.9. Schematic illustration of water quality system in Yodo River basin, Japan

The problem is thus reduced to how to allocate the BOD input to be discharged to each river region so as to optimize the following objectives: water quality at the outlet of the Yodo river, and operational costs of waste treatment plants in the three tributaries. The control variables s_1, s_2 and s_3 in the optimization problem are defined as:

s_1: treatment ratio of the waste water (BOD) from the treatment plant in the Katsura River of the upper region;

s_2: same in the Uji River of the upper region;

s_3: same in the Yodo River of the lower region.

Subject to a physical constraint:

$$0.45 < s_1, s_2, s_3 < 1.0 \tag{7.47}$$

The following three objective functions are defined as components of the multi-objective function $U(J_1, J_2, J_3)$:

1. Treatment cost at the upper region;

$$J_1\,(s_1, s_2) = 287.6 + 2296\,(s_1 - 0.45)^2 + 404.5\,(s_2 - 0.45)^2 \tag{7.48}$$

2. Treatment cost in the lower region;

$$J_2\,(s_3) = 1050 + 10035\,(s_3 - 0.45)^2 \tag{7.49}$$

3. BOD concentration at the outlet of the Yodo River;

$$J_3\,(s_1, s_2, s_3) = 36.0 - (8.05\,s_1 + 1.04\,s_2 + 24.0\,s_3) \tag{7.50}$$

The 'interactive procedure' used in this example is based on relaxation of the constraint, that is, on transforming the minimization of the utility $U(J_1, J_2, J_3)$ on the 'Parato' surface into that of the $U(J_1, J_2, J_3(J_1, J_2))$.

Table 7.1 displays the whole process of the IRM (Iterative Relaxation Method) for obtaining the 'trade-off' solution of the water quality control problem (Nakayama et al., 1980). This decision maker has put the primary weight on water quality (BOD) with the stress on a relatively high ratio of water treatment (s_1 = 0.947, and s_3 = 0.973) in Katsura and Yodo rivers, and a lesser ratio of BOD treatment in the Uji river (s_2 = 0.814).

7.7 CONCLUSION

In this chapter, the following applications of the modern control theory to the water quality problems are illustrated:

1. Identification of a pollution source function from the point monitoring data;
2. Optimal impulse control of the in-stream aeration; and
3. Multi-objective optimization of water quality management in a river basin system.

Table 7.1. Decision making process by IRM

J_2	J_3	J_1	S_1	S_2	S_3	T_2	T_3	M_2-T_2	M_3-T_3
3000	7.0	695.84	0.853	0.746	0.891	0.624	230.10		+
3000	5.0	infeasible							
3000	6.0	959.26	0.967	0.829	0.891	0.799	294.82	–	0
3500	6.0	637.11	0.823	0.723	0.944	0.514	212.50	+	
3200	6.0	814.77	0.908	0.786	0.913	0.674	260.99	–	
3300	6.0	749.94	0.879	0.764	0.923	0.617	244.45	0	+
330	5.0	1025.69	0.992	0.847	0.923	0.781	309.17	–	0
3500	5.0	881.05	0.936	0.806	0.944	0.671	277.22	–	
3700	5.0	760.87	0.884	0.768	0.963	0.577	247.71	0	+
3700	4.0	1042.32	0.998	0.852	0.963	0.728	312.43		–
3700	4.5	893.42	0.941	0.810	0.963	0.653	280.07		+
3700	4.2	980.25	0.975	0.835	0.963	0.694	298.23	–	0
3800	4.2	907.70	0.947	0.814	0.973	0.647	283.31	0	0

The theoretical frameworks for these problems differ in scale and complexity. In recent years, a variety of the optimal control methods and algorithms have become available. Several analytical tools from systems theory provide a basic method of attack on a fairly large variety of water quality control problems (Beck, 1980). However, lack of effective engineering means to abate and monitor the discharges of pollutants, in the sense of real-time regulation and operation, has limited the practical applicability of modern control theory to a rather narrow area such as optimal operations at sewage treatment plants. The need for instruments that control and monitor a variety of variables and conditions in space and time present further difficulities.

Recently, concern has grown over issues that are the ecological (eutrophication of river water), public health (carcinogenic or toxic risks of drinking water) and social (social costs of siting and constructing sewage plants) ones. This concern will lead to further development of an integrated framework of control theory in both socio-economic and ecological contexts (Ikeda, 1987).

APPENDIX 1

From the variation principle (Bryson & Ho, 1975), one obtains the Hamiltonian H as

$$\mathrm{H}(S, U, \mu) = G(S, U) + \mu\,\{\mathrm{F}(t)S + W + U\} \qquad (7.\mathrm{A}1)$$

where $\mu(t)$ is an adjoint variable that satisfies

$$\partial\mu/\partial t = \partial H(S, U, \mu)/\partial S = Q^t(S - S^*) + F(t)^t\mu \qquad (7.\mathrm{A}2)$$

The necessary condition for the minimum of the functional (Equation 7.14) is that the first derivative of H with respect to U is equal to zero, as in the case of a Lagrangian optimization scheme. The sufficient condition for the optimality is the convexity of the Hamiltonian H with respect to S and U that guarantees the existence of a minimum for Hamiltonian H:

$$H_u = \partial H / \partial U = Q_2 U + \mu = 0 \tag{7.A3}$$

$$H_{ss} = \partial^2 H / \partial S^2 > 0, H_{uu} = \partial^2 H / \partial U^2 > 0 \tag{7.A4}$$

One can derive the optimal control $U^o(t)$ as a function of the adjoint variable $\mu(t)$ and the state variable $S(t)$ by solving (7.A2) and (7.A3)

$$U^o(t) = - Q_2^{-1} \mu(t) \tag{7.A5}$$

where the superscript (– 1) denotes the inverse of the matrix.

APPENDIX 2

Even if conditions (1) and (2) are met in a physical sense, condition (3) of 'continuous dependence' on the data does not necessarily hold. Because of the noise involved in the observation data, $\tilde{S}(t)$ on the left hand side of the integral equation could not match the smoothing effect of the kernel K in the right hand side of the integral equations. This fact leads to an important problem in that a small change in the data $\tilde{S}(t)$ may trigger a drastic change in the solution $z(s)$. Small errors in the observed data $\tilde{S}(t)$, which always occur in practice, may produce tremendous estimation errors in the solution $z(s)$, even in an approximate sense (Lavrentiev, 1967).

APPENDIX 3

According to the theory of 'Quasi Variational Inequality' by Bensoussan and Lions (1975), this impulsive control problem is reduced to finding the auxiliary function $V(t, x)$ that satisfies the following 'Quasi Variational Inequalities (Q.V.I.)' (that are compatible with the Hamiltonian) by Hamilton-Jacobi Variational Principle (for the continuous control problem) (Bensoussan et al., 1976).

$$\partial V(t, S)/\partial t + F(t) S \partial V(t, S)/\partial S + g(t, S) \geq 0 \tag{7.A6}$$

$$V(t, S) \leq \min\, [c(t, S, q) + V(t, S - q)] \tag{7.A7}$$

$$[\partial V(t, S)/\partial t + F(t) S \partial V(t, S)/\partial S + g(t, S)] \cdot$$
$$[V(t, S) - \min\, \{c(t, S, q) + g(t, S - q)\}] = 0 \tag{7.A8}$$

$$V(T, S) = h(T, S) \tag{7.A9}$$

The first and second inequalities are the necessary conditions of the optimality for minimizing J, subject to the water quality (7.A7). The third one is a sufficient condition for optimality, and the last one is the terminal condition.

One can construct an optimal impulse control u based on the following function $V^*(t, S)$ so that the minimum value of the right hand of (7.A7) with respect to the impulsive control q is obtained:

$$V^*(t, S) = \min [c(t, S, q) + V(t, s - q)], 0 < t < T \tag{7.A10}$$

The surface of this function V^* becomes a switching curve on which the impulse control u with the aeration power q^i must operate.

For the simplest case of one air injection, initial state $(0, S_0)$ and no control, the solution $S(t; 0, S_0)$ first proceeds across the surface V^*, and then receives aeration $q = q^1$ at $t = t^1$ toward the final state $S(T)$ that satisfies the terminal condition (7.A9). For the case of a linear system with a quadratic performance function J, the switching curve V^* can be obtained by means of the iterative analytical procedure (Koivo & Phillips, 1975; Shima, 1978).

Let $M(t, S)$ be the performance function J without aeration q for the interval (t, T):

$$M(t, S) = h(T, S(T)) + \int_o^T g(t, S(t))dt \tag{7.A11}$$

Differentiation of the function $M(t, S)$ along the solution $S(t)$ demonstrates that M satisfies the Q.V.I. conditions (7.A8) and (7.A9):

$$\partial M(t, S)/\partial t + F(t, S)\, \partial M(t, S)/\partial S + g(t, S) = 0 \tag{7.A12}$$

$$M(T, S(T)) = h(T, S(T)) \tag{7.A13}$$

Then, the following iterative procedure allows one obtain the switching surface $V^*(t, S)$ where one can switch over for the aeration (Shima, 1978) to achieve the minimum of the performance function J:

(1) Obtain an impulsive control q that minimizes

$$N(t, S, q) = \min [c(t, S, q) + M(t, S–q)] \tag{7.A14}$$

where the first term of the right hand side corresponds to the cost function c for the impulse control q, and the second term is the perfomance function (7.A11) without any impulse control.

(2) Define the curve G consisting of the two performance functions, without and with impulsive control q

$$G(t, S) = \{M(t, S) – N(t, S, q)\} \tag{7.A15}$$

and divide it into two positive and negative domains:

$$G^+ = \{(t, S) : G(t, S) > 0 \qquad 0<t<T\}$$
$$G^- = \{(t, S) : G(t, S) > 0 \qquad 0<t<T\} \tag{7.A16}$$

Apparently, one should use the impulse control q in the domain G^+, but not in the domain G^- for attaining a smaller value of the G. However, in order to see the sufficient condition of the optimality, the other possible paths of impulse control q must be assured over the entire domain in the following manner:

(3) Let J_t and $J_{t+\delta t}$ be the following performance functions

J_t: with impulse control q at $p = t$

$J_{t+\delta t}$: without impulse control after $p = t + \delta t$

Then the difference of the two functions $J_t - J_{t+\delta t}$ can be expanded in the following variational equation of the order 0 (δt):

$$J_t - J_{t+\delta t} = Q(t, S) = \partial c(t, S, q)/\partial t + F(t)\, \partial c(t, S, q)/\partial S$$
$$+ \partial M(t, S-q)/\partial t + F(t)\, \partial M(t, S-q)/\partial S + g(t, S) + 0(\delta t) \tag{7.A17}$$

The resulting function $Q(t, S)$ that satisfies the Q.V.I. conditions of the optimally is divided into two domains:

$$Q^+ = \{(t, S) : Q(t, S) > 0\}$$
$$Q^- = \{(t, S) : Q(t, S) < 0\} \tag{7.A18}$$

The surface of the function Q, called 'Free Boundary' can guarantee whether the impulse is optimal or not, that is, whether aerator is to be switched or not.

(4) Repeat steps (2) and (3) until the solution $S(t;0,S(0))$ hits again the surface G (the necessary condition) and the surface Q (the sufficient condition) looking for the next possible switching of the aeration.

REFERENCES

Beck, M. 1980. Forecasting and control of water quality. In E. Wood (ed.), *Real-time forecasting/control of water resource systems*. Pergamon Press.

Bensoussan,A. & J. Lions 1975. New method of impulse control. *Applied Mathematics & Optimization*, 1(4): 289-315.

Bryson, A. & Y. Ho 1975. *Applied Optimal Control*. New York: John Wiley.

Geoffrion, J., J. Dyer & A. Feinberg 1972. An interactive approach for multi-criterion optimization. *Management Sci.*, 19: 359-368.

Gromiec, M., D. Loucks & G. Orlob 1983. Stream quality modeling. In G. Orlob (ed.), *Mathematical modeling of water quality*. New York: John Wiley.

Haimes, Y., W. Hall & H. Freedman 1975. *Multiobjective optimization in water resources systems*. Amsterdam: Elsevier.

Ikeda, S., S. Miyamoto & Y. Sawaragi 1974. Identification method in environmental systems and its application to water pollution. *Int. J. of Systems Sci.*, 5(8): 707-723.

Ikeda, S. 1987. Economic-Ecological Models in regional total systems, In L. Braat (ed.), *Economic-Ecological Modeling*, Amsterdam, North-Holland.
Ito, K. 1966. *Theory of partial differential equations* (in Japanese).
James, A. (ed.) 1978. *Mathematical models in water pollution control.* New York: John Wiley.
Keeney, R. & H. Raiffa 1976. *Decisions with multiple objectives – Preferences and value trade-offs.* New York: John Wiley.
Koivo, A. & G. Phillips 1975. Optimal estimation and control of polluted stream variables. *Proc. of 6th IFAC Congress, Boston.*
Lavrentiev, M. 1967. Some improperly posed problems of mathematical physics. R. Sacker (translated). Springer-Verlag.
Matsuoka, Y., H. Sumitomo & S. Iwai 1977. An estimation of pollution sources with monitoring of river water quality. *Proc. of IFAC Symp. on Environmental Systems Planning, Design, and Control, Paper A-1.2, Kyoto.*
Nakayama, H., T. Tanino & Y. Sawaragi 1980. An interactive method in multicriteria decision making. *IEEE Trans. SMC*, 10(3): 163-169.
Phillips, D. 1962. A technique for the numerical solution of certain integral equations of the first kind. *J. of Assoc. Comp. Mach.*, 9: 84-97.
Rao, C. & S. Mitra 1971. *Generalized inverse of matrices.* New York: John Wiley.
Rinald, S., R. Scocini-Sessa, H. Stehfest & H. Tamura 1979. *Modeling and control of river quality.* New York: McGraw-Hill.
Sawaragi, Y. & S. Ikeda 1975. The methodology of water quality monitoring – Prediction and detection of pollution sources. *Proc. of Human Environ. Sci. Conf., Japan Science Council, Kyoto.*
Sawaragi, Y., S. Ikeda, H. Nakayama, T. Tanino & C. Okuyama 1978. An interactive optimization method for a water resource problem. *Proc. of IFAC Symp. on Environmental Systems, Planning Design and Control, Kyoto.* Pergamon Press.
Shima, M. 1978. A study of optimal impulse control problems – Systems and control. *Japan Soc. for Instr. & Control*, 22(10): 624-632.
Tihonov, A. 1963. Solution of incorrectly formulated problems and the regularization method. *Dokl. Akad. Nauk. SSSR*, 161: 559-562.
Tihonov, A. & V. Arsenin 1977. *Solution of ill-posed problems* (English translation). New York: John Wiley.
Wierzbicki, A. 1980. The use of reference objectives in multiobjective optimization. In G. Fankel & T. Gal (eds), *Multiple criteria decision making*, 468-486. Springer-Verlag.

Waste disposal in the ocean 8

GERHARD H. JIRKA
DeFrees Hydraulics Laboratory, School of Civil and Environmental Engineering, Cornell University, Ithaca, New York, USA
JOSEPH H.-W. LEE
Department of Civil and Structural Engineering, University of Hong Kong, Hong Kong

8.1 INTRODUCTION

8.1.1 *The problem*

Man has long used the ocean as a dumping ground for municipal and industrial wastes. Viewed as a whole, the dilution capacity of the ocean is surely enormous, and it may seem to a newcomer to the subject that a 'water pollution' problem can hardly exist. Because chemical, physical and biological processes progressively degrade and transform most discharged materials and assimilate them into naturally occurring material cycles, one might conclude that the global ocean is indeed an ideal receiving environment for waste effluents (excepting highly stable and persistent materials).

Nonetheless, significant 'water pollution' problems may occur on a local or regional scale, as high concentrations of the released substances can cause considerable harm and impact. For example, domestic wastes issuing from the open end of a submerged discharge pipe may be poorly diluted, resulting in foul odors, visible sewage slicks, and high bacterial concentrations at the water surface in the vicinity of the discharge. Typically, environmental regulations apply on that local level and specify concentration limits, or standards, that may not be exceeded (e.g., a target fecal coliform count for protection of bathing beaches, or a minimum dissolved oxygen level for protection of fisheries).

The role of the hydraulic engineer is to prevent, or at least mitigate, any harmful local effects and to provide a smooth and efficient transition to the global large scale degradation and transformation processes. The hydraulic engineer has two principal means to achieve these objectives: First, through the choice and design of a discharge structure the engineer can control the amount of initial mixing and dilution that takes place within a short time (typically within minutes to an hour) after the effluent enters the ocean. The recent development of highly efficient mixing devices, namely multiport diffusers, provides considerable flexibility in achieving very high initial dilutions (well in excess of 100). Secondly, through proper siting strategies and ambient transport predictions the hydraulic engineer can assure that the discharged material is effectively and continuously removed from the discharge locality. The engineer is frequently cooperating with oceanographers in conducting these assessments.

Clearly, these two tasks are intimately connected with each other. For example, a submerged discharge may have been designed for a high initial dilution; however, if it is located in a semi-enclosed bay or an estuary with weak through-flow or flushing, it will never achieve that dilution and will produce high local concentrations. The choice of another discharge location, e.g., outside the bay or further offshore, may be the only remedy in such a situation. From a fluid mechanical viewpoint these two tasks involve the analysis of near-field processes and of far-field processes that affect the mixing and transport of the pollutant effluents.

The near-field (also called active dispersal region or initial mixing region) is defined as the region close to the discharge and is dynamically affected by the discharge. The momentum and the buoyancy of the discharge modify the ambient flow pattern (mean and turbulent), and the discharge generates its own mean velocity and turbulence field. Phenomena such as jet diffusion and entrainment, and buoyant acceleration and spreading occur in the near-field and affect pollutant distribution and concentration levels. The spatial scale of the near-field is of the order of 10 meters (up to perhaps several kilometers for very large installations) with time scales of the order of minutes to one hour.

The far-field (also called passive dispersal region) refers to the much larger region outside the near field in which the dynamic effects of the discharge are unimportant and the diluted waste field is carried by ambient advective currents and diffused by ambient turbulent processes. The spatial scale in coastal far field processes ranges up to several tens of km (perhaps 100 km) while time scales of interest range from several hours to days, including tidal variations. Of course, within these increasing scales the long-term transformation and degradation processes (which govern the ultimate environmental impact on the global scale) may become increasingly important depending on the characteristics of the effluent.

8.1.2 *Summary*

This review is structured into four main sections. The near field mixing characteristics of effluent discharges are summarized in Sections 8.2, 8.3 and 8.4, while predictive techniques for far field dispersal processes are discussed in Section 8.5.

The three major types of discharge into coastal receiving water are by means of a single submerged pipe outfall (Section 8.2), a multiple port (multiport) submerged diffuser outfall (Section 8.3), and a surface discharge by means of a canal at the shoreline (Section 8.4). All of these discharges produce jet-like motions that are modified in various ways by buoyancy, ambient geometry, and ambient currents. The emphasis in these three sections is to provide a classification of the behaviour of these discharges and to present simple methods for predicting their mixing capacity (near field dilution). The major classification consists of a subdivision into stable and unstable discharge conditions. This distinction is important, as shallow coastal environments often do not allow the establishment of distinct jet-like flow (similar to a chimney plume in the atmosphere!); interaction with water surface and bottom can lead to an unstable flow breakdown so that mixing occurs over the entire water column. The review is limited to simple predictive models – usually based on the jet integral technique – that are useful in preliminary designs. Such design procedures are illustrated with several examples. The various references listed provide more detailed results or more elaborate predictive techniques (e.g. those based on a complete solution of the turbulent field equations). Problems related to the internal hydraulics of multiport diffusers (e.g. sea water intrusion) are not covered herein; a discussion of these topics has been given by Wilkinson (1984) and Charlton (1985).

The review of predictive methods for far field transport processes (Section 8.5) also stresses the application of simple models that are readily applied for approximate analysis. These include plume diffusion models, flushing estimates and box models for bays and estuaries. Even an experienced analyst will usually apply such simple models in order to decide on the need for more detailed numerical models before proceeding further. A short summary of numerical far field models, including computational aspects, concludes this section.

In summary, this review is intended (1) as an introduction to the fluid mechanical processes governing waste disposal into the coastal ocean, (2) as a collection of, and a guide to, simple first-order analysis of discharge design and waste field prediction, and (3) as a reference to more advanced treatments (in the form of research articles or numerical models) of the topic. For a general fluid mechanical background on environmental mixing processes the reader is referred to Fischer et al. (1979), Rodi (1980) and Holley & Jirka (1986).

8.2 SINGLE PORT SUBMERGED DISCHARGES

Figure 8.1a shows a schematic diagram of a submerged single port outfall. In its simplest form, this single outlet can be just the open end of a pipe laid on the sea floor. The buoyant waste effluent discharges at a velocity U_o, relative density difference $\Delta\rho_o/\rho_a$, and reduced gravitational acceleration $g' = (\Delta\rho_o/\rho_a)\,g$, from an orifice of diameter D. Provided that the jet Reynolds number U_oD/ν, in which ν = kinematic viscosity, exceeds about 2000 – a condition that is always met in practice – the flow will be turbulent; thus, the dynamics of the mixing is governed mainly by the kinematic momentum flux, $M = U_o^2\pi D^2/4$, and kinematic buoyancy flux, $J = U_o g' \pi D^2/4$ of the discharge. The volume flux $Q = U_o\pi D^2/4$ plays a

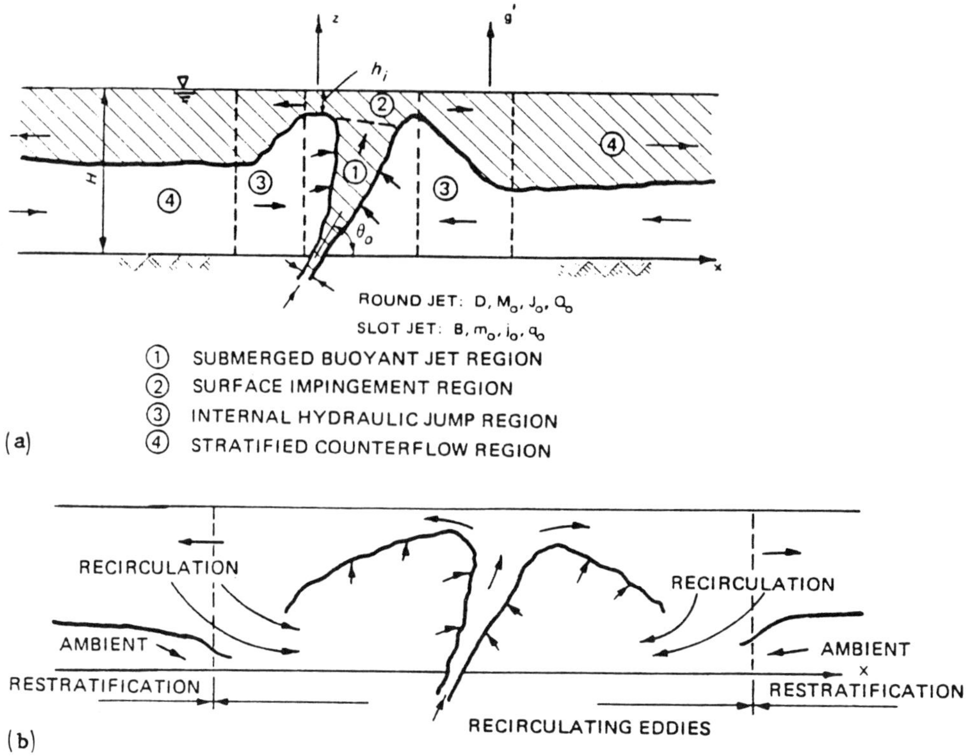

Figure 8.1. Submerged buoyant jet discharging into water of finite depth under stagnant conditions. a) deep water discharge with stable discharge configuration. b) shallow discharge with unstable recirculating zone. These stability conditions are applicable for the round jet as well as for the slot jet

minor dynamic role and becomes important only for very shallow depths. Close to the point of discharge, the high velocity shear induces turbulent entrainment, and the source fluid mixes with the uncontaminated ambient water as it rises, gaining vertical momentum by buoyant acceleration. A spreading layer forms at the free surface.

If c_o is the concentration of a conservative substance in the discharge, then the local dilution at a given point can be defined in terms of its concentration $c(x, y, z, t)$ as $S = c_o/c$. Frequently used dilution values are the centerline dilution $S_c = c_o/c_c$ in which c_c = concentration at centerline or a flux-averaged dilution $\bar{S}$. The unknown jet trajectory, the concentration and velocity field are determined by two important governing parameters: a densimetric Froude Number, $F_o = U_o/\sqrt{g'D}$, is the ratio of inertia to buoyancy forces, and the relative depth z/D. For $F_o \sim 1$, the discharge behaves like a pure source of buoyancy (plume), and for $F_o \to \infty$, the discharge resembles a pure momentum jet. For a buoyant jet, it can be shown (Fischer et al., 1979) that the flow is jet-like near the discharge, and plume-like at large z/D where the initial source kinetic energy has been dissipated.

8.2.1 *Discharge stability*

The physical picture depicted in Figure 8.1a is defined as a deep water condition in which a distinct buoyant jet rises to the surface and dilution occurs because of turbulent jet entrainment. Under shallow water conditions, however, the discharge momentum may be sufficiently strong to cause a dynamic breakdown (instability) of the buoyant jet motion and create a local recirculation zone (Fig. 8.1b). A stability analysis by Lee & Jirka (1981) yielded the following approximate criterion for the deep water condition:

$$\frac{H}{D} > 0.22\,F_o \tag{8.1}$$

(where H = depth of receiving water) with apparently little sensitivity to the discharge angle (Jirka 1982a). The dynamic distinction into deep and shallow water conditions is important as entirely different techniques must be used for prediction of concentration. Sewage effluents with typically small and highly buoyant flows operate in the deep water condition. In contrast, high velocity, weakly buoyant cooling water discharges in shallow coastal zones often fall in the unstable condition.

Example 1: Determine the discharge condition for a) a vertical sewage discharge with jet velocity $U_o = 1$ m/s, $D = 0.15$ m, $\Delta\rho_o/\rho_a = 0.025$, and $H = 15$ m, and b) a cooling water discharge with $U_o = 5$ m/s, $D = 0.6$ m, $\Delta\rho_o/\rho_a = 1.8 \times 10^{-3}$ (corresponding to a temperature excess of about 8°C), and $H = 5$ m.

Solution: The two governing dimensionless parameters can be calculated to be: a) $F_o = 5.2$ and $H/D = 100$; by Equation 8.1, the discharge is stable; b) $F_o = 49$, $H/D = 8.3$; the high jet momentum will induce a recirculating motion over the entire water depth. The details of that motion will depend on the discharge angle θ_o.

8.2.2 *Deep water discharges in stagnant homogeneous ambient fluid*

Buoyant jet mixing for the deep water condition has been extensively studied (e.g. Fischer et al., 1979). For the stagnant condition, integral jet models have been developed, verified, and applied with great success. The models typically are based on the assumption of self-similar Gaussian profiles for velocity and concentration in the jet cross-section. The governing equations of mass and momentum conservation can then be integrated across the jet to yield a set of ordinary differential equations in terms of characteristic variables. Turbulent closure is achieved by either the entrainment hypothesis (Morton et al., 1956), in which the entrainment velocity at the periphery of the jet is related to a characteristic local jet velocity via an entrainment coefficient, or by the assumption of a constant local jet spreading angle (Abraham, 1963). Predictions of the centerline trajectory, velocity and concentration (dilution) are obtained by numerical integration of the resulting parabolic equations along the buoyant jet axis. Such models have been presented by Abraham (1963), Fan & Brooks (1969), Cederwall (1968), and others. A clear exposition of the basic technique and the treatment of the initial zone of flow establishment has been given by Brooks (1972), Jirka et al. (1975) and Holley & Jirka (1986).

Figure 8.2a shows a numerical solution of the normalized centerline dilution S_c/F_o, as a function of z/DF_o; it can be seen that the dilution at a given depth is dependent on the jet densimetric Froude number close to the discharge, but at large z/D all jets tend to behave like plumes. Integral jet model predictions such as that displayed in Figure 8.2 compare well with experimental data over a wide range of conditions. They are routinely used in environmental hydraulic design.

Of special interest is the limiting case of a vertical point plume. For this case, the centerline dilution at elevation z above the discharge has been given by Rouse et al. (1952):

$$S_c = 0.11\,(z/D)^{5/3}\,F_o^{-2/3} = 0.11\,J^{1/3}\,z^{5/3}\,/Q \tag{8.2}$$

For the case of a horizontal buoyant jet, a useful dilution formula that fits Figure 8.2a was given by Cederwall (1968):

$$S_c = 0.54\,F_o\left(0.38\,\frac{z}{D\,F_o} + 0.66\right)^{5/3} \qquad \text{for } z/D \geq 0.5\,F_o \tag{8.3}$$

The role of the main variables can be seen clearly in Equations 8.2 and 8.3 which can be used to provide first estimates of minimum dilution for many deep water discharges. For small F_o and large z/D, Equation 8.3 reduces to Equation 8.2. For round jets, the flux-weighted dilution $\bar{S}$ is approximately 1.7 times the centerline dilution.

As a surface layer is typically formed, not much dilution occurs beyond the base of the surface wastewater field (Fig. 8.1a). The additional mixing above this level serves only to even out concentration gradients. The upper layer thickness for the stagnant condition can be shown to be approximately 0.1 to 0.15 of H (Lee & Jirka, 1981). Equation 8.2 or 8.3 can be applied to calculate the initial dilution at the surface with the depth reduced by 10 to 15%, $z = 0.85\text{-}0.9\,H$; alternatively, the numerical calculation can be taken up to an elevation such that the upper layer thickness equals the local plume radius (Wallis, 1977).

In recent years, 'length scale analysis' has been increasingly used in the solution of buoyant jet problems. These length scales can be formed by dimensional reasoning from the source fluxes; each length scale can be given a clear physical interpretation. For example, $l_Q = Q/\,M^{1/2} = (\pi/4)^{1/2}\,D$, and is proportional to the jet diameter; it can be interpreted as the characteristic distance within

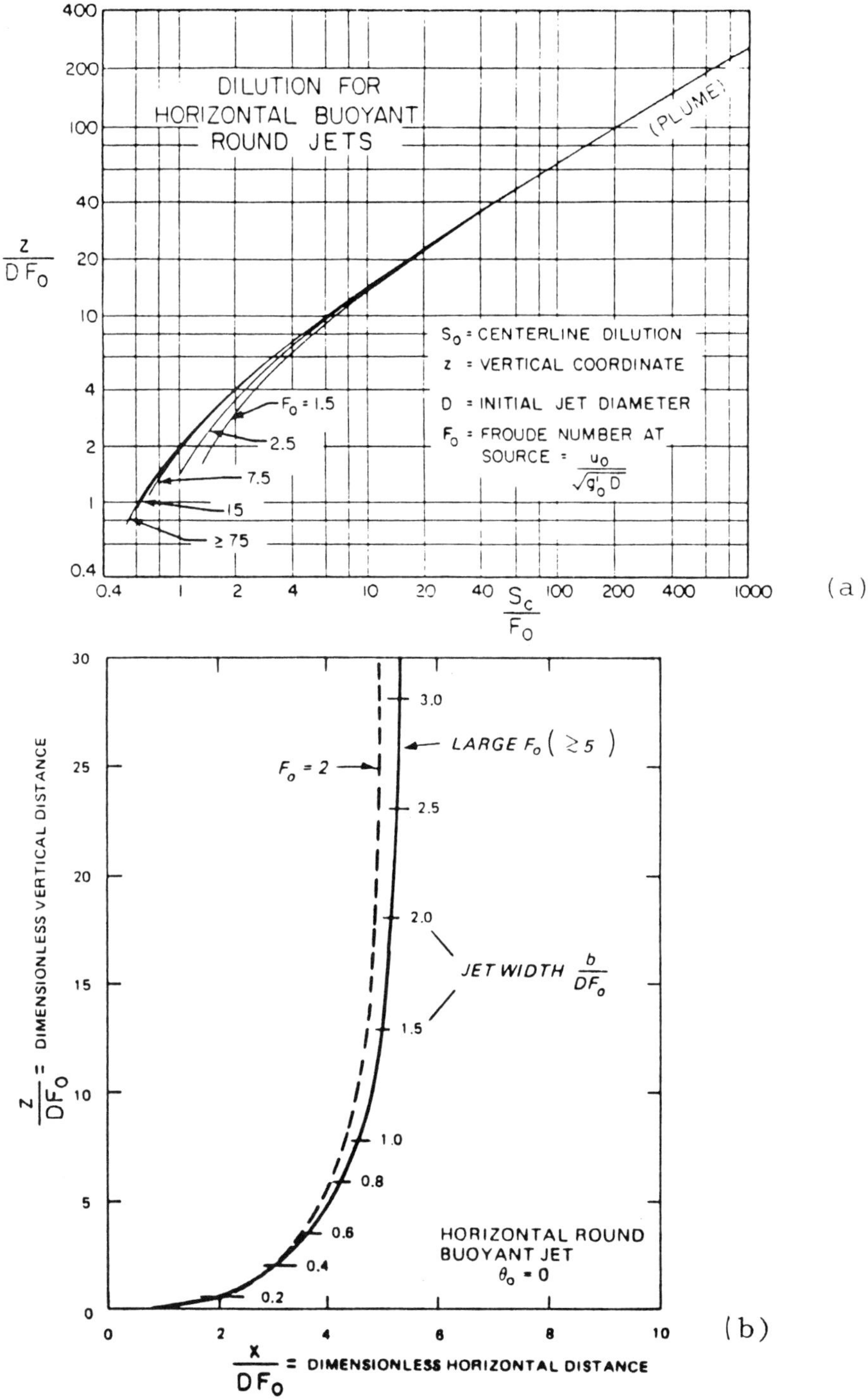

Figure 8.2. Normalized solution diagrams for round buoyant jet discharged horizontally in stillwater: a) centerline dilution as a function of vertical distance (after Brooks, 1980). b) trajectory and jet width (after Holley & Jirka, 1986)

which the source geometry is important. $l_M = M^{3/4} / J^{1/2}$ can be interpreted as the distance within which the initial momentum effect is important; for $z \ll l_M$, behavior is jet-like (momentum-dominated), and for $z \gg l_M$ behavior is plume-like (buoyancy-dominated). Equations 8.1 and 8.3 can also be expressed equivalently in terms of length scales. For example, since $l_M = (\pi/4)^{1/4}\, DF_o$, the definition of near field stability can be written simply as $H > 0.23\, l_M$. The reader is referred to Fischer et al. (1979) for further details.

8.2.3 *Shallow water discharges*

For high momentum discharges in shallow water resulting in complicated recirculating flows, Equations 8.2 and 8.3 cannot be used. For the vertical discharge, Lee & Jirka (1981) give a flux-weighted (bulk) dilution that characterizes the local recirculation cell:

$$S = 0.9 \left(\frac{H}{D}\right)^{5/3} F_o^{-2/3} \tag{8.4}$$

Compared to the plume solution (Equation 8.2), Equation 8.4 predicts higher dilutions which reflect the bulk mixing over the local recirculating cell. More important, Equation 8.4 gives an upper bound on the dilution achievable by high momentum discharges in shallow water (for a pure momentum jet in deep water the corresponding dilution increases with jet momentum as $S = 0.25\, M^{1/2} H/Q$).

8.2.4 *Deep water discharge with ambient stratification*

In coastal locations the water column is frequently density-stratified. Such stratification may be caused by temperature differences (due to heating of the upper layers) or by salinity differences (due to river inflows). The ambient density structure, $\rho_a(z)$, may show a continuous variation that can be approximated by a linear density gradient, $d\rho_a/dz = \text{constant}$. In other instances, the ambient density may undergo a relatively sharp transition (pycnocline) that separates a lighter upper layer from a denser lower layer. In any of these situations the waste effluent may become trapped at a level z_{max} below the free surface, $z_{max} < H$.

For a linear density gradient the maximum height of rise and the centerline dilution at this terminal level are given by Morton et al. (1956), Hino (1964) and Brooks (1972):

$$z_{max} = 3.98\, J^{1/4} \left(-\frac{g}{\rho_1}\frac{d\rho_a}{dz}\right)^{-3/8} \tag{8.5}$$

$$S_c = 0.071\, \frac{J^{1/3} z_{max}^{5/3}}{Q} \tag{8.6}$$

where $\rho_1 = \rho_a\ (z = 0)$. For a nonlinear stratification, Brooks also presented an expedient graphical procedure to obtain an approximate solution based on the assumption that the ambient density variation is linear from the source level up to the unknown terminal level.

8.2.5 *Effects of ambient currents*

In the presence of an ambient current u_a (Fig. 8.3), the buoyant jet is deflected by pressure form drag as the current flows around the jet, and by the entrainment of ambient water having crossflow momentum. In principle, the mixing is governed by the momentum flux and the buoyancy flux of the discharge, as well as by the forced entrainment from the current. Since the momentum flux is a directed quantity the orientation of the discharge port relative to the ambient crossflow should also be important. In actual practice, however, most ocean discharges are governed mainly by the buoyancy flux and are little affected by the momentum flux and, hence, by the orientation of the discharge. Over the past two decades, various mathematical models have been formulated and refined (Fan, 1967; Abraham, 1971; Hirst, 1972; Schatzmann, 1981; Jirka & Fong, 1981; Delvigne, 1980; Chu, 1974, 1975; Frick, 1984). Unfortunately, model validations are limited to high momentum discharges in relatively weak crossflows.

Using a length scale approach, Wright (1977) analyzed laboratory data for a vertical buoyant jet in a crossflow and obtained the following empirical correlations for the minimum (centerline) dilution in the weakly deflected stage (buoyancy dominated near field, bdnf) and the strongly deflected stage (buoyancy

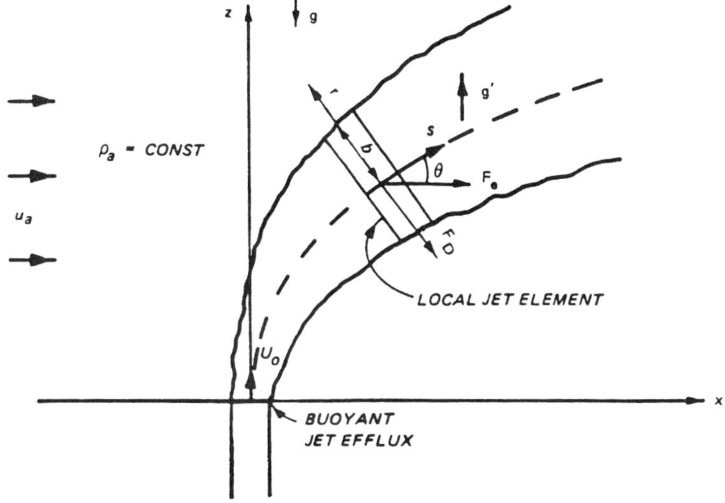

Figure 8.3. Submerged round jet in crossflow. The buoyant jet is deflected by a pressure force F_D and an entrainment force F_e

dominated far field, bdff)*:

bdnf:

$$S_c = 0.27 \frac{J^{1/3} z^{5/3}}{Q} \qquad \frac{z\, u_a^3}{J} < 1 \tag{8.7}$$

bdff:

$$S_c = 0.40 \frac{u_a z^2}{Q} \qquad \frac{z\, u_a^3}{J} > 1 \tag{8.8}$$

Chu (1979) reanalyzed the comprehensive laboratory data of Fan and obtained a similar relationship.

Field investigations of initial dilution using fluorescent dye and radioactive tracers are reported by Agg (1972, 1978), Vigliani et al. (1980), Bennett (1981), Bettess & Munro (1981), and Toms & Botes (1986). Laboratory experiments of horizontal heated jets in an unstratified crossflow were also conducted (Hydraulics Research Station, 1977). A recent analysis and interpretation of all available field and laboratory data of initial dilution of a buoyant jet in an unstratified crossflow (with the port oriented horizontally across the flow) has confirmed the validity of the length scale approach. The following equations for the estimation of initial dilution of a horizontal round jet in a crossflow have been proposed (Lee & Neville-Jones, 1987a):

bdnf:

$$S_m = 0.31 \frac{J^{1/3} H^{5/3}}{Q} \qquad \frac{H\, u_a^3}{J} < 5 \tag{8.9}$$

bdff:

$$S_m = 0.32 \frac{u_a H^2}{Q} \qquad \frac{H\, u_a^3}{J} > 5 \tag{8.10}$$

in which H refers to the total water depth available above the discharge and S_m denotes the minimum dilution observed at the surface. The statistical variability of initial dilution is also discussed (Lee & Neville-Jones, 1987b).

Example 2: For the deep water sewage discharge in Example 1, estimate the initial dilution for a) a vertical discharge, and b) a horizontal discharge. Also estimate the initial dilution in unstratified moving water for a design current of $u_a = 0.05$ m/s and 0.3 m/s, respectively.

*In this section 'near' and 'far' field refer to the plume-dominated (weakly deflected) and the current-dominated (bent over) phases of jet mixing, respectively. The reader is cautioned that this terminology is different from the customary usage of near and far field to distinguish the region of active mixing close to the discharge point, and the region dominated by ambient processes further downstream (as described in Section 8.1).

Solution: The parameters are $F_o = 5.2$ and $H/D = 100$. Accounting for a surface layer of 0.1 H thickness, Equation 8.2 gives

$$S_c = 0.11 \times (90)^{5/3} (5.2)^{-2/3} = 66$$

for a vertical discharge, and Equation 8.3 gives

$$S_c = 0.54\,(5.2)\left[\frac{0.38\,(90)}{5.2} + 0.66\right]^{5/3} = 76$$

for a horizontal discharge. To find the dilution in a current, first compute the buoyancy flux

$$J = (1)\,\frac{\pi\,(0.15)^2}{4}\,(9.81)\,(0.025) = 4.3 \times 10^{-3}\ \mathrm{m^4/s^3}$$

For $u_a = 0.05$ m/s, $Hu_a^3/J = 0.4$, therefore Equation 8.9 can be used to compute the bdnf dilution

$$S_m = 0.31\,(4.3 \times 10^{-3})^{1/3}\, 15^{5/3}\,/0.0177 = 261$$

For $u_a = 0.3$ m/s, $Hu_a^3/J = 93$, use Equation 8.10 to obtain

$$S_m = 0.32\,(0.3)\,15^2/0.0177 = 1220$$

Note that the initial dilution is increased considerably even in the presence of a weak current. In field experiments the lowest measured surface dilution often exceeds the predicted still water dilution by a factor of 2 to 3 times. In coastal locations with adequate currents, Equations 8.9 and 8.10 give more realistic predictions of dilution. The still water dilution should be used for conservative design.

When the ambient crossflow is stratified. Wright's experiments (1977) have shown that mixing up to the maximum equilibrium height of rise z_{max} is little affected by the stratification. For a point plume in linearly stratified fluid, z_{max} is given by:

$$z_{max} = 1.8 \left(\frac{u_a}{\varepsilon^{1/2}}\right)^{2/3} \left(\frac{J}{u_a^3}\right)^{1/3} \qquad (8.10a)$$

where

$$\varepsilon = \left(-\frac{g}{\rho_1}\frac{d\rho_a}{dz}\right)$$

is a stratification parameter. The dilution can then be estimated by using Equation 8.8 applied at z_{max}. Field data support for Equation 8.10a has not been reported.

8.2.6 *Detailed mathematical models*

The foregoing equations provide first estimates of dilutions and associated

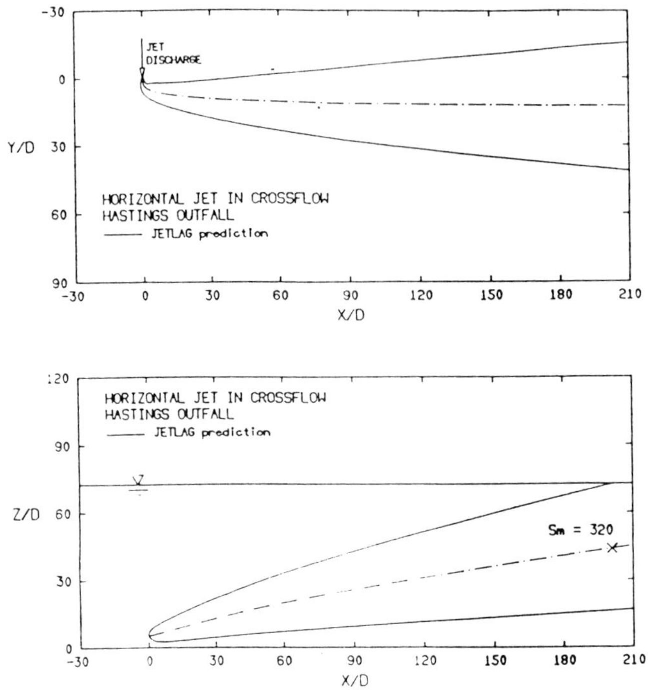

a)

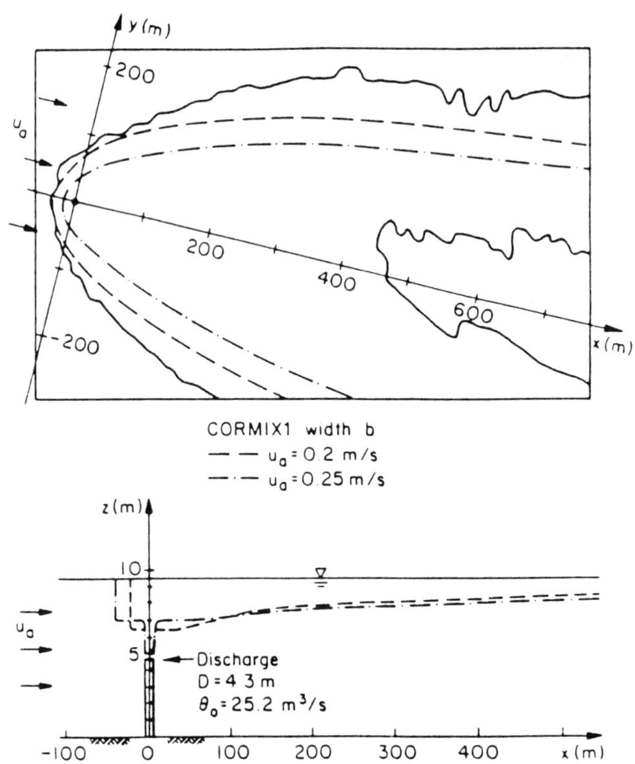

b)

Figure 8.4. Examples of mathematical model predictions for submerged discharges: a) three-dimensional submerged jet trajectory predicted with integral model of JETLAG (Lee & Cheung, 1990), H/D = 68, F_o = 6.1 and u_a/U_o = 0.23. b) prediction of submerged jet behaviour in confined water depth, resulting in buoyant spreading along water surface and upstream intrusion; expert system CORMIX1 (Doneker & Jirka, 1990)

concentration levels for simple discharge geometries and ambient conditions. Mathematical models are however needed for some situations. Even in stagnant water, numerical models are required if the discharge geometry is relatively complex (e.g. inclined discharges). For example, the United States Environmental Protection Agency (USEPA) model UPLUME (Mullenhoff et al., 1985) is an elaborate computer implementation of Abraham's (1963) theory. For the more complicated case of buoyant jet in a current, integral jet models (e.g., the USEPA model UDKHDEN, which is based on Hirst's (1972) formulation) are often used in practical applications if both discharge momentum and buoyancy are important, if the ambient density profile cannot be adequately approximated as a linear variation or if the ambient velocity profile is non-uniform. Figure 8.4a shows an example calculation for a horizontal sewage outfall discharge into a perpendicular tidal current, with a three-dimensional jet trajectory (Lee & Cheung, 1990). Rodi (1980) has presented examples of buoyant jet solutions computed by more sophisticated numerical models that solve for the turbulent kinetic energy and dissipation rate and that do not rely on a self-similarity assumption.

In general, integral jet models do not account for the dynamic effect of boundaries; many flow configurations, such as plume attachment to the sea bottom or buoyant spreading of surface layer, cannot be predicted. In addition, the models tend to be specialized, and are often used beyond their range of applicability. In view of the diverse hydrodynamic flow patterns that a waste discharge can exhibit, a micro-computer based expert system has been developed to facilitate initial mixing analyses (Doneker & Jirka, 1990; Jirka & Doneker, 1991). The system (CORMIX1) incorporates a broad knowledge base for flow classification; complex features such as near field instability, upstream intrusion, and buoyant spreading are included. Predictions of flow geometry and dilution are then provided by a sequence of relatively simple sub-models based on length scale analysis and basic experimental data. The overall mixing behavior of the majority of waste discharges can be satisfactorily simulated by such a system, which does not require of the user a detailed understanding of model applicability and selection. Figure 8.4b shows an example simulation of the San Onofre Nuclear Power Station thermal discharge by CORMIX1.

8.3 DISCHARGES FROM SUBMERGED MULTIPORT DIFFUSERS

Large flows of waste effluents are commonly discharged through multiport diffusers. This type of diffuser is a linear structure with a (usually large) number of ports mounted on, or connected through risers to, a submerged pipe laid on the ocean floor. The nozzles can be oriented in various ways, both in plan and in section, to effect particular types of mixing patterns and to take advantage of the direction of the prevailing ambient current. A multiport diffuser is an efficient mixing device, capable of rapidly diluting the effluent within a short distance.

Three genetic types of nozzle/port arrangements – the unidirectional, alternating, and staged diffuser – are shown in Figure 8.5. In the unidirectional diffuser, all the jets point in the same direction perpendicular to the diffuser axis. In a staged diffuser, the jets are arranged in tandem, and point in the same direction parallel to the diffuser axis. Both the unidirectional and staged diffusers have a net momentum and a preferred direction, while in the alternating diffuser, the jets discharge in opposite directions alternately and, hence, the flow from the diffuser has no net horizontal momentum. To prevent bottom scour the jets are usually oriented upwards at a small angle above the horizontal.

Of course, there are variations on the basic theme for each of the three diffuser types. For example, there may be double or triple nozzle arrangements (with a small internal angle) for both unidirectional or staged diffusers, and the port orientation angle β may differ somewhat from the nominal value, 90° or 0°, respectively. Or, in the case of the alternating diffuser, there may be multiple port assemblies for each riser with several ports arranged in a circular fashion. Furthermore, alternating diffusers for thermal discharges in shallow water may have a variable port orientation along the diffuser axis to control instabilities and horizontal circulations (for details, see Jirka, 1982b). Another special case of an alternating diffuser is given by a vertical discharge from all ports.

Besides the general arrangement, the pertinent diffuser variables are the diffuser flow Q, ambient water depth H, diffuser length $L_D = Ns$, where s = port spacing, N = number of ports, jet velocity U_o, and port area $D^2\pi/4$, where D = port diameter.

The criterion for near field instability of a buoyant diffuser discharge in a stagnant ambient fluid is given by Jirka & Harleman (1973, 1979) as:

$$\frac{m}{j^{2/3} H} = \frac{0.54}{(1+\cos^2\theta_o)^2} \tag{8.11}$$

where $m = qU_o$ is momentum flux per unit diffuser length, $q = Q/L_D$ is the unit discharge, $j = g'q$ is the unit buoyancy flux, and θ_o the vertical angle above horizontal. In addition to discharge momentum, ambient current is often another destabilizing factor, and a complete stability diagram can be found in Jirka (1982b).

Application of Equation 8.11 to various discharge situations yields the following practical distinctions: Municipal and industrial wastewater diffusers, in which the effluent has essentially the density of freshwater, typically operate in the stable regime; in contrast, 'thermal diffusers' as used for the once-through cooling of large steam-electric power plants usually have much higher flow rates with smaller density differences (corresponding to the temperature rise in the condensers) and result in unstable flows. These differences and associated examples are discussed in more detail by Jirka (1982b).

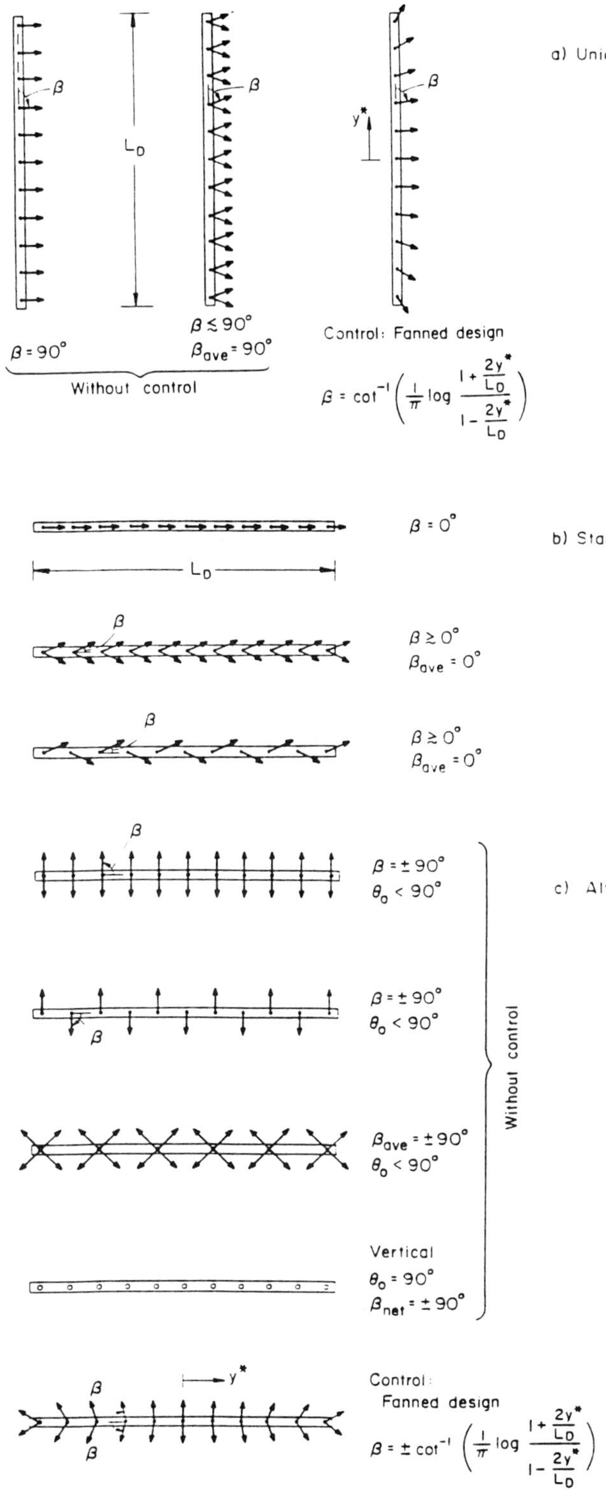

Figure 8.5. The three major diffuser types with different nozzle (port) orientations (plan view)

8.3.1 *Deep water diffusers*

For stable buoyant discharges in deep water, the nozzle arrangement is less crucial. The dilution of the wastewater field is primarily governed by the buoyancy flux per unit length of diffuser and the depth of discharge. Initially, the effluent leaves the diffuser ports as a series of round plumes (Fig. 8.6) and the equations for a round buoyant jet can be used for prediction. After travelling some distance, adjacent plumes merge with each other to form a rising curtain, and the flow is essentially two-dimensional. As with the single port discharge, the buoyancy force is the major driving force for the mixing; the initial momentum is important only within the region a few meters from the port. A general integral model that enables one to compute the jet characteristics of the initial three-dimensional round plume region, the transition to an essentially two-dimensional flow, and the subsequent plane plume can be formulated (Koh & Fan, 1970). An example calculation is shown in Figure 8.7. Under typical discharge conditions, it can be shown (Brooks, 1972; Wallis, 1977) that for depths in excess of about 40 m, initial dilution predictions for the same buoyancy flux per unit diffuser length agree to within 15%, regardless of the details of the individual ports.

One of the most important equations for dilution is the equation for a line plume in a stagnant unstratified ocean (Rouse et al., 1952):

$$S_c = 0.38\frac{j^{1/3} H}{q} \tag{8.12}$$

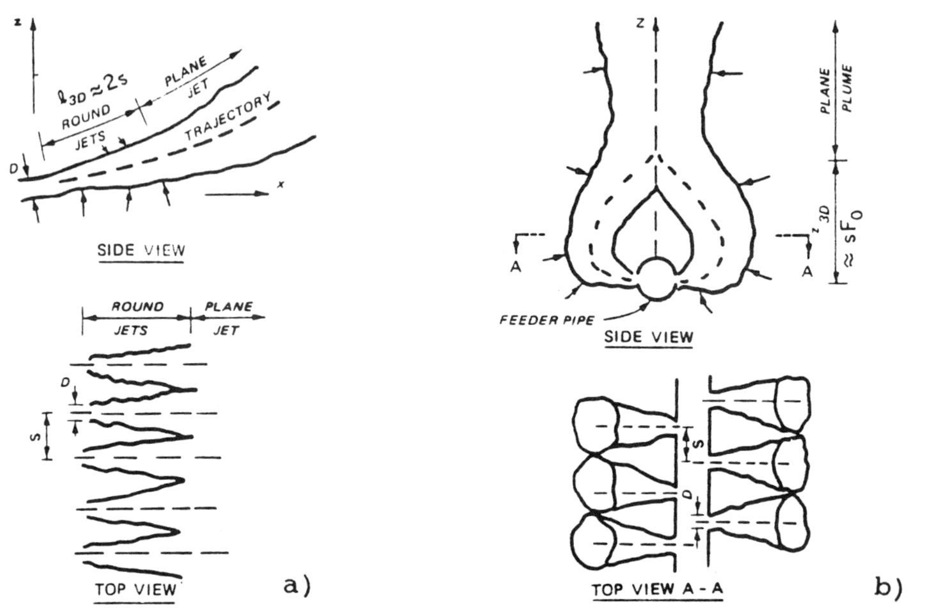

Figure 8.6. Interference of individual round jets from multiport diffuser discharges forming two-dimensional (slot) jets or plumes. a) Jets merging in unidirectional diffuser. b) Jets merging in alternating diffuser

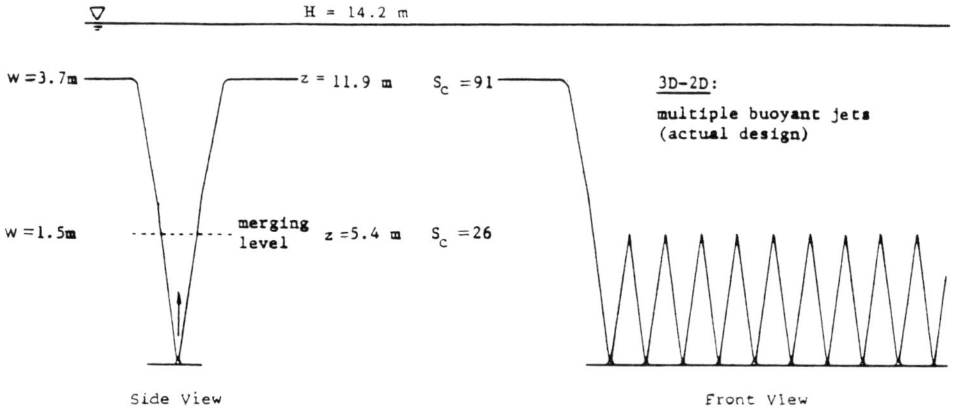

Figure 8.7. Discharge prediction for the Kwun Tong (Hong Kong) submarine sewage outfall. Total discharge flow is 1.95 m^3/s through 275 vertical ports of 0.125 m diameter. The ambient water is unstratified, and the discharge density difference is $\Delta\rho_o = 0.025\ g/cm^3$. The local centerline dilution S_c and total jet width $w = 2\sqrt{2}\,b$ are shown at the merging level and at the impingement level

For a given flow Q, the unit discharge q and unit buoyancy flux j are inversely proportional to the diffuser length L_D, and Equation 8.12 suggests that a higher dilution is obtained by increasing the length of the diffuser. For a line plume, the minimum dilution can be multiplied by a factor of $\sqrt{2}$ to give the average dilution.

The overall behavior of a submerged diffuser plume in deep water is shown in Figure 8.8. For a line plume in uniform stagnant ambient (Fig. 8.8a), laboratory experiments show that the wastewater field thickness to be about 30% of the depth. This result is also predicted by a model which couples the dynamics of the rising buoyant jet and the subsequent gravitational spreading at the surface (Koh, 1983).

It has been demonstrated both theoretically and experimentally (Fischer et al., 1979) that maximum mixing can be achieved with closely spaced ports that allow some interference of adjacent jets. In relatively shallow coastal waters of typical depth 5-15 m, however, it is often the case that, given practical considerations (e.g. in order to maintain a minimum jet velocity and minimum diameter), multiport diffusers are designed to minimize interference of adjacent plumes. In such cases, the required spacing is about $H/3$. The equations given in the previous section on single port discharges also apply for these non-interferring plumes.

For the same entrainment coefficients as for a uniform stagnant ambient, the following equations for a line plume in a linearly stratified ambient (Fig. 8.8b) can be obtained from integral models:

$$z_{max} = 2.84\, j^{1/3} \left(-\frac{g}{\rho_a} \frac{d\rho_a}{dz} \right)^{-1/2} \qquad (8.13)$$

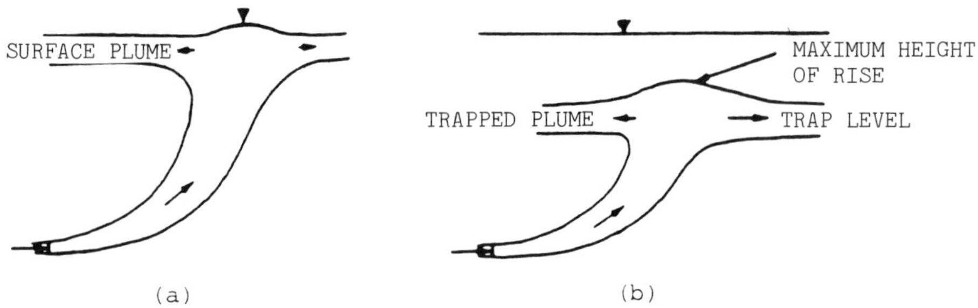

Figure 8.8. Behavior of line plumes from submerged multiport diffuser in deep water. a) uniform ambient with plume rise to water surface. b) stratified ambient with internal trapping of plume

$$S_c = 0.31 \frac{j^{1/3} z_{\max}}{q} \tag{8.14}$$

In a linearly stratified ambient, the spreading layer is found to occupy about 40-50% of the rise height. For computing bulk dilutions, one must allow for the thickness of the wastewater field. Simple models to account for blocking in the presence of an ambient current can be found in Fischer et al. (1979). Recent laboratory experiments of plane buoyant jets in a linearly stratified fluid (Wallace & Wright, 1984) give revised coefficients of 3.6 and 0.24 for Equations 8.13 and 8.14 respectively. Related studies (Lee & Cheung, 1986) with inclined plane jets confirm these revised values for a plume. Two interesting features of these plane buoyant jet studies are: a) for jet-like discharges, the spreading layer occupies almost the entire depth of rise, b) the minimum dilution predicted by an integral model at the maximum height of rise compares favorably with the minimum dilution in the spreading layer. These experiments were carried out in a flume of finite size in which the ambient stratification undergoes continuous change. More experimental and field studies are needed to clarify the connection between these small scale experiments and prototype performance.

Roberts (1979, 1980) studied the mixing of a line source of buoyancy in an ambient current, and found that the shape of the flow field and the dilution is determined by the parameter $F = u_a^3/j$. F measures the ratio of the ambient current velocity to the buoyancy-induced velocity. For $F < 0.1$, the minimum surface dilution S_m is little affected by the current and is given by:

$$S_m = 0.27 \frac{j^{1/3} H}{q} \tag{8.15a}$$

Compared with Equation 8.12, the smaller dilution coefficient reflects the effect of blocking of the surface layer. For higher crossflow, $F > 0.1$, however, the entrainment is dominated by the crossflow, and the alignment angle γ between the diffuser line and the current direction is important. As shown in Figure 8.9, higher

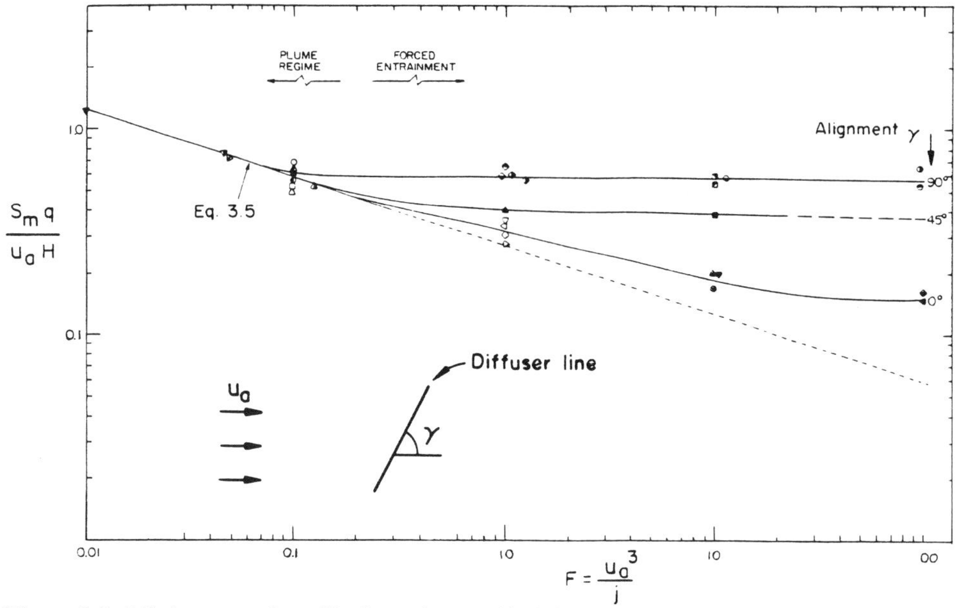

Figure 8.9. Minimum surface dilutions observed in laboratory study for deep water multiport diffuser in crossflow as a function of alignment angle γ (after Roberts, 1979)

dilution results for a perpendicular alignment, $\gamma = 90°$, in which the maximum amount of flow is intercepted while the parallel alignment, $\gamma = 0°$, gives the lowest dilution. For $F \approx 100$, the perpendicular alignment results in a dilution that is 4 times greater than that for the parallel alignment.

The dilution and rise height predicted by a computer model of multiport coastal wastewater outfalls were compared to field observations by Wallis (1981). Reasonable agreement is found for slack or near slack conditions, although in most cases the measured dilutions are somewhat higher than predicted. In a detailed investigation of in-situ dilution for the Alyeska diffuser, Alaska, both the minimum dilution and the trap level of the effluent were measured in the presence of ambient stratification for near stagnant conditions. Based on this data, a comparative evaluation of six selected initial dilution models was made (Alam et al., 1982). Realistic predictions were again confirmed for near stagnant conditions. Roberts' predictions for line plumes in a current also appear to be supported by some field measurments reported by Hendricks (1977).

Figure 8.9 has also been extrapolated to treat a line source in a stratified current in the model ULINE (Muellenhoff et al., 1985). The model can be used to provide estimates of the trap level and thickness of the waste field in a tidal current. Experiments to study the dilution and wastefield formation of a line plume in a steady linearly density-stratified current have recently been reported by Roberts (1989). The results show that mixing behavior is determined by the crossflow parameter F, buoyancy flux j, and the ambient density gradient. For all diffuser alignments γ, an increase in current speed results in increase of dilution, and decrease of rise height and wastefield thickness. As in an unstratified current, maximum mixing is achieved with the perpendicular alignment. However, compared with the unstratified case, the dilution is less sensitive to diffuser orientation.

For $F < 0.1$, the maximum height of rise and the minimum dilution in the wastefield are found to be independent of γ, and approximately the same as given by Equations 8.13 and 8.14, with revised coefficients of 2.6 and 0.37 respectively. For $0.1 < F < 100$, the rise height and minimum dilution for a perpendicular alignment are:

$$z_{\max} = 2.5 \frac{j^{1/2}}{u_a^{1/2}} \left(-\frac{g}{\rho_1} \frac{d\rho_a}{dz} \right)^{-1/2} \tag{8.15b}$$

$$S_m = 0.4 \frac{j^{1/3} z_{\max} F^{1/6}}{q} (2.19 F^{1/6} - 0.52) \tag{8.15c}$$

Equation 8.15b is applicable for $z_{\max} < H$. Similar equations for other alignments are presented in Roberts (1989). Equation 8.15b and 8.15c are based on experiments performed with only one rather strong ambient stratification; the results have not been validated against field data.

Example 3: Estimate the maximum discharge per unit diffuser length that can be released while achieving an initial dilution of 150 in 50 m of stagnant unstratified ocean. What would be the dilution for a perpendicular diffuser alignment in an ocean current of 0.1 m/s?. For this current, also estimate the height of rise and minimum dilution in the wastewater field for a linear density stratified ambient, with $-d\rho_a/dz = 0.2\ \mathrm{kg/m^4}$. The ambient density is 1025 $\mathrm{kg/m^3}$ at discharge level, and the initial relative density deficit is 0.025.

Solution: With H = 50 m, S = 150, and accounting for the surface layer, with Equation 8.15a to give

$$q^{2/3} = \frac{(0.27)\,(9.806 \times 0.025)^{1/3}\, 50}{150}$$

therefore

$$q = 0.0133\ \mathrm{m^2/s}$$

Many ocean sewage diffusers are characterized by unit discharges in the order of 0.01 $\mathrm{m^2/s}$. With $u_a = 0.1$, $F = 0.1^3/(0.0133 \times 9.806 \times 0.025) = 0.31$. From Figure 8.9, one obtains $Sq/u_aH = 0.58$, to compute a surface dilution

$$S = 0.58\,(0.1)\,(50)/0.0133 = 218$$

With the given ambient density gradient, the stratification parameter is

$$\left(-\frac{9.806}{1025}\times 0.2\right) = 1.91\times 10^{-4}\, s^{-2}$$

Using Equation 8.15b to estimate the rise height (top of waste field)

$$z_{\max} = 2.5\,\frac{(0.0133\times 9.0806\times 0.025)^{1/2}}{(0.1)^{1/2}}\quad (1.91\times 10^{-4})^{-1/2} = 32.7 \text{ m}$$

which is less than the total water depth. The minimum centerline dilution in the waste field is (Equation 8.15c)

$$S_m = 0.4\,\frac{(9.806\times 0.025)^{1/3}\,(32.7)}{(0.0133)^{2/3}}\quad (0.31)^{1/6}\,[2.19\,(0.31)^{1/6} - 0.52] = 154$$

Surface waters are protected by the submergence of the wastefield, at the expense of a lower initial dilution.

Example 4: Consider a sewage diffuser of length L_D and discharge flow Q. Compare the initial dilution achieved by 2 alternative designs in stillwater: a) N non-interfering plumes spaced at $s = H/3$, and b) a large number of closely spaced ports simulating a line source.

Solution: For design a), Equation 8.2 gives

$$S_1 = \frac{0.11\, g'^{1/3} H^{5/3}}{(Q/N)^{2/3}} = (0.11)3^{2/3}\quad \frac{g'^{1/3} H L_D^{2/3}}{Q^{2/3}}$$

For a line plume. design b), Equation 8.12 gives

$$S_2 = 0.38\,\frac{g'^{1/3} H L_D^{2/3}}{Q^{2/3}}$$

Therefore, $S_1/S_2 = (0.11)\,3^{2/3}/0.38 = 0.6$. Given a sufficiently large depth (say 40 m) for the multiple plumes to form a true line plume, better mixing is obtained by designing the ports to simulate a line source of buoyancy. However, in shallow coastal waters of typical depth 5 to 10 m, other practical considerations may limit the number of ports, and the initial dilution may be poorly approximated by the line plume formula. In many cases, satisfactory mixing can be achieved with design a).

8.3.2 *Shallow water diffusers*

If the stability parameter in Equation 8.11 is exceeded, then techniques that are altogether different from those for buoyant jets, given previously for deep water diffusers, must be used. The shallow water diffuser theory is required primarily in the design of submerged cooling water discharges. Thermal diffusers typically discharge large flows (30 to 40 m^3/s per 1000 MW) in shallow depths of perhaps

5 to 10 m, and the mixing ratio required is relatively low, in the order of 10. The near field, with a horizontal extent of the order of L_D, is typically unstable and vertically well-mixed, although the mixed flow usually stratifies in the intermediate and far fields. Further, the diffuser length is typically larger than the depth, $L_D/H \gg 1$. Various nozzle arrangements can produce quite different flow configurations.

As shown in Figure 8.10, both the unidirectional and the staged diffusers produce vertically mixed plumes that flow in the same direction as the jet discharge. As opposed to sewage diffusers, the fluid motion is independent of the buoyancy, and the main governing parameters are the momentum flux per unit diffuser length, m, and the receiving water depth H. The essentially two-dimensional inviscid flow can be viewed as being generated by a line source of momentum.

The flow induced by a unidirectional diffuser is driven by pressure gradients set up by the momentum input. The induced flow separates near the ends of the diffuser into a contracting slipstream, and resembles the flow through a nonlinear actuator disk (such as a propeller). The two-dimensional velocity field can be obtained from an exact inviscid solution (Lee & Greenberg, 1984). The bulk (average) dilution in a current aligned with the momentum input (coflowing diffuser) is given by (Adams, 1982; Lee & Jirka, 1980; Lee, 1985):

$$S = 1/2\left(V + \left(V^2 + \frac{2\,m\,H}{q^2}\right)^{1/2}\right) \tag{8.16}$$

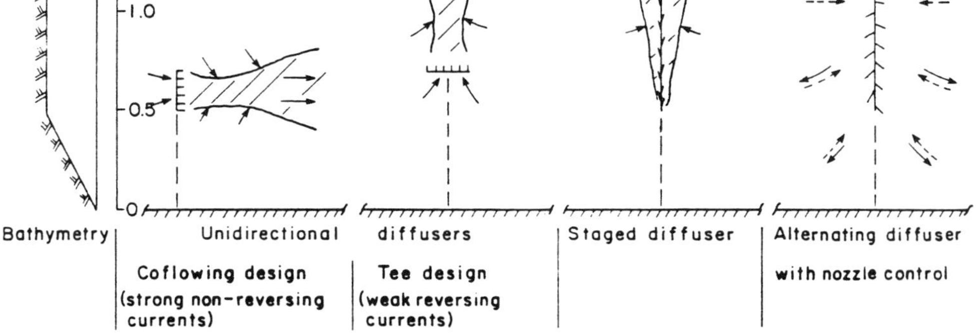

Figure 8.10. Examples of diffuser configuration and flow fields for thermal discharge in a shallow coastal environment under stagnant conditions (from Jirka, 1982). Dimensions are typical for cooling water discharge from a 1000 MW power plant

where $V = u_a H / q$. Equation 8.16 has been verified with good accuracy against excess temperature measurements and detailed velocity measurements. Predictions of excess temperatures and velocities for the shallow water plume in the intermediate field have also been given (Lee & Jirka, 1980). The unidirectional diffuser is effective in a coflowing current, but gives poor mixing performance in an opposing current. If the discharge is in an offshore direction, the net momentum imparted to the receiving water directs the thermal effluent away from sensitive nearshore areas of biological importance. However, in a perpendicular longshore current, experiments show that the mixing performance of such a 'Tee-diffuser' depends on the ratio of the ambient to discharge momentum flux, m_a / m, where $m_a = u_a^2 H$. For $m_a / m < 0.1$, the dilution is similar to that given by Equation 8.16; however, the dilution deteriorates rapidly for larger values of m_a / m. Analysis over a wide range of data (Adams & Stolzenbach, 1976) leads to an empirical factor r_s, which gives the dilution relative to the still water dilution (Equation 8.16 with $V = 0$):

$$r_s = \left(1 + 5\frac{m_a}{m}\right)^{-\frac{1}{2}} \tag{8.17}$$

The dilution reduction is caused by the interaction of the individual jets; the pressure distribution set up by the ambient current limits the quantity of water which can enter from behind the diffuser.

Because of the large induced flows involved, the unidirectional diffuser has to be placed sufficiently far offshore to provide enough space for the entrainment flow from behind the diffuser. With a shoreline boundary located at a separation distance x_s from and parallel to the diffuser line (Tee design in Fig. 8.10), theory and experiments (Fig. 8.11) show that significant reduction of induced flows can result if the separation distance is less than 0.5 L_D (Adams et al., 1982; Lee, 1984).

The jets in a staged diffuser usually point at a small angle to the diffuser axis with a predominantly offshore momentum. The induced flow has a boundary layer geometry (Fig. 8.10) and can be modelled as a line source of momentum continuously imparted to the receiving water over the diffuser length. A model based on an entrainment hypothesis (Lee, 1980; Almquist & Stolzenbach, 1980; Brocard, 1980) gives the bulk dilution as:

$$S = \frac{V}{2} + \left[\left(\frac{V}{2}\right)^2 + 0.2\frac{mH}{q^2}\right]^{\frac{1}{2}} \tag{8.18}$$

For stagnant ambient Equation 8.18 reduces to $S = 0.45\,(mH)^{\frac{1}{2}}/q$. An alternate theory based on momentum superposition (Jirka, 1982b) gives a stagnant dilution coefficient of 0.67 instead of 0.45. The staged diffuser combines good offshore transport of diffuser plume during stagnant conditions with a good capture of longshore current flow from either direction. Also, the mixing performance improves as the current strength increases.

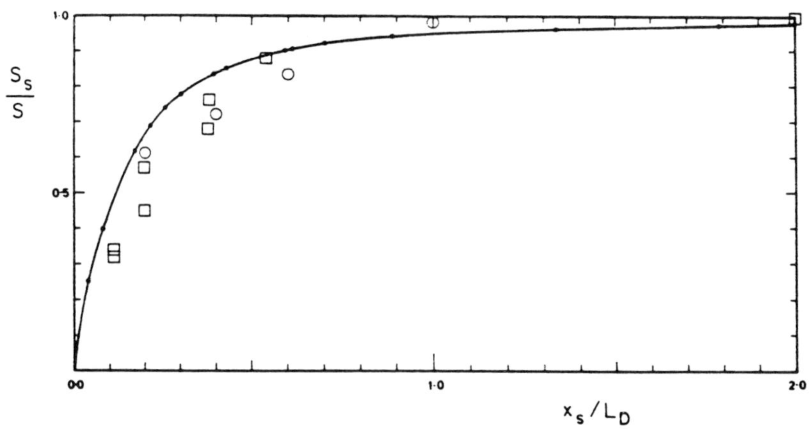

Figure 8.11. Effect of limited separation distance x_s/L_D between diffuser line and shoreline. The bulk dilution S_s is reduced relative to the unconstrained bulk dilution S as predicted by Equation 8.16 (with $V=0$) (from Lee, 1984)

Unlike the previous two designs, the alternating diffuser does not impart any net horizontal momentum. A stratified counterflow region is generated outside the unstable recirculation zone. The bulk dilution can be predicted from stratified flow theory as:

$$S = c j^{1/3} H/q \tag{8.19}$$

where the coefficient c = 0.45 – 0.55 depends on friction effects in the counterflow. Note the similarity of Equation 8.19 to the line plume solution, Equation 8.12. The dilution in an ambient current can be estimated by vectorial addition of the stagnant water dilution and the ambient flow dilution. Jirka (1973, 1982b) has demonstrated that the use of 'fanned out' alternate jet discharges, in which the individual jets have a variable angle with respect to the diffuser line, helps prevent strong induced horizontal circulations and ensures a stably stratified intermediate field.

The equations given for bulk dilution for each type of diffuser provide a basis for determining the required diffuser length in terms of a given depth, plant heat rejection rate, condenser temperature rise, discharge velocity, and maximum permissible near field excess temperature. For the stagnant condition, the unidirectional design gives the best mixing capacity, whereas the alternating diffuser induces the weakest currents. Characterized by their symmetrical performance with respect to ambient current direction, the staged and alternating diffusers are generally more appropriate in tidal waters. The Tee-diffuser can be used if the magnitude of the ambient current is small. A useful guide to thermal discharge prediction techniques for coastal application has been given by the British Hydromechanics Research Association (Miller & Brighouse, 1984). As an interesting case study, the detailed behavior of the thermal field in the vicinity of the submerged discharge at the Sizewell power station at different stages of the tide is discussed in that publication.

Example 5: Consider a 170 m long diffuser consisting of $N = 12$ ports discharging heated cooling water at an excess temperature of 20°C. Inclined at 10° above the horizontal, the jets discharge at 3.5 m/s into coastal water of 10 m depth. The jet diameter is 1.25 m, and the ambient temperature is 26°C, resulting in a relative density difference of 0.0082. The design ambient current is 0.1 m/s. Determine the near field dilution for three port arrangements: a) coflowing diffuser, b) staged diffuser, and c) alternating diffuser.

Solution: $N = 12$, $D = 1.25$ m, $L_D = 170$ m, $H = 10$ m, $U_o = 3.5$ m/s, $\Delta\rho_o/\rho =$ 0.0082. The discharge per unit diffuser length is $q = (12)\,\pi(1.25)^2\,3.5/(4\times 170) =$ 0.30 m²/s, $m = q\,(3.5)$, and $j = q\,(9.81)\,0.0082$. Therefore from Equation 8.11 the stability parameter can be calculated for the stagnant condition:

$$\frac{(3.5)\,(0.30)^{1/3}}{(9.801\times 0.0082)^{2/3}\times 10} = 1.262$$

while the limiting value for instability is $0.54/(1+\cos^2(10°))^2 = 0.14$. Therefore, as expected, this thermal diffuser operates in the unstable range.

a) *Unidirectional diffuser*: For $u_a = 0.1$ m/s, the ratio of ambient crossflow to discharge flow per unit length is $V = (0.1)\ 10/0.30 = 3.3$. Equation 8.16 then gives

$$S = 1/2\left(3.3 + \left(3.3^2 + \frac{2\times 0.30\times 3.5\times 10}{(0.30)^2}\right)^{1/2}\right) = 9.4$$

for the coflowing diffuser. A maximum surface excess temperature of 20/9.4 = 2.1°C is hence expected.

b) With an *alternating diffuser*, Equation 8.19 gives the stillwater dilution:

$$S_o = \frac{0.5\,(9.806\times 0.0082)^{1/3}\,10}{(0.30)^{2/3}} = 4.8$$

The dilution in the presence of a current is then found by vectorial addition:

$$S = (S_o^2 + V^2)^{1/2} = (4.8^2 + 3.3)^{1/2} = 5.8$$

giving an excess temperature of 3.4°C.

c) Using a *staged diffuser*, the stagnant dilution is given by Equation 8.18

$$S_o = (0.2\times 0.30\times 3.5\times 10)^{1/2}/0.30 = 4.8$$

With an ambient current, the dilution is increased to (Equation 8.18):

$$S = \frac{3.3}{2} + \left[\left(\frac{3.3}{2}\right)^2 + 4.8^2\right]^{1/2} = 6.7$$

with an excess temperature of 3.0°C.

8.3.3 *Detailed mathematical models*

The initial mixing relations presented in the foregoing are applicable only to simplified discharge and ambient conditions; more comprehensive models have to be used for complex situations. The USEPA models UDKHDEN and UMERGE account for the merging of the individual round buoyant jets to form a line plume, but are limited to deep receiving water and to a perpendicular diffuser alignment. In shallow receiving water these models are not flux conserving and may give unrealistic predictions (Jirka, 1991).

A general expert system (CORMIX2) for the hydrodynamic classification and initial mixing analyses of multiport diffuser discharges has recently been reported (Akar & Jirka, 1990; Jirka & Akar, 1991). CORMIX2 addresses the full range of discharge geometries and ambient conditions, and predicts flow configurations ranging from internally trapped plumes, buoyant plumes in uniform density layers with or without shallow water instabilities, and sinking (negatively-buoyant) plumes. Boundary interaction, upstream intrusion, buoyant spreading and passive diffusion in the far field (see Section 8.5) are also considered. Figure 8.12 shows an example prediction for a staged multiport thermal diffuser plume in shallow water and its comparison with data from a hydraulic model study.

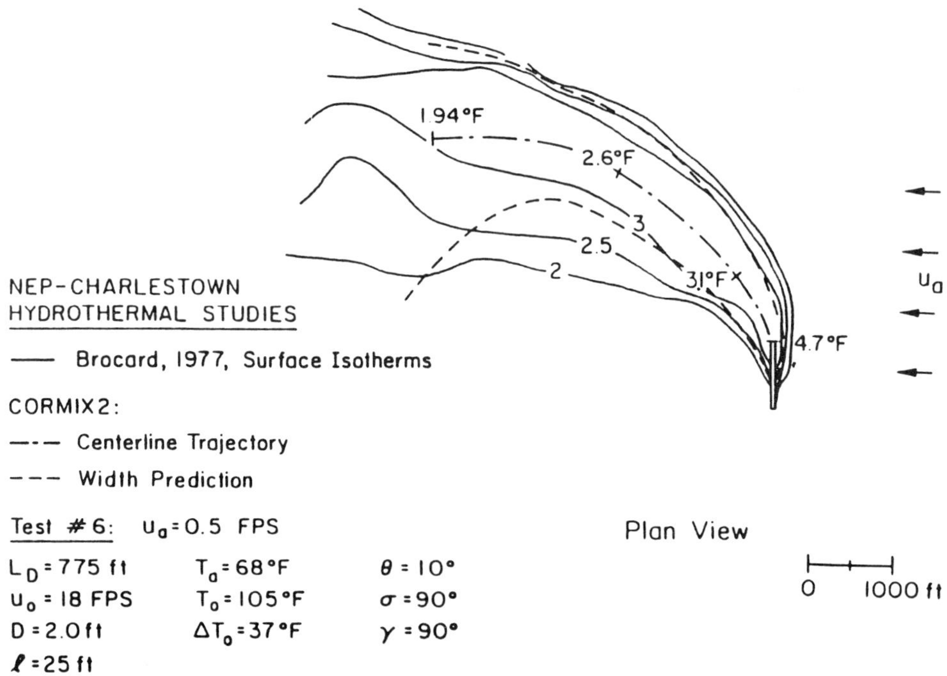

Figure 8.12. Plume pattern (top view) predicted by expert system CORMIX2 for a staged multiport thermal diffuser in perpendicular alignment in shallow coastal water. The plume is vertically fully mixed. Laboratory data by Brocard (1977) is also shown for comparison

8.4 SURFACE DISCHARGES

Perhaps the most common, and often the most economical, approach to waste disposal is to discharge effluents through an existing or artificial open conduit at the shoreline. Many power stations discharge heated condenser cooling water in the form of such surface jets. Nevertheless, surface discharges often have the disadvantage of limited near-field dilutions achievable, and of pollutant build-up directly at the shoreline in the form of a plume that may attach to the shore in the presence of strong cross-currents. Such shoreline impacts may be undesirable from an ecological viewpoint.

Mathematical, laboratory, and field studies of buoyant surface jets have been carried out by many investigators. A summary of properties of surface jets is presented in Jirka et al. (1981). A more fundamental discussion on the fluid mechanics of buoyant surface jets and plumes has recently been given by Chu and Jirka (1986).

Stagnant water: As shown in Figure 8.13, a three-dimensional buoyant surface jet in stagnant water is characterized by discharge channel width $2b_o$, depth h_o, discharge velocity U_o, and reduced gravitational acceleration g'. Two regions of distinct dynamic behavior can be discerned. Near the discharge point, the flow resembles a three dimensional jet in which buoyancy plays a minor role, jet

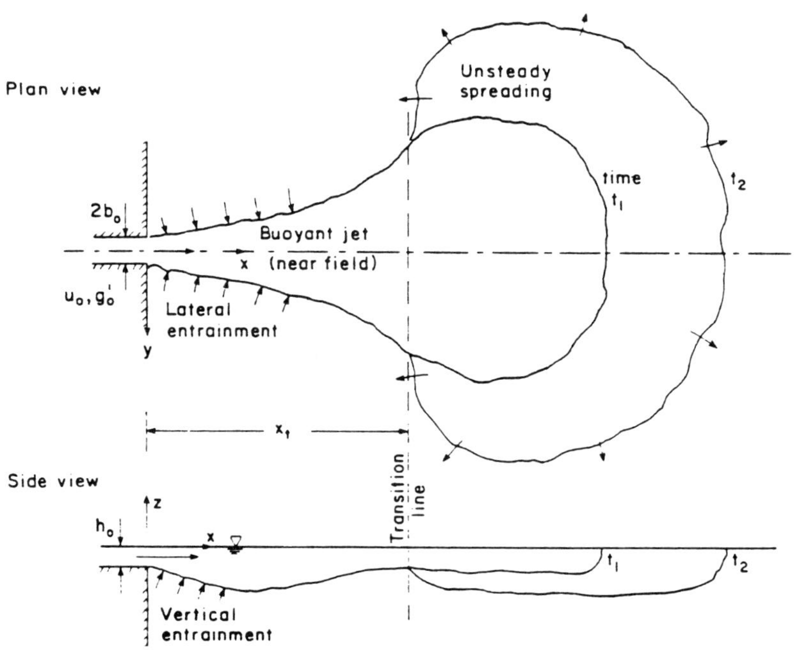

Figure 8.13. Structure of buoyant surface jet in stagnant deep receiving water (after Jirka et al., 1981)

growth is approximately linear, and dilution is achieved by both vertical and lateral entrainment. The mixing results in decreasing local densimetric Froude numbers along the jet trajectory; the effect of buoyancy becomes increasingly important; and, beyond a transitional distance x_t, the flow field is governed entirely by gravitational spreading in all directions. Within the present discussion this unsteady flow field is labelled as an 'intermediate field' as it provides an intermediate stage between the near field and the passive far field mixing processes. The buoyant forces in the intermediate field also damp vertical entrainment and negligible mixing occurs beyond x_t. The important dimensionless parameters are the densimetric Froude number based on a characteristic length scale $l_o = \sqrt{h_o b_o}$, $F'_o = U_o / \sqrt{g' l_o}$, and the aspect ratio $A = h_o / b_o$. A number of mathematical models for the surface jet (Stolzenbach & Harleman, 1971; Shirazi & Davis, 1974; Prych, 1972) have been formulated. In general, for deep water conditions, model predictions agree reasonably well with temperature and velocity measurements. The following equations for the bulk properties of the surface jets are derived from an integral jet model (Stolzenbach & Harleman, 1971) and have been verified against data from various sources.

The maximum jet thickness, defined by the depth at which the pollutant concentration becomes negligible, is given by:

$$\frac{h_{max}}{l_o} = 0.42\, F'_o \tag{8.20}$$

and is located at a longitudinal position

$$\frac{x_{max}}{l_o} = 5.5\, F'_o \tag{8.21}$$

The transition to the intermediate field occurs at:

$$\frac{x_t}{l_o} = 12\, F'_o A^{-0.2} \tag{8.22}$$

The mixing process beyond the transition distance x_t is rather inefficient, and in particular, the centerline dilution reaches a stable plateau. The stable centerline and bulk dilution for $F'_o > 3$ is given by:

$$S_c = F'_o \tag{8.23}$$

$$S = 1.4\, F'_o \tag{8.24}$$

Equation 8.22 to 8.24 provide a starting point for transient passive field calculations, as it is often acceptable to neglect the intermediate field spreading.

Shallow water effects: If discharged in shallow depths, the vertical confinement of the fluid layer leads to higher induced velocities and reduced hydrostatic pressures for a given entrainment demand, and leads to bottom attachment of the deepening jet. Further, the mixing capacity is reduced. Experiments with buoyant jets show that for small h_{max}/H, good agreement with deep water predictions is obtained, while for large h_{max}/H induced concentration levels are higher. A simple criterion for shallow water conditions was given by Jirka et al. (1981):

$$\frac{h_{max}}{H} > \frac{3}{4} \tag{8.25}$$

where h_{max} is the maximum jet thickness predicted by Equation 8.20 and H is the receiving water depth. Most cooling water discharges exhibit some degree of bottom attachment. The dilution reduction factor for shallow water conditions can be estimated by an empirical correlation:

$$r_s = \left(\frac{0.75}{h_{max}/H}\right)^{0.75} \quad \text{for} \ \frac{h_{max}}{H} > 0.75 \tag{8.26}$$

Ambient currents: In a strong ambient current u_a, the surface jet may become attached to the shoreline boundary, resulting in recirculation in the lee of the jet. For many applications, the criterion for shoreline attachment of a perpendicular surface discharge shown in Figure 8.14 can be used. Based on the ambient current

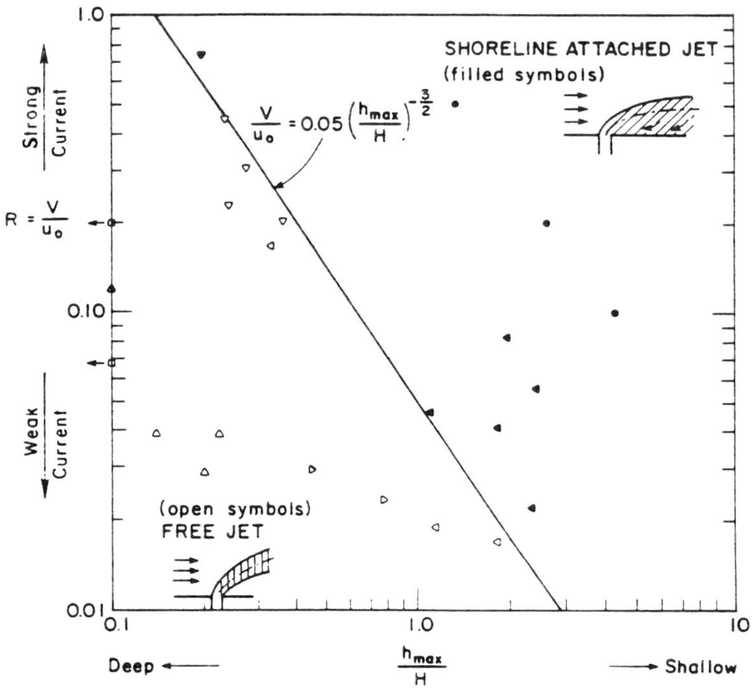

Figure 8.14. Shoreline attachment for buoyant surface jets as function of velocity ratio, u_a/U_o, shallowness, h_{max}/H (after Jirka et al., 1981)

ratio $R = u_a/U_o$ and the shallowness factor h_{max}/H, this figure classifies observations into two categories. As expected, the combination of a large crossflow in shallow water is conducive to attachment. Experimental results on shoreline-attached buoyant jets were summarized in the form of design nomograms by Schatzmann & Naudascher (1980).

A more extensive classification diagram that also includes the behavior of highly buoyant discharges (low Froude number), such as natural rivers entering the ocean, has been presented by Chu & Jirka (1986). Discharges of the highly buoyant category may, in fact, not display the jet-like near-field at all, the result being a strong buoyant spreading against the direction of the ambient current ('upstream intrusion') and the formation of a curved 'sharp front' separating ambient and effluent water.

A summary of surface jet applications for thermal discharges from power plants is given by Miller & Brighouse (1984). Integral jet models presented in the foregoing have also been incorporated in an expert system (Jones & Jirka, 1991) for the classification and predictions of surface discharges.

Example 6: A coastal power station discharges 64 m^3/s of heated condenser cooling water (relative density difference 0.002) into a tidal estuary of mean depth 10.5 m. A 12 m wide surface channel along the shoreline is used. The tidal range is 3 m and the depth of the channel is 4 m at mean sea level. Compute the near-field mixing parameters of interest for the surface jet at high tide and low tide.

Solution: a) *High tide*: $h_o = 5.5$ m, $H = 12$ m, $U_o = 64/(12 \times 5.5) = 0.97$ m/s, $l_o = \sqrt{6 \times 5.5} = 5.75$ m, with an aspect ratio of $5.5/6 = 0.92$. The surface jet densimetric Froude number is

$$F'_o = \frac{0.97}{(9.81 \times 0.002 \times 5.75)^{1/2}} = 2.9$$

Equation 8.20 then gives $h_{max} = 0.42 \times 2.9 \times 5.75 = 7$ m.

Since $h_{max}/H = 7/12 = 0.58 < 0.75$ (Equation 8.25), the jet discharges in deep water, with a maximum penetration at (Equation 8.21)

$$x_{max} = 5.5 \times 2.9 \times 5.75 = 91 \text{ m},$$

and the transition to far field occurs at (Equation 8.22)

$$x_t = 12 \times 2.9 \times (0.92)^{-0.2} \times 5.75 = 203 \text{ m}$$

The centerline and bulk dilution in the stable buoyancy dominated region are (Equation 8.23 and 8.24):

$$S_c \approx 3 \qquad \text{and} \qquad S = 1.4 \times 3 = 4.2$$

b) *Low tide*: $h_o = 2.5$ m, $H = 9$ m, $U_o = 2.1$ m/s, $l_o = 3.9$m, and $F'_o = 7.73$.

Therefore, the maximum depth of surface jet predicted by deep water theory is $h_{max} = 0.42 \times 7.7 \times 3.9 = 12.6$ m. Since $h'_{max}/H = 1.4$, strong bottom attachment is expected. Using Equation 8.26, the reduction in dilution is given by a factor of $(0.75/1.4)^{0.75} = 0.63$, and the bulk dilution is $S = 1.4 \times 7.73 \times 0.63 = 6.8$.

The difference in dynamic behavior at different stages of the tide should be noted. In general, narrow elongated jet like patterns are observed at low tide, and gravitational spreading with wide thermal fronts are dominant at high water (see Chu & Jirka, 1986).

8.5 FAR FIELD MIXING AND TRANSPORT

After the near field mixing phase is complete, the dynamic (momentum and buoyancy) source inputs of the waste effluent have been sufficiently dispersed or dissipated; further mixing and transport of the effluent are dominated by passive far field effects. The ambient advective transport and turbulent diffusion in the ocean environment control the far field mass distribution. (Occasionally residual buoyancy effects persist and they may modify somewhat the far field flow field or limit vertical diffusion; for large scales of interest, however, these phenomena can usually be neglected in engineering analyses).

The required detail of a far field calculation can vary considerably from case to case. It depends on the complexity of the coastal ocean environment, the availability of data, and finally, the severity of the pollution problem. For an open coastal environment with a prevailing current structure a simple plume model (Section 8.5.1) may suffice if a far field calculation is needed at all. For estuaries or semi-enclosed bays, a flushing analysis (Section 8.5.2) may be required to ascertain the net flow-through and potential long-term accumulation of pollutants. Finally, in coastal environments or tidal networks of complex topography and current structure, numerical transport models (Section 8.5.3) have to be employed for solution of the convective diffusion equation in order to predict the far-field pollutant distribution. The velocity field in these transport models may have to be calculated from a separate circulation model or be obtained from detailed field measurements.

In any of these far-field modeling approaches, one can describe through appropriate source/sink terms any of the ultimate transformation processes that the pollutant may be subject to. These include biological or chemical reactions, coagulation and settling processes, adsorption mechanisms at the sediment interface, and exchange processes at the air-water interface.

8.5.1 *Simple plume models*

Beyond the initial mixing zone governed by buoyant jet mixing, a pool of wastewater is formed either at the surface or at the level of submergence under stratification conditions. The wastewater could be brought shoreward either by lateral turbulent diffusion in a predominantly alongshore current or directly by wind generated onshore currents. A common water quality objective is to prevent high concentrations of pollutants from reaching coastal recreational waters or shell fisheries.

For a line diffuser normal to a current, the initial width of the wastewater field can be taken to be equal to the diffuser length; the depth of the field in moving stratified water can be estimated on the basis of continuity considerations and accounting for partial blocking above the base of the field (Brooks, 1972):

$$h = z_{max} \{(QS/u_a L_D z_{max})/(1 + QS/u_a L_D z_{max})\} \qquad (8.27)$$

where S, the average dilution, and z_{max} are obtained from the near-field solution, Equation 8.13 and 8.14, respectively. Equation 8.27 is expected to hold for $SQ/(u_a L_D z_{max}) \leq 2$. For a point source, the initial half-width and depth of the wastewater field can be estimated as 0.3 H for the bdnf (Equation 8.9) or as $(SQ/\pi u_a)^{1/2}$ for the bdff (Equation 8.10). These dimensions are needed as initial conditions for transport calculations in the far field. Further examples of the coupling of near field solutions to far field plume models are contained in Adams et al. (1975) and Jirka et al. (1986).

In still water, the surface field with its residual buoyancy spreads in an unsteady fashion. Much recent research is devoted to the dynamics of the intermediate field in which buoyancy plays a significant role (e.g. Koh, 1983). However, under high dilution conditions in moving water, the wastewater field is essentially advected passively in a tidal current. Vertical mixing is damped by buoyancy, so that the subsequent dilution is mainly due to horizontal mixing by the turbulent eddies. A review of instantaneous plume prediction equations in ocean waters is given by Okubo (1962). Investigations have shown (e.g. Csanady, 1973; Fischer, 1979) that this dilution process is in general non-Fickian; the eddy diffusivity depends on the scale of diffusion. A 4/3 power law is often adopted for engineering calculations (Okubo, 1971):

$$E = \alpha\, l^{4/3} \qquad (8.28)$$

with $\alpha = 0.001$ to $0.1\ \text{cm}^{2/3}$ /s, depending on the scale l (in cm) of the pollutant distribution. A more narrow range, $\alpha = 0.01$ to $0.02\ \text{cm}^{2/3}$ /s, is an adequate representation over a wide data range.

Brooks (1960) gives a widely used method of estimating the subsequent dilution of a wastewater field due to lateral mixing by oceanic turbulence. A line source is assumed with initial conditions determined by the initial mixing phase as discussed above; the horizontal transport of a pollutant layer of constant thickness in a steady uniform current normal to the source is then formulated. By assuming a 4/3 power dependence of the eddy diffusivity on the local plume width, Equation 8.28, and a first order decay with rate constant k, the centerline concentration in the plume is given by:

$$\frac{c_{max}}{c_i} = \left\{ \operatorname{erf} \left[\frac{3/2}{\left(1 + \frac{8E_i t}{l_i^2}\right)^3 - 1} \right] \right\}^{1/2} e^{-kt} \tag{8.29}$$

where t = travel time = x/u_a, E_i = diffusivity corresponding to the initial width of the plume l_i, and c_i = initial concentration.

Subject to the global flushing constraints outlined in Section 8.5.2, Equation 8.29 is useful in giving a conservative estimate of the subsequent dilution and elucidating the relative importance of horizontal diffusion and decay processes. It is, however, limited in two respects: a) nearshore tidal currents are unsteady, and b) a wind-generated onshore surface current is usually accompanied by a compensatory offshore bottom current; Munro & Mollowney (1974) have shown that in shallow coastal waters vertical mixing in such a counter current system can lead to substantial additional reductions in concentration. This result is supported by detailed observations of bacterial distributions at several sea outfalls (Gameson, 1982; Munro, 1984).

The above restrictions can be removed by using general predictive mathematical models (e.g. Koh & Fan, 1970; Adams et al., 1975) based upon similar physical reasoning. By noting that concentration distributions in the horizontal plane are usually Gaussian, the method of moments is applied, and the integrated form of the advective diffusion equation solved numerically as a parabolic initial value problem. The passive diffusion in a nonuniform steady current and the time dependent diffusion in a uniform unsteady current can be treated. Various realistic features (e.g. the return of previously diluted sewage over the outfall and different source conditions) can be readily simulated by superposition methods. However, these methods (as opposed to the general numerical models discussed in Section 8.5.3) are limited to a uniform current field and only allow very simple coastal boundary conditions (e.g. a straight coastline). More recently, Chin & Roberts (1985) developed a Lagrangian random-walk model that directly uses data from spatially distributed, continuously recording current meters to simulate far field dispersion. The concept of 'visitation frequency' has also been suggested and applied to estimate the probability of the wastefield reaching a particular coastal location (Csanady & Churchill, 1987; Koh, 1988; Roberts, 1986).

Example 7: An ocean outfall running parallel to the shore 4 km away discharges 2 m^3/s of sewage through multiple ports in a diffuser 400 m long. The velocity of the shoreward ocean current is 0.15 m/s, and the initial dilution of the wastewater in the surface wastewater field is 85. The coliform bacteria concentration of the effluent is 5×10^7 counts/100 mL. Estimate a) the subsequent dilution due to lateral mixing of a conservative pollutant at the shore, and b) the coliform bacteria concentration expected at the shore if the mortalilty of bacteria is characterized by a T_{90} (time for inactivation of 90% of organisms) of 4 hours.

Solution: a) The eddy diffusivity corresponding to the initial width of the waste-water field is $E_i = 0.01\ (400 \times 10^2)^{4/3}\ \mathrm{cm^2/s} = 1.37 \mathrm{m^2/s}$ and the travel time is

$$t = x/u_a = \frac{4000}{(0.15) \times 3600} = 7.4\ \mathrm{h}$$

such that

$$1 + \frac{8E_i t}{l_i^2} = 2.83$$

Equation 8.29 then gives for the conservative case ($k = 0$)

$$\frac{c_{\max}}{c_i} = \left\{ \mathrm{erf}\left(\frac{3/2}{22.67 - 1} \right) \right\}^{1/2} = 0.28$$

The subsequent dilution by horizontal mixing is about 3.

b) With the first order decay assumed for coliform bacteria, the expected bacteria concentration is (k = 13.8 per day) $c_{\max} = 5 \times 10^7/85 \times 0.28 \times e^{-kt} = 2.3 \times 10^3/100$ mL with a concentration reduction factor of about 70 due to bacterial mortality alone. It can be shown that in the far field subsequent dilution is relatively slow for long diffusers which effect good initial mixing and vice versa. Bacterial natural mortality (often accelerated by solar radiation) is the most important factor for concentration reduction. Consequently, outfalls are often sited so as to allow for adequate travel times to shoreline regions to be protected.

8.5.2 *Pollutant flushing in estuaries and coastal bays*

The circulation processes in an estuary are often complicated, and driven by tides (barotropic), salinity gradients (baroclinic), and wind. For waste discharges located in estuaries or semi-enclosed bays, it is often of interest to determine the residence period of any pollutant released in the estuary – i.e. the average time that any conservative pollutant particles stay in the region of interest. In addition, as each flood tide brings back some of the pollutant discharged into the ocean on the previous ebb, it is also important to determine the tidal exchanges at the ocean mouth. In general, the long term average conditions of pollution in a coastal water are governed by these global considerations that usually cannot be computed a priori. Although two dimensional numerical transport models are commonly used in lieu of traditional tidally averaged or slack tide water quality models, it is useful to outline some practical methods of predicting estuarine pollutant concentrations. When used in conjunction with field data, these models give reliable first estimates of the flushing times and of the changes in water quality to be expected after the installation of a proposed waste discharge. Often, these preliminary calculations allow alternatives to be screened and indicate whether further detailed pollutant transport calculations are required. For an introduction to the subject, the reader is referred to Fischer et al. (1979) and Dyer (1973).

8.5.2.1 *Tidal prism models*

The classical modified tidal prism method due to Ketchum (1951) divides the estuary into a number of segments of length equal to the local average tidal excursion. Each segment is assumed to be completely mixed at high tide. Defined in this manner, the low tide volume of each segment is then equal to the high tide volume of the next landward segment. Given mean tidal data, the estuary geometry and the river flow, a segmentation can be obtained by starting at the river end. Once the segmentation is established, the pollutant (freshwater) concentration and the flushing time can be calculated from the low and high water volumes of each segment. The method is simple and appears to reproduce observed salinity distributions in well-mixed estuaries when the number of segments is large. Although the method possesses certain internal inconsistencies, it can give reasonable answers, especially in questions of a comparative nature, even in the absence of any measurements – the only required inputs are a nautical map and a representative tidal range. Based on mass balance considerations, Dyer (1973) has proposed a more rigorous approach which removes the theoretical difficulties. An interesting adaptation of the tidal prism method for the case of no freshwater runoff has also been reported (Kuo, 1976).

8.5.2.2 *Tidally-averaged segmented ('box') models*

A more common engineering tool is the 'box' or segmented model. The area of interest is divided into segments within each of which the pollutant is assumed to be well-mixed. Across each segment boundary, the model allows for a net advective flow (corresponding to the freshwater flow) as well as a 'tidal exchange' flow that circulates mass between adjacent segments over a tidal cycle.

Figure 8.15 illustrates a self-explanatory mass balance that can be formulated for each segment; non-conservative transformations can be easily incorporated. If the equations are applied to calculate salinity, as a tracer, the tidal exchanges can be inferred from observed salinities. This type of model is usually applied for tidally averaged conditions, although it can also be used for slack tide models (Lee & Choi, 1986) in which case the salinities measured at slack tide should be used. In this method all the hydrodynamic exchanges are completely determined by the salinity distribution which mirrors the mixing processes. The exchange flows are given by:

$$Q_{n,\,n+1} = \frac{Q_{f,\,n}\, s_n}{s_{n+1} - s_n} \tag{8.30}$$

where $Q_{f,\,n}$ = the freshwater flow landward of transect n, and s_n, s_{n+1} are the observed salinities in segments n and $n + 1$. The flushing time or residence period of river water or any released pollutant in a particular segment is given by:

$$T_n = \frac{V_n f_n}{Q_{f,\,n}} \tag{8.31}$$

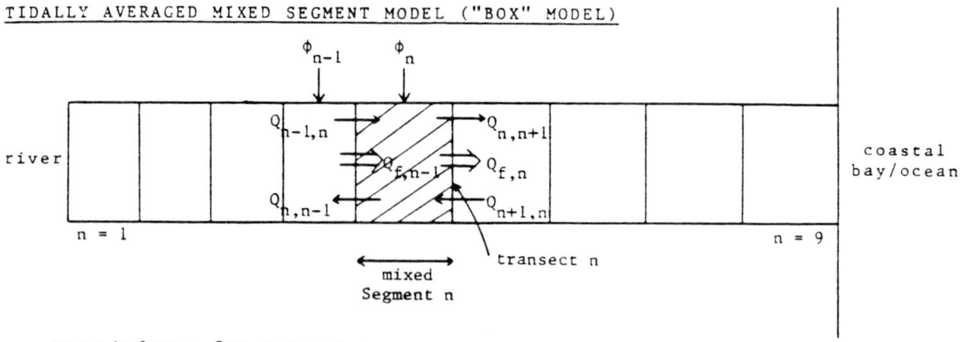

Mass balance for segment n:

$$V_n c_n^{k+1} = V_n c_n^k - \underbrace{k_d V_n c_n^k}_{\text{decay}} + \underbrace{Q_{f,n-1} c_{n-1}^k - Q_{f,n} c_n^k}_{\text{freshwater flushing}}$$

$$+ \underbrace{Q_{n,n-1}(c_{n-1}^k - c_n^k) + Q_{n,n+1}(c_{n+1}^k - c_n^k)}_{\text{tidal mixing}} + \phi_n$$

c_n^k = tidally averaged concentration in segment n in tidal cycle k

$Q_{f,n}$ = freshwater runoff per tide landward of transect n

$Q_{n,n+1}$ = 'exchange' flow per tide between segment n and n+1 = $Q_{n+1,n}$

ϕ_n = pollutant loading per tide in segment n

Figure 8.15. Definitions and formulation of segment ('box') model for tidally averaged conditions

where V_n is the mean volume of the segment and f_n is the freshwater concentration, $f_n = 1 - s_n/s_s$ in which s_s = ocean salinity. Required inputs are salinity, estuary geometry and river flow, and these quantities are often known. There are no undetermined or undefined hydrodynamic coefficients. On the other hand, the method is restricted to the saline portion of an estuary. In estuaries with major river inflows, a significant region may exist upstream of the region of salinity intrusion in which tidal effects occur but the method is not directly applicable.

Figure 8.16 shows an application of the method to study the global flushing characteristics of Tolo Harbour, Hong Kong. The harbour was divided into 7 segments, and the exchange flows were calculated from mean freshwater flows and salinities. It can be seen that the exchange flow increases towards the ocean, and pollutant flushing is relatively slow; it takes more than 40 days for a pollutant released in Tai Po Hoi (Segment 3) to get out of the harbour.

This essentially one dimensional approach is suited for well-mixed estuaries. It can also be generalized to include vertical exchanges between layers in a partially

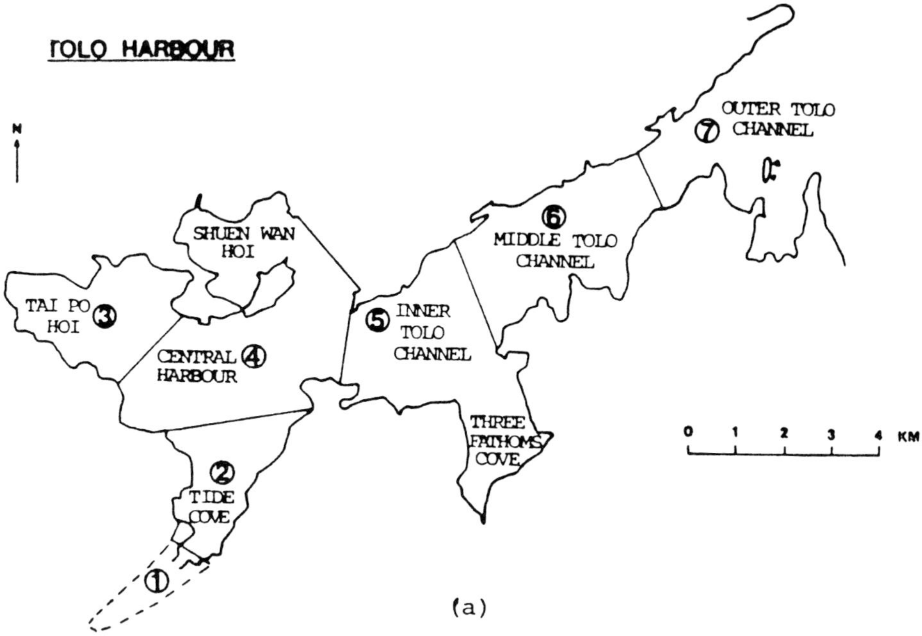

(a)

Segment n	Observed Mean Salinity o/oo	Observed Freshwater Flow $10^5 m^3/day$	Computed Residence Period days	Computed Exchange Flows $10^5 m^3/day$
1	27.0	1.8	5.1	Q_{12} = 19.4
2	29.5	3.17	6.9	Q_{24} = 93.5
3	28.6	1.37	20.4	Q_{34} = 20.6
4	30.5	6.15	14.6	Q_{45} = 250.1
5	31.2	8.16	7.9	Q_{56} = 340.0
6	32.0	9.27	2.1	Q_{67} = 593.3
7	32.5			

(b)

Figure 8.16. Application of a tidally averaged segment model to Tolo Harbour, Hong Kong: a) segmentation. b) observed salinities and freshwater flows, and computed residence times and exchange flows

mixed estuary (Pritchard, 1969). Comparisons of box model predictions with analytical solutions and refined hydrodynamic numerical model results have been given by Officer (1980). The Water Pollution Research Laboratory has also successfully applied this approach in the study of the Thames estuary (McDowell & O'Connor, 1977).

Although a formal connection can be established between the box model and either tidally averaged models (Stommel, 1953) or slack tide models (O'Connor, 1960), which lump the tidal mixing into dispersion coefficients, the 'box model' approach is conceptually more appealing and less restrictive. In particular, it can be shown that if the salinity of the river segment is assumed to be zero, then the concentration of a conservative pollutant at the discharge point ($x = 0$) is given by:

$$\frac{c(0)}{W} = \frac{f(0)}{Q_f} \tag{8.32}$$

where $c(0)$ = pollutant concentration, W = pollution mass discharge flow, Q_f = freshwater flow, $f(0)$ = freshwater concentration = $1 - s(0)/s_s$, and $s(0)$ = salinity at the discharge point, and s_s = ocean salinity. Upstream of the outfall (any position x), the pollutant concentration is given by the ratio of salinities:

$$\frac{c(x)}{c(0)} = \frac{s(x)}{s(0)} \tag{8.33}$$

Downstream of the discharge point, the pollutant concentration is given by the ratio of freshwater concentrations:

$$\frac{c(x)}{c(0)} = \frac{f(x)}{f(0)} \tag{8.34}$$

Equation 8.32 to 8.34 can also be obtained from the classical solution of Stommel (1953).

If the entire bay is thought of as one segment, the intersegment exchange flow can be meaningfully interpreted as 'a circulation of clean ocean water', as suggested by Fischer et al. (1979); essentially similar approximate prediction methods were arrived at by accounting for both mass and salt conservation.

Example 8: A factory discharges 2 m^3/s of effluent containing 100 ppt of a toxic substance into an estuary with a freshwater flow of 330 m^3/s. The mean salinity at the point of discharge is 18 ppt and the ocean salinity is 32 ppt. Estimate a) the pollutant concentration in the vicinity of the discharge point and b) the concentration at points upstream and downstream of the outfall where the salinity is 5 ppt and 25 ppt respectively.

Solution: Assuming the toxic substance to be conservative, the concentration at the discharge point is

$$c(0) = 2 \times 100 \times (1 - 18/32)/330 = 0.26 \text{ ppt}$$

the concentration at the upstream point is

$$c(x) = 0.26 \times 5/18 = 0.074 \text{ ppt}$$

and the downstream point

$$c(x) = 0.26 \times (1 - 25/32)/(1 - 18/32) = 0.13 \text{ ppt}$$

The concentration $c(0)$ puts a limit on the achievable near field mixing. Even if a diffuser were installed to achieve a yet lower concentration, the limited flushing in this example estuary would still dictate the concentrations!

Calculations based on such a 'box' model provide only order of magnitude estimates of the cross-sectional average concentration to be expected in the estuary. Maximum concentrations may exceed average figures substantially in stratified waters where the pollutant may be confined to the surface layer. Lung & O'Connor (1984) have presented a simple and practical method to compute the two-dimensional tidally-averaged mass transport pattern in a partially-mixed estuary. The circulation in the vertical plane along the estuary is assumed to be governed essentially by a balance of longitudinal pressure gradient and turbulent shear due to vertical variation of horizontal velocity (Officer, 1976). Measured vertical profiles of salinity and, if available, horizontal velocity are used to derive the horizontal velocities in the upper and lower layer along the estuary, as well as the vertical velocity at the interface. The method appears to reproduce salinity and velocity observations, and has been applied in conjunction with multi-component water qualtiy models to a number of estuaries. Although the tidally-averaged transport is consistent with the characteristic time scales of interest of many water quality parameters, many questions important in engineering design cannot be answered by such an approach, e.g. the effect of accidental/transient releases. To obtain more detail on spatial and temporal changes, one inevitably turns to a numerical model.

8.5.3 *Numerical modelling of pollutant transport*

The solution of the advective diffusion equation governing mass transport with appropriate near field source conditions and ambient boundary conditions determines the far field pollutant distribution. In general, this equation is transient and three-dimensional. Given typical shallow coastal environments, simplification to two horizontal dimensions, or, with channel-like geometries, even to one dimension are customary, implying some averaging over the missing dimension.

Knowledge of the detailed advective velocity field is an important element for the solution of the transport equation. The velocity field may be obtained from a separate numerical calculation of the equations of motion (a so-called 'circulation model'), or from field surveys, or occasionally from a physical scale model. In recent years, increasingly powerful computer models have become available.

Circulation models: The tidal circulation can be computed by a finite difference (FD) or finite element (FE) model which solves the vertically averaged equations of motion and continuity. There is a vast literature on various numerical solutions of shallow water circulation problems. Many examples of circulation models can be found in Leendertse (1967, 1972, 1975), Cheng (1976), Abbott (1982), Fischer (1981), and Benque et al. (1982). In general, the accuracy of the prediction is to a great extent dependent on the accuracy of the open boundary condition. Most of the models in industrial use are finite difference codes which use Alternating Direction Implicit (ADI) schemes on a staggered grid. However, in the past decade, a number of finite element models which can better handle complex geometries have been proposed (e.g. Wang & Connor, 1975; Walters & Cheng, 1980; Gray, 1977). A comprehensive critical review of finite element circulation models has been given by Gray (1982). As an example, Figure 8.17 shows a finite element grid for Massachusetts Bay and the computed flood and

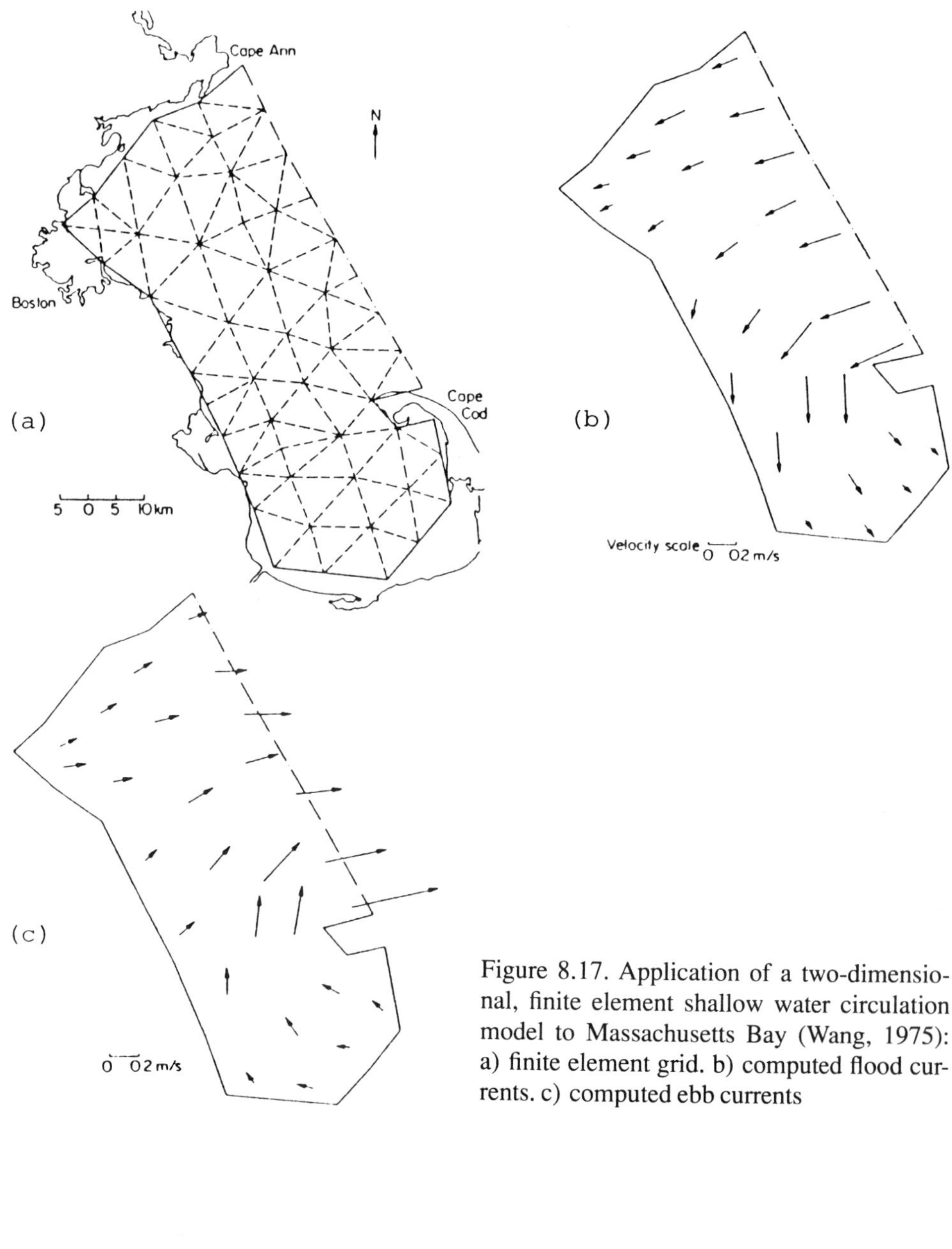

Figure 8.17. Application of a two-dimensional, finite element shallow water circulation model to Massachusetts Bay (Wang, 1975): a) finite element grid. b) computed flood currents. c) computed ebb currents

ebb current field. Most finite element models are limited in two respects: a) many simulations exhibit serious node to node oscillations that are caused by the poor propagation of the short wave components, and b) most models are dictated by the Courant stability criterion, thus rendering the method uneconomical in many coastal computations. More recently, successful attempts at overcoming some of the earlier difficulties have been reported (Peraire et al., 1986; Li et al., 1986). A review of recent advances in the numerical hydrodynamic modelling of estuaries and coastal regions has also been presented by Blumberg (1985).

Knowledge of the detailed velocity field can aid greatly towards understanding the pollutant transport. For example, the computed velocity field can be time averaged over the tidal cycle to give residual currents which would be helpful in siting outfalls. The velocity time history at selected points can also be used to construct progressive vector diagrams that give a useful picture of the fate of pollutant from an outfall.

Transport models: Given the velocity field, the pollutant concentration field is typically obtained by solving the Eulerian convective diffusion equation in one or two dimensions. Vertical averaging is applied over the entire (shallow) water depth, or in case of deep stratified bodies over an upper layer above a pycnocline. The required parameters are shear flow dispersion coefficients that express the combined effects of velocity shear and vertical (and horizontal for one-dimensional models) diffusion. These coefficients are often obtained from drogue studies and dye tracing exercises for representative spring and neap cycles. In general, as stressed by Harleman (1974), the more detailed the description of the advective field, the less sensitive are the results to the exact value of the dispersion coefficients. For example, dispersion coefficients in a tidally averaged model are necessarily much larger than true dispersion coefficients in a real time model that simulates the intratidal advective transport – in order to account for the information lost in the averaging process. Also required is the open boundary condition that may have to be obtained by field measurement. Essentially the same type of models can be used to predict salinity intrusion.

The pollutant transport in coastal areas is typically dominated by advection, with coastal current velocities in the order of 0.1 to 1 m/s. Great care has to be exercised in the numerical solution of such problems. The solution may exhibit serious numerical dispersion and/or contain excessive numerical dissipation. These numerical difficulties can be traced to the discrete approximation of the convective term. Considerable research has been devoted to the solution of such advection-dominated problems (e.g. Hughes, 1979; Pinder & Gray, 1977). A split-step characteristic-based finite difference scheme (Holly & Preissmann, 1977) that uses Hermitian cubic interpolation functions gives good results. A detailed comparative analysis and discussion of the numerical behavior of different schemes has been given by Sobey (1984).

As an example, Figure 8.18 shows a finite element solution of the concentration due to a continuous source (wastewater outfall) at $x = 0$ in a tidal flow and subject

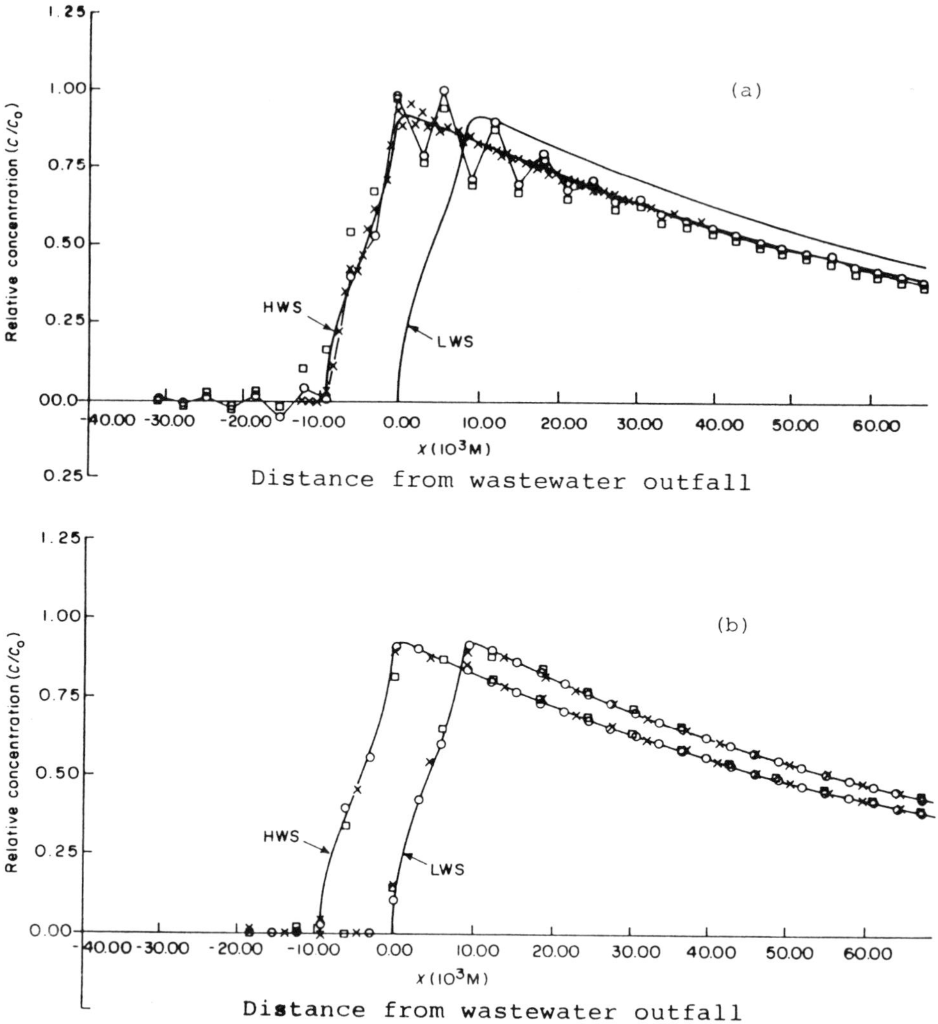

Figure 8.18. Example of numerical oscillations in one-dimensional mass transport calculation in an advection-dominated problem with continuous source in tidal flow. a) finite element solution (symbols) with linear basis functions at High Water Slack (HWS) and Low Water Slack (LWS). Solid line indicates exact solution. Note the strong numerical oscillations around the source at $z = 0$. b) solution with Hermitian cubic basis functions. Both solutions use the Crank Nicolson scheme in time. c_o = reference concentration resulting from full mixing of pollutant load with river flow (Li & Lee, 1985)

to a first order decay (Li & Lee, 1985). With linear elements (Fig. 8.18a), the numerical solution exhibits serious oscillations near $x = 0$, where the source furnishes a supply of sharp concentration gradients. This problem can be eliminated by local grid refinement, or by the use of higher order basis functions (Fig. 8.18b). The figure shows that the pollutant distribution is transported up and downstream of the outfall over a distance equal to the tidal excursion. Upstream of the outfall, advective transport governs the concentration while freshwater flushing and pollutant decay are dominant downstream.

An example of a numerical transport model computation is shown in Figure 8.19a for the heated discharge from a power station into a coastal bay. Concurrent temperature observations in the field are shown for comparison in Figure 8.19b. The quality of agreement is typical for such computations and is attributable to a number of factors, including deficiencies in model formulation, incomplete

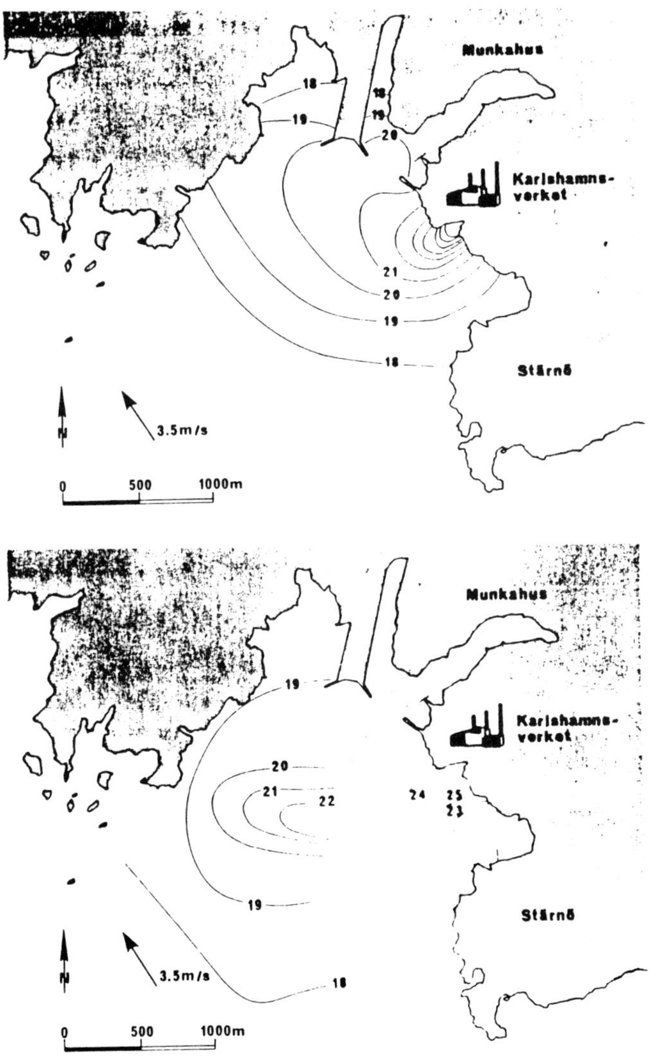

(a)

(b)

Figure 8.19. Comparison of computed and observed surface temperatures in the vicinity of a thermal power plant discharge off the Swedish coast (after Vasseur et al., 1980). a) Calculated temperature at the surface at 1400, 2 September 1974. b) Measured temperature at the surface at 1300-1620, 2 September 1974

knowledge of boundary conditions and environmental forcing functions, and, finally, inaccuracies in the field data.

As an alternative to the above Eulerian approach, Lagrangian models which have a minimal numerical diffusion can be gainfully applied in certain cases. Fischer (1975, 1977) has proposed a Lagrangian method based on the use of streamtubes. In coastal waters where the open boundary condition may not be satisfactorily known, models of this kind, which capture the essential dispersion mechanisms, may yield good answers using only representative alongshore velocity measurements. As previously mentioned, superposition models (e.g. Munro, 1975; Adams et al., 1975) which track the advection of pollutant patches with the mean flow and calculate the diffusion and decay of each patch separately according to a selected diffusion law can be applied along relatively straight coastlines with simple bathymetry. Details of comparisons of Eulerian and Lagrangian models can be found in Sobey (1984). More recently, combined Eulerian-Lagrangian methods have been developed which link a convection step (without diffusion) with a subsequent diffusion step (without convection). These methods appear to be stable, exhibit minimal numerical diffusion, and are cost-effective. These developments for ocean dispersion modeling have been summarized by Adams & Baptista (1986).

Finally, much progress in the numerical modelling of coastal hydrodynamics and water quality has been made in recent years. An idea of these developments and their applications can be gained from the many contributions presented at a recent conference on environmental hydraulics (Lee & Cheung, 1991).

8.6 REFERENCES

Abbott, M.B. & J.A. Cunge 1982. Engineering applications of computational hydraulics. Pitman, 1.

Abraham, G. 1963. Jet diffusion in stagnant ambient fluid. *Delft Hydraulics Laboratory*, Publ. No. 29.

Abraham, G. 1971. The flow of round buoyant jets issuing vertically into ambient fluid flowing in a horizontal direction. *Delft Hydraulics Laboratory*, Publ. No. 81.

Adams, E.E. 1982. Dilution analysis for uni-directional diffusers. *J. Hydraulics Div., ASCE*, 108(HY3): 327-342.

Adams, E.E. & A.M. Baptista 1986. Ocean dispersion modeling. In N.Cheremisinoff (ed.), *Encyclopedia of Fluid Mechanics*, 6. Complex Flow Phenomena, Gulf Publishing Co.

Adams, E.E. & K.D. Stolzenbach 1976. Comparison of alternative diffuser designs for the discharge of heated water into shallow receiving water. *Proc. Conf. on Waste Heat Management and Utilization, U. of Miami*, 1.

Adams, E.E., K.D. Stolzenbach & D.R.F. Harleman 1975. *Near and far field analysis of buoyant surface discharges into large bodies of water.* Tech. Rep. 205, MIT Parsons Laboratory.

Adams, E.E., W. Skamarock & R.V. Nothaft 1982. Shoreline effects on mixing of tee-diffuser. *J. Hydraulics Div., ASCE*, 108(HY10): 1232-1238.

Agg, A.R. 1978. *Investigations of sewage discharges to some British coastal waters, Chapter 6, Initial dilution*. Water Research Centre, Tech. Rep. 99.

Agg, A.R. & A.C. Wakeford 1972. Field studies of jet dilution of sewage at sea outfalls. *Institution of Public Health Engineers Journal*, 71: 126-149.

Akar, P.J. & G.H. Jirka 1990. CORMIX2: *An expert system for hydrodynamic mixing zone analysis of conventional and toxic submerged multiport diffuser discharges*. Tech. Rep., DeFrees Hydraulic Laboratory, School of Civil and Environmental Engineering, Cornell University (also published by US Environmental Protection Agency, Tech. Rep., Environmental Research Lab., Athens, Georgia).

Alam, A.M.Z., D.R.F. Harleman & J.M. Colonell 1982. Evaluation of selected initial dilution models. *J. of Environmental Engineering, ASCE*, 108(EE 1): 159-186.

Almquist, C.W. & K.D. Stolzenbach 1980. Staged multiport diffusers. *J. of Hydraulics Div., ASCE*, 106(HY2): 285-302.

Bennett, N.J. 1981. Initial dilution: A practical study on the Hastings Long Sea Outfall. *Proc. Inst. of Civil Engrg.*, Part I, (70): 113-122.

Benque, J.P., A. Hauguel & P.L. Viollet 1982. *Engineering Applications of Computational Hydraulics*, Pitman, 2.

Benque, J.P., J.A. Cunge, J. Feuillet, A. Hauguel & F.M. Holly 1982. New method for tidal current computations. *J. of Waterway, Port, Coastal and Ocean Division, ASCE*, 108(WW3): 396-417.

Bettess, R. & D. Munro 1981. *Field measurement of initial dilution at a sea outfall*. Technical Report 161, Water Research Centre.

Blumberg, A.F. & L.Y. Oey 1985. Modeling circulation and mixing in estuaries and coastal oceans. *Advances in Geophysics*, 28a: 525-547.

Brocard, D.N. 1980. Discussion of 'Staged Multiport Diffusers', *ASCE*, 106(HY10): 1723-1725.

Brooks, N.H. 1960. Diffusion of sewage effluent in an ocean current. *Proceedings of the 1st International Conference on Waste Disposal in the Marine Environment, Pergamon*: 246-267.

Brooks, N.H. 1972. *Dispersion in hydrologic and coastal environments*. W.M. Keck Laboratory of Hydraulics & Water Resources, California Institute of Technology, Rep. KH-R-29.

Brooks, N.H. 1980. Synthesis of stratified flow phenomenon for design of outfalls. *Proc. 2nd Int. Symposium on Stratified Flows, Trondheim, Norway*, 2: 809-831.

Cederwall, K. 1968. *Hydraulics of marine waste disposal*. Hydraulic Div., Chalmers Institute of Technology, Goteborg, Sweden, Report No. 42.

Charlton, J. 1985. Sea outfalls. In P. Novak (ed.), *Developments in hydraulic engineering*, Elsevier, 3: 79-127.

Cheng, R.T.S. 1976. Modeling of hydraulic systems. *Advances in Hydroscience*, 11: 208-284.

Chin, D.A. & P.J.W. Roberts 1985. Model of dispersion in coastal waters. *J. of Hydraulic Engineering, ASCE*, 111(1): 12-27.

Chu, V.H. 1975. Three dimensional forced plumes in a laminar crossflow. *Proc. 15th IAHR Congress*, 4: 31-39.

Chu, V.H. 1978. Discussion of 'Mean behavior of buoyant jets in a crossflow'. *J. of Hydraulics Div., ASCE*, 104(HY2): 318-320.

Chu, V.H. 1979. L.N. Fan's data on buoyant jets in a crossflow. *J. of Hydraulics Div., ASCE*, 105(HY5): 612-617.

Chu, V.H. & M.B. Goldberg 1974. Buoyant forced plumes in crossflow. *J. of Hydraulics Division, ASCE*, 100(HY9): 1203-1214.

Chu, V.H. & G.H. Jirka 1986. Surface buoyant jets and plumes. In N. Cheremisinoff (ed.), *Encyclopedia of Fluid Mechanics*, 6. Complex Flow Phenomena, Gulf Publishing Co.

Csanady, G.T. 1973. *Turbulent diffusion in the environment*. D. Reidel.

Csanady, G.T. & J.H. Churchill 1987. Environmental engineering analysis of potential dumpsites. In T.P. O'Connor (ed.), *Ocean processes in marine pollution*, 2: 21-31.

Delvigne, G.A.L. 1980. *Round buoyant jets with three-dimensional trajectory in ambient flow*. Delft Hydraulics Laboratory, Pub. No. 228.

Doneker, R.L. & G.H. Jirka 1990. CORMIX1: *An expert system for hydrodynamic mixing zone analysis of conventional and toxic submerged single port discharges*. Tech. Report., DeFrees Hydraulic Laboratory, School of Civil and Environmental Engineering, Cornell University (also published by US Environmental Protection Agency, Tech. Rept. EPA/600/3-90/012, Environmental Research Lab., Athens, Georgia.

Dyer, K.R. 1973. *Estuaries: A physical introduction*. Wiley.

Dyer, K.R. & P.A. Taylor 1973. A simple, segmented prism model of tidal mixing in well-mixed estuaries. *Estuarine and Coastal Marine Science*, 1: 411-418.

Fan, L.N. 1967. *Turbulent buoyant jets into stratified or flowing ambient fluids*. W.M. Keck Laboratory of Hydraulics and Water Resources, California Institute of Technology, Rep. No. KH-R-15.

Fan, L.N. & N.H. Brooks 1969. *Numerical solutions of turbulent buoyant jet problems*. W.M. Keck Laboratory of Hydraulics and Water Resources, California Institute of Technology, Rep. No. KH-R-18.

Fischer, H.B. 1975. Numerical modeling of dispersion in estuaries. In A.L.H. Gameson (ed.), *Discharge of sewage from sea outfalls*: 371-380. Pergamon.

Fischer, H.B. 1977. A numerical model of the Suisan Marsh, California. *Proc. 17th IAHR Congress, Baden-Baden*, 1: 425-432.

Fischer, H.B., E.J. List, R.C.Y. Koh, J. Imberger & N.H. Brooks 1979. *Mixing in inland and coastal waters*. Academic Press.

Fischer, H.B. (ed.) 1981. *Transport models for inland and coastal waters*. Academic Press.

Frick, W.E. 1984. Non-empirical closure of the plume equations. *Atmospheric Environment*, 18(4): 653-662.

Gameson, A.L.H. 1982. *Investigations of sewage discharges to some British coastal waters, Chapter 5, Bacterial distributions, Part 4*. Tech. Rep. 176, Water Research Centre.

Gray, W.G. 1977. An efficient finite element scheme for two-dimensional surface water computations. In W.G. Gray et al. (eds), *Finite elements in water resources*. Pentech.

Gray, W.G. 1982. Some inadequacies in finite element models as simulators of two-dimensional circulation. *Advances in Water Resources*, 5(3): 171-177.

Harleman, D.R.F. & M.L. Thatcher 1974. Longitudinal dispersion and unsteady salinity intrusion in estuaries. *La Houille Blanche*, 29(1): 25-33.

Hendricks, T.J. 1977. *In-situ measurements of initial dilution*. Southern California Coastal Water Research Project Report: 41-51.

Hino, M. 1964. Limit of smoke ascent in a calm inversion layer of atmosphere, *J. Tech. Laboratory, CRIEPI*, vol. 14, no. 1, 9-44.

Hirst, E. 1972. Buoyant jets with three-dimensional trajectories. *J. of Hydraulics Division, ASCE*, 98(HY11): 1999-2014.

Holley, E.R. & G.H. Jirka. 1986. *Mixing and solute transport in rivers*. Tech. Rep. E-86-11, US Army Engineers Waterways Experiment Station, Vicksburg, Mississippi.

Holly, F.M. & A. Preissman 1977. Accurate calculation of transport in two dimensions. *J. of Hydraulics Div., ASCE*, 103(HY11): 1259-1277.

Hughes, T.J.R. (ed.) 1979. *Finite element methods for convection dominated flows*. American Society of Mechanical Engineers, AMD, 34.

Jirka, G.H. 1982a. Turbulent buoyant jets in shallow fluid layers. In W. Rodi (ed.), *Turbulent buoyant jets and plumes*. Pergamon Press.

Jirka, G.H. 1982b. Multiport diffuser for heat disposal: A summary. *J. of Hydraulics Div., ASCE*, 108(HY12): 1425-1468.

Jirka, G.H. 1991. Use of mixing zone models in estuarine waste load allocations. In R.B. Ambrose & J.L. Martin (eds), *Part III of Technical Guidance Manual for Performing Waste Load Allocations, Book III: Estuaries*. US Environmental Protection Agency, Tech. Rep., Environmental Research Lab., Athens, Georgia (in print).

Jirka, G.H. & P.J. Akar 1991. Hydrodynamic classification of submerged multiport diffuser discharges. *J. Hydraulic Engineering, ASCE*, 117(9): 1113-1128.

Jirka, G.H. & R.L. Doneker 1991. Hydrodynamic classification of submerged single port discharges. *J. Hydraulic Engineering, ASCE*, 117(9): 1095-1112.

Jirka, G.H. & H.L.M. Fong 1981. Vortex dynamics and bifurcation of buoyant jets in crossflow. *Journal of Engineering Mechanics Division, ASCE*, 107(EM3): 479-499.

Jirka, G.H. & D.R.F. Harleman 1973. *The mechanics of submerged multiport diffusers for buoyant discharges in shallow water*. R.M. Parsons Laboratory, Tech. Rep. 169, MIT.

Jirka, G.H. & D.R.F. Harleman 1979. Stability and mixing of vertical plane buoyant jets in confined depth. *Journal of Fluid Mechanics*, 94: 275-304.

Jirka, G.H., G. Abraham & D.R.F. Harleman 1975. *Assessment of techniques for hydrothermal impact prediction*. R.M. Parsons Laboratory, Tech. Rep. 203, MIT.

Jirka, G.H., E.E. Adams & K.D. Stolzenbach 1981. Properties of buoyant surface jets. *J. of Hydraulics Div., ASCE*, 107(HY11): 1467-1488.

Jirka, G.H., J. Colonell & D. Jones 1986. Outfall mixing design in shallow coastal waters under arctic ice cover. *J. Marine Technology Society*, 20(6): 44-53.

Jones, G.R. & G.H. Jirka 1991. *CORMIX3: An expert system for the analysis and prediction of buoyant surface discharges*. Tech. Rep., DeFrees Hydraulic Laboratory, School of Civil and Environmental Engineering, Cornell University (also to be published by US Environmental Protection Agency, Tech. Rep., Environmental Research Lab., Athens, Georgia).

Ketchum, B.H. 1951. The flushing of tidal estuaries. *Sewage and Industrial Wastes*, 23(2): 198-208.

Koh, R.C.Y. 1983. Wastewater field thickness and initial dilution. *J. of Hydraulics Div., ASCE*, 109(HY9) 1232-1240.

Koh, R.C.Y. 1988. Shoreline impact from ocean waste discharges. *J. of Hydraulic Engr., ASCE*, 114(4): 361-376.

Koh, R.C.Y. & N. Brooks 1975. Fluid mechanics of wastewater disposal in the ocean. *Ann. Rev. Fluid Mech.*, 7: 187-211.

Koh, R.C.Y. & L.N. Fan 1970. *Mathematical models for the prediction of temperature distributions resulting from the discharge of heated water into large bodies of water*. US EPA Rep. 16130 DWO 10/70.

Kuo, A.Y. 1976. A model of tidal flushing for small coastal basins. *Proc. EPA Conf. on Environmental Modeling and Simulation, Ohio*, 543-547.

Lee, J.H.W. 1980. Near field mixing of staged diffuser. *J. Hydraulics Div., ASCE*, 106(HY8): 1309-1324.

Lee, J.H.W. 1984. Boundary effects on a submerged jet group. *J. of Hydraulic Research*, 22(4): 199-216.

Lee, J.H.W. 1985. Comparison of two river diffuser models. *J. of Hydraulic Engineering, ASCE*, 111(HY7): 1069-1078.

Lee, J.H.W. & V. Cheung 1986. Inclined plane buoyant jet in stratified fluid. *J. of Hydraulic Eng., ASCE*, 112(HY7): 580-589.

Lee, J.H.W. & V. Cheung 1990. Generalized Lagrangian model for buoyant jets in a current. *J. of Environmental Engineering, ASCE*, 116(6): 1085-1106.

Lee, J.H.W. & Y.K. Cheung (eds) 1991. *Proceedings of the International Symposium on Environmental Hydraulics, Hong Kong*, Balkema.

Lee, J.H.W. & K.W. Choi 1986. Slack tide oxygen balance model. *J. of Environmental Engineering, ASCE*, 112(HY5): 985-991.

Lee, J.H.W. & M.D. Greenberg 1984. Line momentum source in shallow inviscid fluid. *J. Fluid Mech.*, 145: 287-304.

Lee, J.H.W. & G.H. Jirka 1980. Multiport diffuser as line source of momentum in shallow water. *Water Resources Research*, 16(4): 695-708.

Lee, J.H.W. & G.H. Jirka 1981. Vertical round buoyant jet in shallow water. *J. of Hydraulics Div., ASCE*, 107(HY12): 1651-1675.

Lee, J.H.W. & P. Neville-Jones 1987a. Initial dilution of horizontal jet in crossflow. *Journal of Hydraulic Engineering, ASCE*, 113(5): 615-629.

Lee, J.H.W. & P. Neville-Jones 1987b. Design of sea outfalls – Prediction of initial dilution. *Proc. Inst. of Civil Engrs.*, Part I, 82: 981-994.

Leendertse, J.J. 1967. *Aspects of a computational model for long period water wave propagation*. Rand Corporation Memorandum RM-5294-PR.

Leendertse, J.J. 1972. *A water quality simulation model for well mixed estuaries and coastal seas, (IV), Jamaica Bay Tidal Flows*. Rand Corporation R-1009-NYC.

Leendertse, J.J. & S.K. Liu. 1975. *A three-dimensional model for estuaries and coastal seas*, (II), *Aspects of computation*. Rand Corporation Rep. R-1764-OWRT.

Li, C.W. & J.H.W. Lee 1985. Continuous source in tidal flow – A numerical study of the transport equation. *Applied Mathematical Modelling*, 9(4): 281-288.

Li, C.W., J.H.W. Lee & Y.K. Cheung 1986. Mathematical model study of tidal circulation in Tolo Harbour, Hong Kong: Development and verification of a semi-implicit finite element scheme. *Proc. Institution of Civil Engineers*, Part II, 81: 569-592.

Lung, W.S. & D.J. O'Connor 1984. Two-dimensional mass transport in estuaries. *J. of Hydraulic Engineering, ASCE*, 110(10): 1340-1357.

McDowell, D.M. & B.A. O'Connor 1977. *Hydraulic behaviour of estuaries*. MacMillan.

Miller, D.S. & B.A. Brighouse 1984. *Thermal discharges*. British Hydromechanics Research Association.

Morton, B.R., G.I. Taylor & J.S. Turner 1956. Turbulent gravitational convection from maintained and instantaneous sources. *Proc. Roy. Soc.* A234: 1-23.

Muellenhoff, W.P., A.M. Soldate, D.J. Baumgartner, M.D. Schuldt, L.R. Davis & W.E. Frick 1985. *Initial mixing characteristics of municipal ocean discharges*. US Environmental Protection Agency, Corvallis Environmental Research Laboratory, Corvallis, Oregon.

Munro, D. 1975. Observed and predicted coliform distributions near a sea outfall. In A.L.H. Gameson (ed.), *Discharge of sewage from sea outfalls*: 353-362. Pergamon Press.

Munro, D. 1984. Sewage dilution between outfall and shore. *Public Health Engineer*, 12(2): 113-114.

Munro, D. & B.M. Mollowney 1974. The effect of a bottom countercurrent and vertical diffusion on coliform concentrations at the shore due mainly to wind-induced onshore currents. *Rapports et Proces-verbaux des Reunions Conseil International pour l'Exploration de la Mer*, 167: 37-40.

O'Connor, D.J. 1960. Oxygen balance of an estuary. *J. of Sanitary Engineering Div., ASCE*, 86(SA3): 35-55.

Officer, C.B. 1976. *Physical oceanography of estuaries*. Wiley.

Officer, C.B. 1980. Box models revisited. In P. Hamilton & K.B. MacDonald (eds), *Estuarine and wetland processes*: 65-114. Plenum.

Officer, C.B. & J.H. Ryther 1977. Secondary sewage treatment versus ocean outfalls. *Science*, 197: 1056-1060.

Okubo, A. 1962. A review of theoretical models for turbulent diffusion in the sea. *J. of the Oceanographic Society of Japan, 20th Anniversary Volume*.

Okubo, A. 1971. Oceanic diffusion diagrams. *Deep Sea Research*, 18: 789-802.

Pearson, E.A. 1975. Conceptual design of marine waste disposal. In A.L.H. Gameson (ed.), *Discharge of sewage from sea outfalls*: 403-413. Pergamon Press.

Peraire, J., O.C. Zienkiewicz & K. Morgan 1986. Shallow water problems: A general explicit formulation. *Int. J. for Numerical Methods in Engineering*, 22(3): 547-574.

Pinder, G.F. & W.G. Gray 1977. *Finite element simulation in surface and subsurface hydrology*. Academic Press.

Pritchard, D.W. 1969. Dispersion and flushing of pollutants in estuaries. *J. of Hydraulics Div., ASCE*, 95(HY1): 115-124.

Prych, E.A. 1972. A warm water effluent analysed as a buoyant surface jet. *Serie Hydrologi. Sveriges Meteorologiska och Hydrologiska Inst., Stockholm*, no. 21.

Roberts, P.J.W. 1979. Line plume and ocean outfall dispersion. *J. of Hydraulics Div., ASCE*, 105(HY4): 313-331 (see also discussion by G.H. Jirka, HY12: 1573-1575).

Roberts, P.J.W. 1980. Ocean outfall dilution: Effect of currents. *J. of Hydraulics Div., ASCE*, 106(HY5): 769-782.

Roberts, P.J.W. 1986. The use of current data in ocean outfall design. *Water Science & Technology*, 18(11): 111-120.

Roberts, P.J.W., W.H. Snyder & D.J. Baumgartner 1989. Ocean outfalls I: Submerged wastefield formation. *J. of Hydraulic Engineering, ASCE*, 115(1): 1-25.

Rodi, W. 1980. *Turbulence models and their applications in hydraulics*. IAHR.

Rouse, H., C.S. Yih & H.W. Humphreys 1952. Gravitational convection from a boundary source. *Tellus*, 4: 201-210.

Schatzmann, M. 1981. Mathematical modeling of submerged discharges into coastal waters. *Proc. 19th IAHR Congress, New Delhi*, 3: 239-246.

Schatzmann, M. & E. Naudascher 1980. Design criteria for cooling water outlet structures. *J. Hydraulics Div., ASCE*, 106(HY3): 397-408.

Shirazi, M.A. & L.R. Davies 1974. *Workbook and thermal plume prediction – Vol. 2, Surface Discharges*. Report No. EPA R2-72-0056, USEPA Corvallis, Oregon.

Sobey, R.J. 1984. Numerical alternatives in transient stream response, *J. of Hydraulic Engineering, ASCE*, 110(6): 749-772.

Stolzenbach, K.D. & D.R.F. Harleman 1971. *An analytical and experimental investigation of surface discharges of heated water*. R.M. Parsons Laboratory, Tech. Rep. 135, MIT.

Stommel, H. 1953. Computation of pollution in a vertically mixed estuary. *Sewage and Industrial Wastes*, 25(9): 1065-1071.

Toms, G. & W.A.M. Botes 1986. Dye studies of initial dilution and the applicability of the stagnant water dilution. *Water Science & Technology*, 18(11): 189-197.

Vasseur, B., L. Funkquist & J.B. Paul 1980. *Verification of a numerical model for thermal plumes*. Swedish Meteorological and Hydrological Institute, Tech. Rep. ISSN 0347-7827, Norkopping.

Vigliani, P.G., B. Sclavi, R. Olivotti, A. Visconti & G.F. Tartaglia 1980. An investigation of the initial dilution of sewage discharged through submarine diffusers. *Prog. Wat. Tech.*, 12(1): 321-337.

Wallace, R.B. & S.J. Wright 1984. Spreading layer of two-dimensional buoyant jet. *J. of Hydraulic Engineering, ASCE*, 110(6): 813-828.

Wallingford Hydraulics Research Station 1977. *Horizontal outfalls in flowing water*, Hydraulics Research Station Rep. EX 763.

Wallis, I.G. 1977. Initial dilution with deep water diffusers. *J. Water Pollution Control Federation*, 49(7): 1621-1626.

Wallis, I.G. 1981. Verification of ocean outfall performance prediction. *J. of Environmental Engineering, ASCE*, 107, EE2: 421-425.

Walters, R.A. & R.T. Cheng 1980. Accuracy of a estuarine hydrodynamic model using smooth elements. *Water Resources Research*, 16(1):187-195.

Wang, J.D. & J.J. Connor 1975. *Mathematical modeling of near coastal circulation*. Parsons Laboratory Tech. Rep. 203, MIT.

Wilkinson, D.L. 1984. Purging of saline wedges from ocean outfalls. *J. of Hydraulic Engineering, ASCE*, 110(12): 1815-1829.

Wright, S.J. 1977. Mean Behavior of Buoyant Jets in a Crossflow. *J. of Hydraulics Div., ASCE*, 103(HY5): 499-513 (subsequent closure 1978(HY9): 1359-1360).

Fish protection 9

N. RAJARATNAM
Department of Civil Engineering, University of Alberta, Edmonton, Alberta, Canada
C. KATOPODIS
Freshwater Institute, Department of Fisheries and Oceans, Winnipeg, Manitoba, Canada

9.1 INTRODUCTION

General problems of fish protection in streams, rivers and lakes include artificial destratification in reservoirs, placement of multiple level outlets in dams, control of instream flows, control and reduction of gas supersaturation and protection measures at inlets to such structures as canals, pump stations for process industries, thermal power stations and turbines in hydroelectric power stations. The protection measures at these inlets include track-racks, screens of different kinds (travelling screens, resolving drum screens, plane inclined screens), louver diverters with a bypass and velocity caps on intake structures. These measures for the protection of fish have been discussed well by Clay (1961) and in the recent ASCE manual on the design of water intake structures for fish protection (1982). Reviews of environmental impacts of stream and lake regulation are available in Baxter & Glaude (1980), Davies & Walker (1986) and others. This chapter is confined to a discussion of the hydraulics of another class of structures used for fish protection in the aquatic habitat namely fishways including culverts with fish passing arrangements; the latter are referred to simply as culvert fishways.

In river systems, fish migrations involve completing a cycle of movements, both downstream and upstream (Northcote, 1984; Watt, 1974). The migratory sequence depends on the migrant's life stage, location and type of migration. Migration at a specific life stage may involve both downstream and upstream movements. Characteristically though, downstream migration is a feature of early life stages (eggs, larvae, fry, juveniles) and upstream migration is a feature of late life stages (subadults, adults, spawners and postspawners).

Types of fish migration include movements that involve the habitats for reproduction, feeding and refuge. Reproductive migrations are the well known spawning runs (especially for the salmon) when the mature fish return to the place where they were born to spawn. In these spawning runs, it is important that fish are not delayed or stopped by obstructions. Feeding migrations include movements between areas in stream, lake, and other habitat that are rich in food.

Migrations to seek refuge occur to enhance survival and avoid harmful environmental conditions. For instance, in climates where shallow streams may freeze completely in the winter, fish are forced to seek deep streams or lakes.

Fishways are hydraulic structures built in streams, at outlets from lakes, at diversions, at waterfalls and elsewhere; they enable fish migrating upstream, to move past the obstruction in question. In many countries, if a hydraulic structure poses potential problems for the passage of fish, a fishway is required by law. Similar requirements exist for culverts that are built in streams having fish runs.

Pool-weir fishways are amongst the earliest of fishways built. They were later modified by the addition of one or more orifices (or ports) in the weir walls. Pool-weir fishways with or without orifices are commonly used in European countries among others. Some of the largest fishways of this type are on the dams of the Columbia River system in the western US (Clay, 1961).

In a series of papers (starting in 1909), Denil published the results of his studies on what was then a new type of fishway, now known as the Denil fishway (see Clay, 1961). Some simplified forms of Denil fishways have become common in Canada, the US and Europe. Decker (1967) described the largest Denil, built in Maine (USA) across the St. Croix River, that has a length of 227 m and vertical rise of 15.2 m.

The first vertical slot fishway (with twin jets) was built in 1946 at Hell's Gate on the Fraser River in British Columbia (Canada) to pass salmon (Clay, 1961). It was designed to pass about 20 000 sockeye salmon per hour. Vertical slot fishways have become popular in Canada and the USA (Collins & Gillis, 1985) mainly with a single slot.

For culverts laid on slopes steeper than 0.5% and having lengths greater than about 20 m, some form of baffle-system is usually necessary for fish to pass (Dane, 1978). The offset-baffle system was suggested by McKinley & Webb (1956) and has since been used in several culverts in western Canada and western US. The spoiler baffle system was studied by Engel in 1974. The side-baffle was suggested by Watts (1974). Amongst other forms of baffle systems used in culverts are fish-weirs and fish-baffles, and these have been used in some culverts in Alberta (Canada) (Rajaratnam et al., 1986).

9.2 FISH SWIMMING PERFORMANCE

To build successful fishways, one must understand the behaviour of fish, especially their swimming performance, motivation, how to attract them to the fishway entrance and the periods of their migration and other movements. Zoologists, hydraulic engineers and fluid mechanicians, who have studied how fish swim, both experimentally and theoretically, have divided swimming performance into three broad classes: sustained speed (including cruising speed), prolonged speed and burst (or darting) speed (Blake, 1983). Sustained speed is the speed that a fish can maintain for periods longer than about 200 minutes, without fatigue. This speed is used for cruising, foraging, station holding and for maintaining vertical elevation for negatively buoyant fish. Energy for sustained swimming comes from aerobic pathways (metabolism).

Prolonged swimming speed is a higher speed that a fish can maintain for periods of up to about 200 minutes. Energy comes from both aerobic and anaerobic pathways. A fish might use this speed going through some fishways or culverts.

Burst speed is a high speed that fish can maintain for only short duration of less than 15 seconds. Energy for this activity comes from anaerobic pathways. A fish might use this kind of speed for catching its prey or escaping a predator. Fish appear to use this speed in fishways to get past slots or orifices or to go over weirs. Bell (1973) provides the three classes of swimming speeds for different fish including whitefish, trout and salmon. Fish performance speeds are obtained by testing the fish both in laboratories (in curved channels with maintained water speeds and in closed water tunnels) and in the field (see Blake, 1983; Watts, 1974).

9.3 POOL-WEIR AND POOL-ORIFICE FISHWAYS

A pool-weir fishway consists of a number of pools formed by constructing a series of weirs as shown in Figure 9.1. Water flows from the headwater side to the tailwater region over these weirs. The fish swim over each weir using their burst speeds or, in some cases as for salmon and trout, by jumping over them (Orsborn, 1985). The fish then rest in the pool for a while before passing over the next weir, and so on, till they swim into the upstream region.

The many variations of the pool-weir design include the double weir at Ice Harbour Dam Fishway and weirs with notches. The success of a pool-weir fishway depends on the existence of adequate tailwater and headwater levels. This requirement can be met by the provision of orifices in the weir walls close to the bed of the fishway. In some designs, chutes are attached to weirs.

Pool-weir fishways (with or without orifices) are generally designed with the head drop per pool in the range of 0.2 to 0.3 m even though somewhat larger drops can be allowed for stronger fish like salmon. The bed slopes vary from 1 on 8 to 1 on 10.

Pool-weir fishways differ significantly from classical weirs, which have been studied extensively (Lakshmana Rao, 1975). The main difference is that instead of a single weir, a fishway is composed of a series of weirs that are placed sufficiently close to each other (approximately 3 m apart) that they act hydraulically together rather than as individual weirs. For fishways the overflow has either a 'plunging nappe' or a kind of 'streaming' flow with a transition state in between (Fig. 9.1).

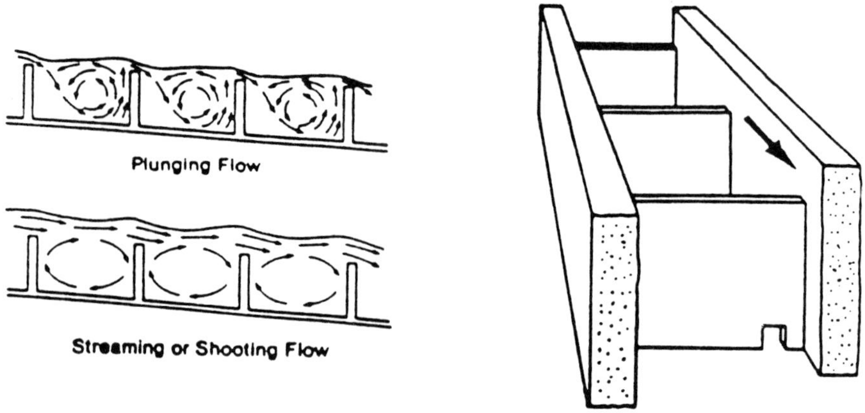

Figure 9.1. Pool-weir fishway

For the plunging nappe, the downstream water level is generally below the crest of the weir immediately upstream. The kinetic energy of the jet is lost by turbulent mixing and diffusion. For the head-discharge relation, the weir behaves as through unsubmerged. In streaming flow, (especially in its fully developed state), a surface jet of water flows over the crests of the weirs and over the recirculating mass of fluid in the pools with an approximately constant thickness. If the streaming flow is not fully developed, a local acceleration occurs over the weir crest with a standing wave below it; the thickness of the surface stream thus varies cyclically along the fishway. The following text contains an approximate analysis of the behaviour of pool-weir fishways.

For fully developed streaming flow over the weirs, neglecting the side wall shear stress (which is believed to be small), the balance of forces acting on the water is

$$d\Delta x\gamma S_0 = \bar{\tau}\Delta x \tag{9.1}$$

wherein d is the depth or thickness of the stream, S_0 is the slope of the fishway (equal to sine of the angle between the bed of the fishway and the horizontal), Δx is an elemental length along the fishway, γ is the weight per unit volume of the fluid and $\bar{\tau}$ is the average turbulent shear stress between the streaming flow and the recirculating mass in the pools. The mean shear stress $\bar{\tau}$ can be expressed as

$$\bar{\tau} = c_f \frac{\rho V^2}{2} \tag{9.2}$$

wherin c_f is a coefficient of fluid friction, ρ is the mass density and V is the mean velocity in the surface stream; a combination of Equations 9.1 and 9.2, after some simplification, leads to an equation for the volumetric flow rate (or discharge) as

$$Q^* = \frac{Q}{bd^{3/2}\sqrt{gS_0}} = \sqrt{\frac{2}{c_f}} \tag{9.3}$$

In Equation 9.3, Q^* is the dimensionless flow rate and b is the width of the fishway. Rajaratnam et al. (1988) found that for streaming flow, Q^* is given approximately by the equation

$$Q^* = 1.5\sqrt{L/d} \tag{9.4}$$

where L is the length of the pool.

For the plunging nappe regime, the experimental observations (of Andrews (1984) and by the authors), indicate that the weirs discharge in the unsubmerged or free-flow mode and the flow equation can be written as

$$Q = bC_d \frac{2}{3}\sqrt{2g}\, h^{3/2} \tag{9.5}$$

where C_d is a coefficient of discharge and h is the head over the weir. Weir crests generally have a thickness of 100 mm, and for design heads of roughly 300 mm, a ratio of head to weir thickness of 3 makes the weirs act as though they were sharp-crested (the lowest value of this ratio needed for the weir to act as sharp-crested is about 1.5 to 1.9). For sharp-crested weirs discharging in the free-flow mode and for the ratio of height of weir to head over weir generally used in fishways (4 to 5), the Rehbock equation (Kandaswamy & Rouse (1957)) yields a value for C_d of 0.605. Equation 9.5 written in a dimensionless form is

$$Q_+ = \frac{Q}{b\sqrt{g}\, h^{2/3}} = \frac{2}{3}\sqrt{2}C_d \tag{9.6}$$

With $C_d = 0.605$, $Q_+ = 0.57$. The preliminary experiments of Rajaratnam et al. (1988) indicate a value of 0.61. The authors' results also indicate that the transition from plunging to streaming flow occurs if

$$\frac{Q}{b\sqrt{g}\, S_0 L^{3/2}} \simeq 0.25 \tag{9.7}$$

where L is the (sloping) length of the pool.

If a pool-weir fishway has single or multiple orifices in the weir walls (Orsborn (1986) contains a summary of orifice sizes used in practice) and if the flow rate is such that no water flows over the weirs, it is referred to as a pool-orifice fishway. Experiments have shown that for the so-called uniform flow for which the central depth y_0 in every pool is the same, and if y_0 is less than the height h_0 of the orifices, the pool-weir fishway operates like a vertical slot fishway. If on the other hand, y_0 is much larger than h_0, then the fishway operates in a submerged orifice mode with the flow from the orifices behaving like submerged jets. In between these two states is a transition state in which the orifices are submerged on the upstream side and unsubmerged on the downstream side. In a pool-orifice fishway, if the flow rate is such that a part of the flow passes through the orifices and the rest over the weirs, then the fishway is referred to as a pool-orifice-weir or pool-weir-orifice fishway.

A method of analysis for pool-orifice fishways, can be developed for a fishway with a number of pools and with an orifice on each wall located near the lower corner all on the same side. If the fishway has a slope of S_0 and the drop per pool is the same for all pools so that one has a kind of uniform flow in the fishway,

$$b_0 h_0 L \gamma S_0 \simeq (b_0 + h_0) L\, c_f \frac{\rho V^2}{2} \tag{9.8}$$

where b_0 is the width of the orifice, h_0 its height and $c_f\, \rho V^2/2$ is once again the average fluid shear stress between the jet and the surrounding recirculating water. In Equation 9.8 the average cross-section of the jet is taken as $(b_0 \times h_0)$ which does not account for the jet expansion; further, the boundary shear stresses on the bed and on the wall are small enough that they can be neglected. Equation 9.8 can be developed to give

$$Q^* = \frac{Q}{\sqrt{g S_0}\, b_o^{5/2}} = \sqrt{\frac{2\,\alpha_*^3}{c_f (1 + \alpha_*)}} \tag{9.9}$$

wherein $\alpha_* = h_0/b_0$. Experience with vertical slot fishways shows that c_f is likely to be approximately constant so that, for a given orifice design, the dimensionless flow rate Q^* is also a constant. The value of this constant would vary with the design since it depends on whether the orifices alternate from one side to the other, are centrally located or otherwise placed. Equation 9.9 has been evaluated by Rajaratnam et al. (1989) for a particular design and in the submerged orifice regime, $Q^* \simeq 2.2$.

For pool-weir-orifice fishways, if the orifices and weirs operate for a given flow, and the flow over the weir is in the uniform surface streaming state, the flow from both the orifice (assumed to be at one corner as in the previous section) and the weir may be added. If Q is the total flow rate,

$$Q = Q_{w*}\, b d^{3/2} \sqrt{g S_0} + Q_{0*} \sqrt{g S_0}\, b_0^{5/2} \tag{9.10}$$

In Equation 9.10, the suffixes w and o refer to, respectively, the weir and orifice.

If the flow over the weir is in the plunging nappe state, the total flow rate Q can be written as

$$Q = Q_+\, b \sqrt{g}\, h^{5/2} + Q_{0*} \sqrt{g S_0}\, b_0^{5/2} \tag{9.11}$$

9.4 DENIL FISHWAYS

The Denil fishway is a straight rectangular flume provided with closely spaced baffles or vanes on the bottom and sides. Of the many Denil types studied, the two commonly adopted types are the 'Alaska Steeppass' with a rather complex system of vanes and the simple Denil that is discussed herein.

A simple Denil fishway, referred to hereafter as the Denil fishway (or Denil) is shown in Figure 9.2; the rectangular flume of width B houses baffles placed at an angle of $\psi = 45°$ with the bed. The clear width of the fishway is b and other details of the baffles are shown in Figure 9.2. In actual installations, the flume is constructed of concrete and the baffles are cut from plywood or metal sheets. For the 'standard' design, $B = 0.56$ m, $b = 0.36$ m, $k = 0.13$ m; the longitudinal spacing of the baffles a is equal to 0.25 m so that $B/b = 1.57$.

In the design suggested by the Institute of Civil Engineers (London) (1942), $B = 0.91$ m, $b = 0.53$ m, $k = 0.11$m, $\psi = 45°$, $a = 0.61$ m and $B/b = 1.71$. Except for the larger value of the spacing a and the smaller k, these designs are nearly the same. Since the simple Denil has been extensively studied, if a bigger b is desired, the simple Denil should be scaled up to a geometrically similar larger design.

Denil fishways can be built with slopes as large as 30% if the length of the fishway is so determined that the fish in the stream can travel through them. The Denil fishways pass fish well because the recirculating mass of fluid between the baffles on the sides and bottom exerts a retarding shear stress (due to momentum exchange) on the active stream flowing down the Denil. If d is the depth of flow above the crest of the baffles, the ratio d/b appears to be the main parameter for predicting the hydraulic performance of the Denil.

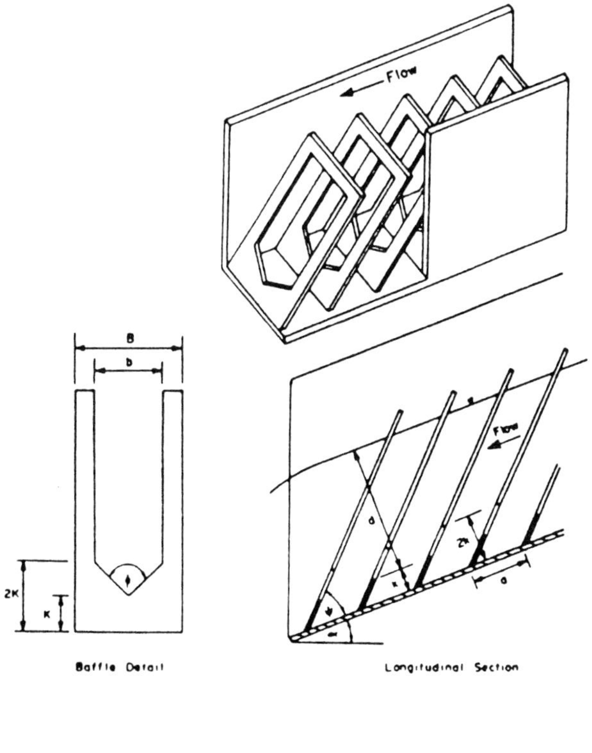

Figure 9.2. Denil fishway

The flow in a Denil fishway is highly turbulent. Preliminary observations with Laser Doppler Anemometer indicate that turbulence intensities in a Denil are many times more than those in a corresponding normal open channel. The flow is established in a relatively short distance, with the flow development length being roughly equal to the depth of flow. Let u be the time-averaged velocity in the longitudinal (x) direction at any height y above the crest of the baffles. Experimental observations have shown that in the central region of the fishway, away from the reaches affected by the entrance and exit, velocity profiles have certain characteristic shapes that depend mainly on the b/d ratio (Rajaratnam & Katopodis, 1984).

If the downstream water levels are such that the depth of flow through the whole fishway is equal to d, then the flow is uniform except near the (water) inlet. If on the other hand, the tailwater is moderately high, then in the downstream part of the fishway a M1-type of backwater curve occurs. If the fishway is submerged deeply, the velocities in the fishway decrease considerably, so that the fish might not be attracted to the fishway. If the tailwater is low, a M2-type of backwater curve will occur in the downstream part of the fishway with approximately critical flow at the downstream end a condition that might form a barrier for fish entrance.

A simple equation exists between the discharge Q and the depth of flow d for the Denil. For uniform flow and a length Δx of the fishway,

$$b\bar{d}\,\Delta x \gamma S_0 = \bar{\tau}_0 (b + 2\bar{d})\,\Delta x \tag{9.12}$$

wherein the width of the active stream in the Denil is b, $\bar{d}$ is the average depth of flow equal to $(d - k/2)$, γ is the specific weight of water and $\bar{\tau}_0$ is the average fluid boundary shear stress between the active stream and the recirculating fluid on the sides and bottom. A combination of Equation 9.2 and 9.12 yields

$$\frac{Q}{\sqrt{g S_0 b^5}} = \sqrt{\frac{2}{c_f}\frac{(\bar{d}/b)}{\sqrt{(2 + b/\bar{d})}}} \tag{9.13}$$

Values of c_f can be calculated from observations of Q, slope S_0 and average depth $\bar{d}$; c_f can be shown to be a function mainly of d/b. Then Equation 9.13 can be rewritten as

$$Q^* = \frac{Q}{\sqrt{g S_0 b^5}} = f(d/b) \tag{9.14}$$

where Q^* is the dimensionless flow rate and f denotes functional dependence. The experimental results show this function to be

$$Q^* = 0.94\,(d/b)^2 \tag{9.15}$$

Equation 9.15 is the flow equation of the simple Denil. Some designers of fishways prefer to use the resistance coefficient n from the Manning equation

$$V = \frac{1}{n} r^{2/3} \sqrt{S_0} \tag{9.16}$$

where r is the hydraulic radius, equal to the area of flow divided by the wetted perimeter. Experimental values of n are as large as 0.15 for $d = 0.2$ m and decrease continuously as the depth increases, falling to about 0.04 for $d \simeq 2$ m.

If the depth averaged velocity in the central plane is V^*, V^*/V has been found to vary mainly with b/d. This velocity ratio increases from about 0.4 at $d/b = 1.0$ to about 1.0 for $d/b = 4.0$. For a Froude number for the uniform flow defined as

$$F^* = \frac{V^*}{\sqrt{gd}} \tag{9.17}$$

F^* was less than one for the whole range of these experiments, even though the slopes were as large as 31.5%. The flow was always subcritical because of the considerable resistance offered to the flow. The backwater curves in Denil have also been investigated (Rajaratnam et al., 1985).

If a Denil fishway of a suitable slope and length is not long enough to reach from the headwater to the tailwater regions, one would need to install a series of them with resting pools in between. The length of a Denil should be based on the swimming speed of the fish.

9.5 VERTICAL SLOT FISHWAYS

Figure 9.3 shows the vertical slot fishway at Hell's Gate, with double slots as well as a single slot. The vertical slot fishway consists of a rectangular channel with a sloping floor (or a stepped floor) that is divided into a number of pools. In this channel, water flows downstream through a vertical slot from one pool to the next one. A jet forms as the water flows through the slot, and the energy is dissipated by jet mixing in the pool. The fishway is designed with a certain value of Δh, the difference in water level between two adjacent pools; the proper value depends upon the burst speed of the fish that would ascend this fishway (Blake, 1983). Usually Δh is the same across every slot and equal to the total head drop from the upstream to the downstream water level divided by the number of slots. In most cases, the width of slot b_0 is 0.305 m. Most of the vertical slot fishways built in Canada and the US are generally patterned after the one-half Hell's Gate model. If an untested design is to be used, the dimensions should be based on hydraulic model studies.

Recently several designs were studied in scale models by Rajaratnam et al. (1986) and the following discussion of their results gives an indication of the flow characteristics of vertical slot fishways. For a vertical slot fishway with a large number of pools and slots, perhaps 10, and for a given discharge Q, the tailwater can be controlled in such a way that the depth of flow in almost all the pools is

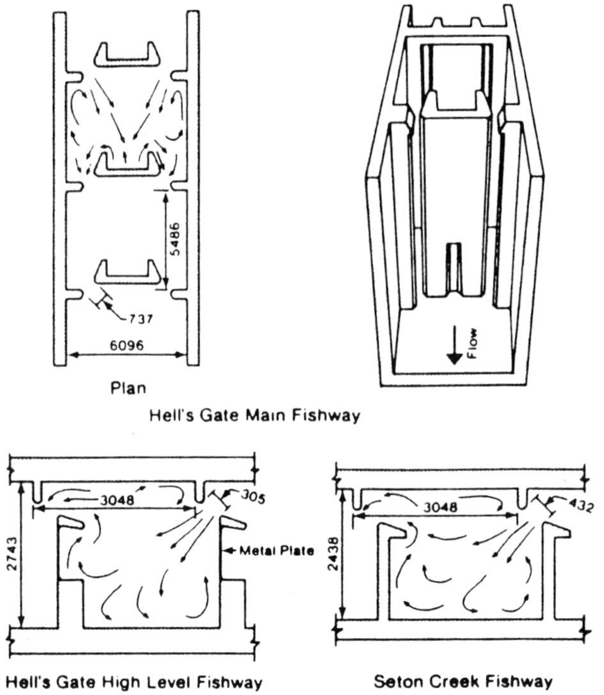

Figure 9.3. Vertical slot fishways (Clay)

approximately the same. Such a flow is essentially uniform for which the drop in pool levels across every slot Δh is the same. In some cases it could increase with x or decrease with x, and a M1 or M2-type backwater curve would occur in the fishway.

For such a uniform flow,

$$b_0 y_0 \, \Delta x \gamma S_0 = \bar{\tau}_0 \, m y_0 \, \Delta x \tag{9.18}$$

wherein b_0 is approximately the width of the stream in the pool, and $\bar{\tau}_0$ is the stress between the stream from the slot and the circulating water mass (bed shear stress on the stream is negligible compared to $\bar{\tau}_0$); $m = 1$ if the stream is bounded on one side by a solid wall and $m = 2$ if bounded by fluid on both sides. Obviously, the length and slope S_0 will have to be adjusted for various designs, but each design would have a characteristic curve in a dimensionless discharge diagram.

If Equation 9.2 is substituted in Equation 9.18, the dimensionless discharge Q^* is given by

$$Q^* = \frac{Q}{\sqrt{g S_0 b_0^5}} = \frac{y_0}{b_0}\sqrt{\frac{2}{m\, c_f}} \tag{9.19}$$

If c_f is a constant, the Q^* is a linear function of y_0 / b_0.

The above analysis describes effectively the experimental results and the functional form of Equation 9.19 is essentially linear as follows

$$Q^* = A\left(\frac{y_0}{b_0}\right) + D \tag{9.20}$$

The values of A and D have been found for several designs. These results as well as a discussion of the structure of the jet and its behaviour in the pools and the effect of backwater can be found in Rajaratnam et al. (1986).

9.6 CULVERT FISHWAYS

To ensure that fish can pass through a culvert without undue or harmful delay, one has to find the flow rate or range of flow rates that would prevail during the periods of fish passage. For spawning runs, a delay at culverts of even a day may be undesirable whereas for certain other types of passage, a delay of several days might be acceptable. Based on these, one can find a fish passing discharge or a range of fish passing discharges for a culvert under consideration. Then for given length and slope of the culvert, the characteristic velocity-length curve of the fish species under consideration would indicate whether special provisions are required to allow fish to pass.

These fish-passing facilities take several forms in practice. They include the offset baffle system, spoiler system, side baffles, fish weirs, fish baffles, the 'stream simulation' approach and others. The offset baffle system, originally developed for box culverts, has been adopted for use in circular culverts. Even though this baffle system has not been studied in detail, it has been used in a number of installations in the western US and Canada. The spoiler system has had limited use. Fish weirs and fish baffles have been used in a number of culverts in Alberta, Canada, but have not been studied in detail. The stream simulation concept has been used in Canada, particularly along the Liard Highway in the Northwest Territories (McKinnon & Hnytka, 1985).

A recent investigation by Rajaratnam et al. (1986) contains design details shown in Figure 9.4 for offset-type culvert fishways (for circular culverts). A functional relation for predicting the flow rate that has been evaluated experimentally can be written as

$$Q^* = \frac{Q}{\sqrt{gS_0 D^5}} = 12.0\,(y_0/D)^{2.6} \tag{9.21}$$

Equation 9.21 describes the experimental observations well for values of y_0/D from about 0.05 to about 0.5. If U is the maximum velocity in the culvert, then

$$\frac{U}{\sqrt{gDS_0}} \simeq 12.8\left(\frac{y_0}{D}\right) \tag{9.22}$$

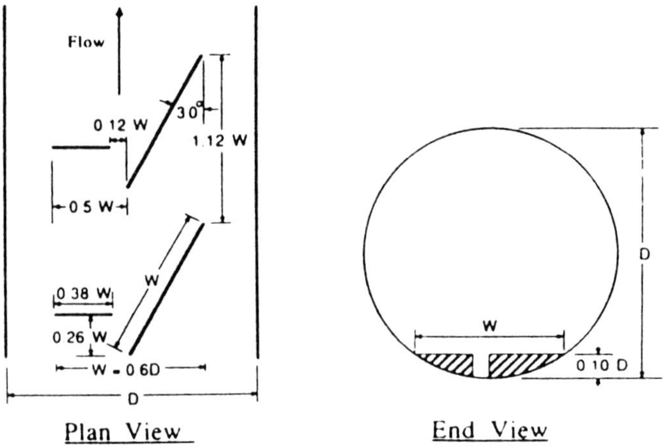

Figure 9.4. Offset baffle system (Engel)

Other types of culvert fishways are currently under investigation. Preliminary design information for these are available in Engel (1974), Katopodis et al. (1978) and Dane (1978).

REFERENCES

ASCE 1982. Design of water intake structures for fish protection. *ASCE*, 163.

Andrew, F.J. 1948. Notes on experiments on fishways (unpublished).

Baxter, R.M. & P. Glaude 1980. *Environmental effects of dams and impoundments in Canada: Experience and prospects*. Canadian Bulletin of Fisheries and Aquatic Sciences, Bulletin 205, Ottawa, 32 .

Bell, M.C. 1973. *Fisheries handbook of engineering requirements and biological criteria*. Corps of Engineers, North Pacific Division, Portland, Oregon, 490 pp.

Blake, R.W. 1983. *Fish locomotion*. Cambridge Univ. Press, London, 208 pp.

Chow, V.T. 1959. *Open-channel hydraulics*. McGraw-Hill Book Co., N.Y., 680 pp.

Clay, C.H. 1961. *Design of fishways and other fish facilities*. Dept. of Fisheries of Canada, Ottawa, 301 pp.

Collins, N.H. & G.F. Gillis 1985. Upstream fish passage, regulated streams-recent advances. *Proc. of Third International Symposium on Regulated Streams, Edmonton, Canada*. NY: Plenum Press.

Dane, B.G. 1978. A review and resolution of fish passage problems to culvert sites in British Columbia, Fisheries and Marine Service TR810, Fisheries and Env. Canada, 126 pp.

Davies, B.R. & K.F. Waker (ed.) 1986. *The ecology of river systems*. Dordrecht, the Netherlands: Doctor W. Junk Publishers.

Engel, P. 1974. *Fish passage faciities for culverts of the Mackenzie Highway*. Canada Centre for Inland Waters, Burlington, Ontario, Canada.

Institution of Civil Engineers 1942. Report of the committee on fish-passes. The Institution Research Committee, London, England, 59 pp.

Katopodis, C., P.R. Robinson & B.G. Sutherland 1978. *A study of model and prototype*

culvert baffling for fish passage. Fisheries & Marine Service TR828, Fisheries and Oceans, Canada.
Katopodis, C . & N. Rajaratnam 1983. *A review and laboratory study of the hydraulics of Denil fishways*. Canadian Tech. Rept. of Fisheries and Aquatic Science 1145, Fisheries and Oceans, Winnipeg, Canada, 181 pp.
Katopodis, C. & N. Rajaratnam 1984. Similarity of scale models of Denil fishways. *IAHR Symposium on Scale Effects, Esslingen, Germany*, 2.8-1 to 6.
Lakhsmana, Rao, N.S. 1975. Theory of weirs. In V.T. Chow (ed.), *HydroScience*, 10: 309-406.
McKinley, W.R. & R.C. Webb 1956. *A proposed correction of migratory fish problems at box culverts*. Washington Dept. of Fisheries Research Papers 1(4).
McKinnon, G.A. & F.N. Hnytka 1985. Fish passage assessment of culverts constructed to simulate stream conditions on Laird River tributaries. *Can. Tech. Rep. Fish. Aquat. Sci.*, 1255: 127 pp.
McLeod, A.M. & P. Nemenyi 1941. *An investigation of fishways*. University of Iowa, Studies in Engineering, Bulletin 24, 63 pp.
McCleave, J.D., G.P. Arnold, J.J. Dodson & W.H. Neil (eds). *Mechanisms of migration in fishes*, 317-355. NY: Plenum.
Orsborn, J.R. 1986. Final project report, Part 1 of *Development of new concepts in fishladder design*. Albrook Hydraulics Lab., Washington State Univ., Pullman, Washington, USA.
Rajaratnam, N., M. Fairbairn & C. Katopodis 1986. *An experimental study of culvert fishways*. Tech. Rept., Dept. of Civil Engrg., Univ. of Alberta, Edmonton, Canada.
Rajaratnam, N. & C. Katopodis 1984. Hydraulics of Denil fishways. *ASCE, J. Hyd. Engrg.*, 110(9): 1219-1233.
Rajaratnam, N., C. Katopodis & S. Lodewyk 1988. Hydraulics of offset baffle culvert fishways. *Canadian Journal of Civil Engrg.*, 15: 1043-51.
Rajaratnam, N., N. Katopodis & A. Mainali 1988. Plunging flows in pool & weir fishways. *ASCE, J. Hydr. Engrg.*, 112(8): 939-944.
Rajaratnam, N., C. Katopodis & A. Mainali 1989. Pool-orifice and pool-orifice-weir fishways. *Cdn. J. of Civil Engrg.*, 15: 774-777.
Rajaratnam, N., C. Katopodis & G. Van der Vinne 1985. M1-type backwater curves in Denil Fishways. *Proc. Canadian Society for Civil Engrg., Hydrotechnical Conf.*, 141-156.
Rajaratnam, N., G. Van der Vinne & C. Katopodis 1986. Hydraulics of vertical slot fishways. *ASCE, J. Hyd. Engrg.*, 112(10).
Watts, F.J. 1974. *Design of culvert fishways*. Water Resources Res. Inst., Univ. of Idaho, Moscow, Idaho, 62 pp.

Subject index